Target and Non-Target Approaches for Food Authenticity and Traceability

Target and Non-Target Approaches for Food Authenticity and Traceability

Editor

Joana S. Amaral

MDPI • Basel • Beijing • Wuhan • Barcelona • Belgrade • Manchester • Tokyo • Cluj • Tianjin

Editor
Joana S. Amaral
Chemical and Biological
Technology
Polytechnic Institute
of Bragança
Bragança
Portugal

Editorial Office
MDPI
St. Alban-Anlage 66
4052 Basel, Switzerland

This is a reprint of articles from the Special Issue published online in the open access journal *Foods* (ISSN 2304-8158) (available at: www.mdpi.com/journal/foods/special_issues/Food_Authenticity_Traceability).

For citation purposes, cite each article independently as indicated on the article page online and as indicated below:

LastName, A.A.; LastName, B.B.; LastName, C.C. Article Title. *Journal Name* **Year**, *Volume Number*, Page Range.

ISBN 978-3-0365-5458-7 (Hbk)
ISBN 978-3-0365-5457-0 (PDF)

Contents

About the Editor

Joana S. Amaral

Joana S. Amaral achieved her degree in Pharmaceutical Sciences at the University of Porto, Portugal, and completed her Ph.D. in Food Chemistry and Nutrition in 2006 at the same institution. She is a Professor at the Polytechnic Institute of Bragança and research member of the Mountain Research Centre (CIMO). She was President of the Food Chemistry Division of the Portuguese Chemical Society (2009–2012) and is currently the Chair of the Food Chemistry Division of the European Chemical Society (EuChemS). Throughout her scientific career, she has been the principal investigator of 6 national and international projects, a member of 26 projects and a COST action, and published over 100 peer-reviewed papers and book chapters (H-index 33). Since 2020, Prof. Amaral is Editor-in-Chief of the section Food Physics and (Bio)Chemistry of the journal *Foods*. Her main line of research focuses on food authentication using DNA-based approaches.

foods

MDPI

Editorial

Target and Non-Target Approaches for Food Authenticity and Traceability

Joana S. Amaral [1,2]

[1] Centro de Investigação de Montanha (CIMO), Instituto Politécnico de Bragança, Campus de Sta. Apolónia, 5301-857 Bragança, Portugal; jamaral@ipb.pt; Tel.: +351-273-383-138
[2] REQUIMTE-LAQV, Faculdade de Farmácia, Universidade do Porto, Rua de Jorge Viterbo Ferreira, 228, 4050-313 Porto, Portugal

Citation: Amaral, J.S. Target and Non-Target Approaches for Food Authenticity and Traceability. *Foods* **2021**, *10*, 172. https://doi.org/10.3390/foods10010172

Received: 11 January 2021
Accepted: 12 January 2021
Published: 16 January 2021

Publisher's Note: MDPI stays neutral with regard to jurisdictional claims in published maps and institutional affiliations.

In the last decade, consumers have become increasingly aware of and concerned about the quality and safety of food, in part due to several scandals that were widely disseminated by the media. Currently, consumers are requesting more information about the food they buy, not only from a nutritional point of view but also regarding origin, safety, traceability, and authenticity. In addition, concerns about environmental and ethical issues are on the rise, with more attention being given to topics such as biodiversity protection, production mode, and food authenticity. The growing demand for higher quality foods, the desire for new experiences associated with delicacy products or foods having particular organoleptic characteristics, together with the increasing willingness to pay more money for such products, provides an overall incentive for the adulteration of premium foods. Moreover, several factors such as international trade, market globalization, long and complex food supply chains, and the booming of e-commerce, further create opportunities for food fraud. While in several cases food adulteration poses no major risk for consumers' health (e.g., mislabeling of geographical origin), in others it can result in health hazards due to toxic or allergenic substances. However, even when health is not jeopardized, food fraud leads to unfair market competition and consumers being deceived. For all these reasons, the issue of food authenticity and food fraud has been receiving increased attention from several stakeholders, including government agencies and policymakers, control labs, producers, industry, and the research community, and different attempts have been made aiming for the definition of these concepts. According to the CEN Workshop Agreement 17369:2019, an authentic food product is "*a food product where there is a match between the actual food product characteristics and the corresponding food product claims; when the food product actually is what the claim says that is*" [1,2]. In the discussion paper on food integrity and food authenticity of the working group of the Codex Alimentarius Commission [3], food fraud is described as "*any deliberate action of businesses or individuals to deceive others in regards to the integrity of food to gain undue advantage*". Moreover, four key elements are identified, namely deliberate intent, deception, financial gain and misrepresentation, which are in line with the European Commission's key criteria to refer to when establishing if a case should be considered as fraud or as non-compliance, namely (i) violation of one or more rules of the European Union agri-food chain legislation as referred to in Article 1(2) of Regulation (EU) 2017/625, (ii) customer deception, (iii) economic gain, (iv) intention [2,4]. Furthermore, different types of food fraud have been described, including substitution, dilution, mislabeling, concealment, and unapproved enhancement, among others [2]. In order to identify, tackle and/or deter fraudulent practices in the agri-food sector, complementary approaches are needed to address this complex issue, including analytical testing and broader strategies such as implementing early warning systems, vulnerability assessments, and intelligence gathering, among which the development of new, fast and advanced analytical methods for checking food authenticity is a central aspect. Thus, several works have been published on the subject with respect to different food matrices, putting in evidence a variety of analytical techniques that can be used for food authentication [2,5–10]. So far, the majority are targeted methods, which look for a pre-defined characteristic or adulterant, thus being focused on the detection of a few selected analytes [11–13]. However, in the last few years, non-targeted methods have increasingly come into focus. These methods do not rely on the analyses of selected individual analytes since the molecules to be detected are not known a priori, but instead aim at studying a global fingerprint that should be as comprehensive as possible [11–13]. This approach can be advantageous when no information about possible adulterants is yet known and/or when unconventional adulterants are added, which would be unlikely to be detected by conventional targeted approaches. Moreover, contrary to targeted methods that frequently need complex and expensive extraction processes, in non-targeted approaches a simple sample preparation is generally performed to get as many matrix components as possible [12]. Despite the many challenges that still need to be overcome, non-targeted methods are becoming increasingly used and their contribution to deterring food fraud, together with targeted methods, is expected to grow in the coming years.

In this regard, this Special Issue aimed at gathering original research and review papers focusing on the development and application of both targeted and non-targeted methodologies to verify food authenticity and traceability. This Special Issue includes eighteen notable contributions, comprising one review paper and seventeen original research papers, these last dealing with the authentication of different foods, including some considered as highly prone to food fraud such as olive oil [14,15], honey [16,17], fish [18–20] and meat [21–24].

Several research articles in this Special Issue reported the application of different analytical techniques including chromatography, spectrometry, and spectroscopy aiming for food authentication. Grazina et al. [18] used a targeted approach to determine nineteen fatty acids by gas-chromatography with flame ionization detection (GC-FID), which were used together with advanced chemometrics to discriminate wild from farmed salmon. Based on seventeen features obtained from the chemical analysis, all the tested approaches, namely principal components analysis (PCA), *t*-distributed stochastic neighbor embedding (*t*-SNE), and seven machine learning classifiers, allowed them to differentiate the two groups (wild vs. farmed). Moreover, five classifiers allowed distinguishing between groups of farmed salmon from different geographical origins. Detecting mislabeling of geographical origin is an issue that has been receiving increasing attention in the last few years, since certified products or those produced in certain regions are frequently associated with a higher price due to their quality and specific characteristics. Analytical testing for identifying the geographical origin of foods is generally considered of high complexity since specifications for agri-food products with geographic indication are frequently based on subjective characteristics such as organoleptic properties [25]. Kim et al. [26] reported the use of hydrophilic and lipophilic metabolite profiling by gas chromatography-mass spectrometry (GC-MS) coupled with orthogonal partial least squares discriminant analysis (OPLS-DA) to differentiate perilla and sesame seeds originating from China and Korea. Furthermore, the authors noticed that glycolic acid was a notable metabolite for discriminating between perilla seeds grown in China and Korea and proposed this compound as being a potential biomarker for such discrimination. Likewise, proline and glycine could be considered potential biomarkers to determine the geographical origin of sesame seeds. The importance of tracing the geographical origin was also addressed in the study of Vukašinović-Pešić et al. [16] on multifloral honeys from different regions of Montenegro. The mineral content determined by inductively coupled plasma-optical emission spectrometry (ICP-OES) and linear discriminant analysis allowed the researchers to distinguish honeys that originated from areas exposed to industrial pollution. A different approach was proposed by Lippoli et al. [17] aiming for the fast authentication of honey's geographical origin. The authors describe the development of a non-targeted method using direct analysis in real time and high resolution mass spectrometry (DART-HRMS) combined with multivariate statistical analysis to discriminate chestnut honey from Portugal and Italy and acacia honey from Italy and China. A non-targeted method coupled with chemometrics was also the approach selected by Barbieri et al. [14] towards the authentication of virgin olive oils. In this study, a classification model was developed based on the raw data from the volatile fraction fingerprint obtained by flash gas chromatography and partial least squares-discriminant analysis (PLS-DA) to predict the commercial category of olive oils (extra virgin, virgin and lampante). The proposed classification model was shown to be robust since it included a high number of diversified samples classified by sensorial analysis ($n = 331$); it was also shown to have good performance, since it was able to correctly classify a high percentage of samples in both cross and external validation. Thus, the proposed approach represents a valid alternative for supporting official sensory panels and increasing the efficiency and fastness of controls, since it could be used as a screening tool allowing for a fast pre-classification of olive oil quality grade, thus supporting the panels by prioritizing the samples or even reducing the number of samples requiring sensory analysis. The comparison of targeted and non-targeted approaches for detecting the adulteration of fresh turkey meat by the fraudulent addition of protein hydrolysates was reported by Wagner et al. [21]. Turkey breast muscles were treated with plant or animal protein hydrolysates (those being produced by enzymatic and acidic hydrolysis and presenting different hydrolyzation degrees—partial or total) and analyzed by traditional high-performance liquid chromatography with ultraviolet-visible detection (HPLC-UV/VIS) targeting ten proteinogenic amino acids and by GC–MS and nuclear magnetic resonance (NMR) spectroscopy as two non-targeted metabolite profiling methodologies. While free amino acids analysis allowed the detection of the injection with fully hydrolyzed proteins, it was not suitable for the detection of food fraud in the case of partial hydrolysates. It was concluded that for lower hydrolyzation degrees, additional compounds originating from protein (such as sugars and the by-products released during hydrolysis) play an important role in the differentiation of nontreated samples and hydrolysate treated ones. Thus, the combination of amino acid profiling and additional compounds can provide stronger evidence for detecting and classifying this kind of adulteration.

The feasibility of using spectroscopic techniques as non-targeted approaches for food authentication was also demonstrated in this Special Issue. Truffles are very expensive mushrooms whose price depends mainly on their species but also on their origin, with the white Piedmont truffle (*Tuber magnatum*) and the black Périgord truffle (*Tuber melanosporum*) being the most valued species. In the paper of Segelke et al. [27] Fourier transform near-infrared (FT-NIR) spectroscopy combined with chemometrics is used to differentiate these truffle species from other species that are less valued but morphologically very similar. Various data pre-processing techniques were evaluated to avoid overfitting and the results compared using several classification models. The results showed the ability to differentiate the expensive white truffle *T. magnatum* from *Tuber borchii* with 100% accuracy, and *T. melanosporum* from *Tuber aestivum* and some species of Chinese black truffles with an accuracy of 99%. Moreover, Piedmont truffles could be differentiated from non-Italian *T. magnatum* truffles with an accuracy of 83%. Therefore, this work demonstrates the potential of FT-NIR spectroscopy as a fast and low-cost authentication tool, not requiring special training for sample preparation and equipment handling, thus being very suited for the industrial screening of samples.

In addition to chemical approaches, several works have been conducted so far describing the development and application of molecular biology techniques for food authentication purposes. These techniques are highly specific and sensitive and are frequently considered as the most suitable tools for the identification of species. Various research papers on the use of DNA-based approaches are also included in this Special Issue, from the comparison of different DNA extraction methods [28] to the use of multiplex polymerase chain reaction (PCR) [23], real-time PCR [19,22,24,29,30], or more advanced techniques such as Digital PCR [31]. Kim et al. [23] proposed the use of a simple qualitative assay based on the use of multiplex PCR to identify three deer species, namely red deer (*Cervus elaphus*), roe deer (*Capreolus capreolus*), and water deer (*Hydropotes inermis*). Three sets of species-specific primers were developed, generating amplicons of different sizes for each species that were then visualized by capillary electrophoresis to increase resolution and accuracy for the detection of the multiple targets. In other works, the specific identification of species was achieved by using real-time PCR. Kim et al. [24] designed new species-specific primers and probe targeting the *cytb* region of donkey (*Equus asinus*) allowing the detection of as low as 0.001% donkey meat in raw and processed meat mixtures made with beef. Velasco et al. [29] reported the development of a real-time PCR based on the use of specific primers and a minor groove binding TaqMan probe targeting the COI (*Cytochrome Oxidase I*) region for the specific authentication of common cuttlefish (*Sepia officinalis*) in seafood products. Commercial samples were also analyzed by FINS (forensically informative nucleotide sequencing) in order to test the reliability of the developed method and guarantee the correctness of the level of mislabeling found in this work (25%). This low-cost method proved to be reliable in the differentiation of this species from other cephalopods and can be very useful for food control authorities, since species from the genus *Sepia* are frequently similar and very difficult to identify after processing because the characteristics for morphological identification are eliminated. Kyriakopoulou and Kalogianni [15] described the development of a new allele-specific real-time PCR to specifically differentiate olive oil from the valuable wild-type *Olea europaea* var *Sylvestris* from the commonly cultivated type *Olea europaea* L. var *Europaea*. Besides being used for species-specific identification, real-time PCR is also reported for quantification purposes [22,29,30]. While Oh et al. [29] estimate the percentage of corn (*Zea mays*) as an added adulterant in turmeric powder (*Curcuma longa*) by using the fluorescent dye SYBRGreen, others propose the use of specific probes [22,30]. Dolch et al. [22] developed two multiplex real-time PCR assays using specific primers and probes, one for the detection and quantification of chicken (*Gallus gallus*), guinea fowl (*Numida meleagris*) and pheasant (*Phasianus colchicus*), and other for quail (*Coturnix japonica*) and turkey (*Meleagris gallopavo*). For each system, three different quantification methods were compared for estimating the relative meat content of these poultry species in meat mixtures. According to the authors, each method had its pros and cons, although the matrix-specific multiplication factors method was the one presenting more accepted recovery rates. By the contrary, in the work of Grazina et al. [30] the ΔCt method was chosen to estimate the percentage of *Ginkgo biloba* in commercial herbal infusions. The proposed normalized real-time PCR system, which required the amplification of the specific target (*G. biloba* ITS1 region) using the novel primer set and TaqMan probe and a reference endogenous gene (nuclear 18S rRNA), exhibited high performance parameters and was successfully validated using blind mixtures. To assess the occurrence of fraud in the swordfish supply chain, Ferrito et al. [20] suggested the use of a different molecular strategy encompassing the PCR amplification of the frequently used barcode COI gene combined with the restriction fragment length polymorphism (RFLP) technique. The COIBar-RFLP procedure was applied on several authenticated reference samples of swordfish (*Xiphias gladius*) and four different shark species to generate species-specific restriction enzyme patterns. Those were further used for the authentication of fresh and frozen commercial swordfish slices, allowing the detection of *Prionace glauca*, *Mustelus mustelus* and *Oxynotus centrina* in slices labeled as *Xiphias gladius*. A different technology, namely digital PCR, is reported in the work of Morcia et al. [31] to identify economically motivated adulteration in the pasta industry by the substitution of *Triticum durum* with cheaper common wheat (*Triticum aestivum*). Moreover, the proposed assay allowed the researchers to track the adulterant down to 3%, which is the critical value established in the legislation as a limit for accidental contamination.

Finally, closing this Special Issue, the review paper by Hassoun et al. [32] discusses the use of different analytical methods for detecting frauds in food products of animal origin, with particular attention being paid to non-targeted spectroscopic detection methods. The advantages, opportunities and challenges associated with the use of spectroscopic techniques are discussed and several application examples are given, covering relevant and recently published works.

Overall, the papers included in the Special Issue "Target and Non-Target Approaches for Food Authenticity and Traceability" put in evidence the global relevance of the topic and the importance of developing different approaches that can be used by control laboratories and governmental agencies to verify and guarantee food authenticity and traceability, allowing agencies to detect and expose eventual food fraud scenarios, and therefore protecting producers and industry from unfair competition as well as increasing consumers' confidence in purchased foods.

Funding: The author acknowledges the Foundation for Science and Technology (FCT, Portugal) for financial support by national funds FCT/MCTES to CIMO (UIDB/00690/2020).

Conflicts of Interest: The author declares no conflict of interest.

References

1. CEN Workshop Agreement (CWA) 17369:2019—Authenticity and Fraud in the Feed and Food Chain—Concepts, Terms, and Definitions. Available online: https://standards.cen.eu/dyn/www/f?p=204:110:0::::FSP_PROJECT,FSP_ORG_ID:68640,2273736&cs=1AE0F1E6D2455306ADD8460579462378C (accessed on 8 January 2021).
2. Goethem, V.; Elliott, C. Forewords & Introduction. In *FoodIntegrity Handbook. A Guide to Food Authenticaticity Issues and Analytical Solutions*; Morin, J.-F., Lees, M., Eds.; Eurofins Analytics France: Nantes, France, 2018; pp. 5–17.
3. Codex Alimentarius—Discussion Paper on Food Integrity and Food Authenticity—Joint FAO/WHO Food Standards Programme. Codex Committee on Food Import and Export Inspection and Certification Systems. Twenty-Fourth Session. Brisbane, Australia, 22–26 October 2018. CX/FICS 18/24/7. Available online: https://www.fao.org/fao-who-codexalimentarius/sh-proxy/en/?lnk=1&url=https%253A%252F%252Fworkspace.fao.org%252Fsites%252Fcodex%252FMeetings%252FCX-733-24%252FWorking%2BDocuments%252Ffc24_07e.pdf (accessed on 8 January 2021).
4. European Commision. 2019 Annual Report, The EU Food Fraud Network and the Administrative Assistance and Cooperation System. Available online: https://ec.europa.eu/food/sites/food/files/safety/docs/ff_ffn_annual-report_2019.pdf (accessed on 8 January 2021).
5. Amaral, J.S.; Mafra, I.; Pissard, A.; Fernández Pierna, J.A.; Baeten, V. Milk and Milk products. In *FoodIntegrity Handbook. A Guide to Food Authenticaticity Issues and Analytical Solutions*; Morin, J.-F., Lees, M., Eds.; Eurofins Analytics France: Nantes, France, 2018; pp. 3–25.
6. Amaral, J.S.; Meira, L.; Oliveira, M.B.P.P.; Mafra, I. Advances in authenticity testing for meat speciation. In *Advances in Food Authenticity Testing*; Downey, G., Ed.; Woodhead Publishing: Duxford, UK, 2016; pp. 369–414.
7. Soares, S.; Amaral, J.S.; Oliveira, M.B.P.P.; Mafra, I. A comprehensive review on the main honey authentication issues: Production and origin. *Compr. Rev. Food Sci. Food Saf.* **2017**, *16*, 1072–1100. [CrossRef] [PubMed]
8. Fernandes, T.J.R.; Amaral, J.S.; Mafra, I. DNA barcode markers applied to seafood authentication: An updated review. *Crit. Rev. Food Sci. Nutr.* **2020**, *25*, 1–32. [CrossRef] [PubMed]
9. Rocha, T.; Amaral, J.S.; Oliveira, M.B.P.P. Adulteration of dietary supplements by the illegal addition of synthetic drugs: A review. *Compr. Rev. Food Sci. Food Saf.* **2016**, *15*, 43–62. [CrossRef] [PubMed]
10. Cunha, S.C.; Amaral, J.S.; Oliveira, M.B.P.P. Authentication of vegetable oils. In *Current Topics on Food Authentication*; Oliveira, M.B.P.P., Mafra, I., Amaral, J.S., Eds.; Transworld Research Network: Kerala, India, 2011; pp. 97–128.
11. Creydt, M.; Fisher, M. Food authentication in real life: How to link non-targeted approaches with routine analytics? *Electrophoresis* **2020**, *41*, 1665–1679. [CrossRef]
12. Cavanna, D.; Righetti, L.; Elliot, C.; Suman, M. The scientific challenges in moving from targeted to non-targeted mass spectrometric methods for food fraud analysis: A proposed validation workflow to bring about a harmonized approach. *Trends Food Sci. Technol.* **2018**, *80*, 223–241. [CrossRef]
13. Ballin, N.Z.; Laursen, K.H. To target or not to target? Definitions and nomenclature for targeted versus non-targeted analytical food authentication. *Trends Food Sci. Technol.* **2019**, *86*, 537–543. [CrossRef]
14. Barbieri, S.; Cevoli, C.; Bendini, A.; Quintanilla-Casas, B.; García-González, D.L.; Gallina-Toschi, T. Flash Gas Chromatography in Tandem with Chemometrics: A Rapid Screening Tool for Quality Grades of Virgin Olive Oils. *Foods* **2020**, *9*, 862. [CrossRef]
15. Kyriakopoulou, C.I.; Kalogianni, D.P. Genetic Identification of the Wild Form of Olive (*Olea Europaea var. Sylvestris*) Using Allele-Specific Real-Time PCR. *Foods* **2020**, *9*, 467. [CrossRef]
16. Vukašinović-Pešić, V.; Blagojević, N.; Brašanac-Vukanović, S.; Savić, A.; Pešić, V. Using Chemometric Analyses for Tracing the Regional Origin of Multifloral Honeys of Montenegro. *Foods* **2020**, *9*, 210.
17. Lippolis, V.; De Angelis, E.; Fiorino, G.M.; Di Gioia, A.; Arlorio, M.; Logrieco, A.F.; Monaci, L. Geographical Origin Discrimination of Monofloral Honeys by Direct Analysis in Real Time Ionization-High Resolution Mass Spectrometry (DART-HRMS). *Foods* **2020**, *9*, 1205. [CrossRef]
18. Grazina, L.; Rodrigues, P.J.; Igrejas, G.; Nunes, M.A.; Mafra, I.; Arlorio, M.; Oliveira, M.B.P.P.; Amaral, J.S. Machine Learning Approaches Applied to GC-FID Fatty Acid Profiles to Discriminate Wild from Farmed Salmon. *Foods* **2020**, *9*, 1622. [CrossRef] [PubMed]

19. Velasco, A.; Ramilo-Fernández, G.; Sotelo, C.G. A Real-Time PCR Method for the Authentication of Common Cuttlefish (*Sepia officinalis*) in Food Products. *Foods* **2020**, *9*, 286. [CrossRef] [PubMed]
20. Ferrito, V.; Raffa, A.; Rossitto, L.; Federico, C.; Saccone, S.; Pappalardo, A.M. Swordfish or Shark Slice? A Rapid Response by COIBar-RFLP. *Foods* **2019**, *8*, 537. [CrossRef] [PubMed]
21. Wagner, L.; Peukert, M.; Kranz, B.; Gerhardt, N.; Andrée, S.; Busch, U.; Brüggemann, D.A. Comparison of Targeted (HPLC) and Nontargeted (GC-MS and NMR) Approaches for the Detection of Undeclared Addition of Protein Hydrolysates in Turkey Breast Muscle. *Foods* **2020**, *9*, 1084. [CrossRef] [PubMed]
22. Dolch, K.; Andrée, S.; Schwägele, F. Comparison of Real-Time PCR Quantification Methods in the Identification of Poultry Species in Meat Products. *Foods* **2020**, *9*, 1049. [CrossRef] [PubMed]
23. Kim, M.J.; Lee, Y.M.; Suh, S.M.; Kim, H.Y. Species Identification of Red Deer (*Cervus elaphus*).; Roe Deer (*Capreolus capreolus*), and Water Deer (*Hydropotes inermis*) Using Capillary Electrophoresis-Based Multiplex PCR. *Foods* **2020**, *9*, 982. [CrossRef]
24. Kim, M.J.; Suh, S.M.; Kim, S.Y.; Qin, P.; Kim, H.R.; Kim, H.Y. Development of a Real-Time PCR Assay for the Detection of Donkey (*Equus asinus*) Meat in Meat Mixtures Treated under Different Processing Conditions. *Foods* **2020**, *9*, 130. [CrossRef]
25. Katerinopoulou, K.; Kontogeorgos, A.; Salmas, C.E.; Patakas, A.; Ladavos, A. Geographical Origin Authentication of Agri-Food Products: A Review. *Foods* **2020**, *9*, 489. [CrossRef]
26. Kim, T.J.; Park, J.G.; Kim, H.Y.; Ha, S.H.; Lee, B.; Park, S.U.; Seo, W.D.; Kim, J.K. Metabolite Profiling and Chemometric Study for the Discrimination Analyses of Geographic Origin of Perilla (*Perilla frutescens*) and Sesame (*Sesamum indicum*) Seeds. *Foods* **2020**, *9*, 989. [CrossRef]
27. Segelke, T.; Schelm, S.; Ahlers, C.; Fischer, M. Food Authentication: Truffle (*Tuber* spp.) Species Differentiation by FT-NIR and Chemometrics. *Foods* **2020**, *9*, 922. [CrossRef]
28. Kaňuková, Š.; Mrkvová, M.; Mihálik, D.; Kraic, J. Procedures for DNA Extraction from Opium Poppy (*Papaver somniferum* L.) and Poppy Seed-Containing Products. *Foods* **2020**, *9*, 1429. [CrossRef] [PubMed]
29. Oh, S.H.; Jang, C.S. Development and Validation of a Real-Time PCR Based Assay to Detect Adulteration with Corn in Commercial Turmeric Powder Products. *Foods* **2020**, *9*, 882. [CrossRef] [PubMed]
30. Grazina, L.; Amaral, J.S.; Costa, J.; Mafra, I. Authentication of *Ginkgo biloba* Herbal Products by a Novel Quantitative Real-Time PCR Approach. *Foods* **2020**, *9*, 1233. [CrossRef] [PubMed]
31. Morcia, C.; Bergami, R.; Scaramagli, S.; Ghizzoni, R.; Carnevali, P.; Terzi, V. A Chip Digital PCR Assay for Quantification of Common Wheat Contamination in Pasta Production Chain. *Foods* **2020**, *9*, 911. [CrossRef] [PubMed]
32. Hassoun, A.; Måge, I.; Schmidt, W.F.; Temiz, H.T.; Li, L.; Kim, H.Y.; Nilsen, H.; Biancolillo, A.; Aït-Kaddour, A.; Sikorski, M.; et al. Fraud in Animal Origin Food Products: Advances in Emerging Spectroscopic Detection Methods over the Past Five Years. *Foods* **2020**, *9*, 1069. [CrossRef]

foods

Review

Fraud in Animal Origin Food Products: Advances in Emerging Spectroscopic Detection Methods over the Past Five Years

Abdo Hassoun [1,*], Ingrid Måge [1], Walter F. Schmidt [2], Havva Tümay Temiz [3], Li Li [4], Hae-Yeong Kim [5], Heidi Nilsen [1], Alessandra Biancolillo [6], Abderrahmane Aït-Kaddour [7], Marek Sikorski [8], Ewa Sikorska [9], Silvia Grassi [10] and Daniel Cozzolino [11]

1 Nofima AS, Norwegian Institute of Food, Fisheries, and Aquaculture Research, Muninbakken 9-13, 9291 Tromsø, Norway; ingrid.mage@Nofima.no (I.M.); heidi.nilsen@nofima.no (H.N.)
2 United States Department of Agriculture, Agricultural Research Service, 10300 Baltimore Avenue, Beltsville, MD 20705-2325, USA; walter.schmidt@usda.gov
3 Department of Food Engineering, Bingol University, 12000 Bingol, Turkey; tumaytemiz89@gmail.com
4 Key Laboratory of Mariculture, Ministry of Education, Ocean University of China, Qingdao 266003, China; l_li@ouc.edu.cn
5 Department of Food Science and Biotechnology, Kyung Hee University, Yongin 17104, Korea; hykim@khu.ac.kr
6 Department of Physical and Chemical Sciences, University of L'Aquila, 67100 Via Vetoio, Coppito, L'Aquila, Italy; alessandra.biancolillo@univaq.it
7 Université Clermont-Auvergne, INRAE, VetAgro Sup, UMR F, 63370 Lempdes, France; abderrahmane.aitkaddour@vetagro-sup.fr
8 Faculty of Chemistry, Adam Mickiewicz University in Poznan, Uniwersytetu Poznanskiego 8, 61-614 Poznan, Poland; sikorski@amu.edu.pl
9 Institute of Quality Science, Poznań University of Economics and Business, al. Niepodległości 10, 61-875 Poznań, Poland; ewa.sikorska@ue.poznan.pl
10 Department of Food, Environmental and Nutritional Sciences (DeFENS), Università degli Studi di Milano, via Celoria, 2, 20133 Milano, Italy; silvia.grassi@unimi.it
11 Centre for Nutrition and Food Sciences, Queensland Alliance for Agriculture and Food Innovation, The University of Queensland, 39 Kessels Rd, Coopers Plains, QLD 4108, Australia; d.cozzolino@uq.edu.au
* Correspondence: abdo.hassoun@nofima.no; Tel.: +47-930-73-906

Received: 3 July 2020; Accepted: 1 August 2020; Published: 6 August 2020

Abstract: Animal origin food products, including fish and seafood, meat and poultry, milk and dairy foods, and other related products play significant roles in human nutrition. However, fraud in this food sector frequently occurs, leading to negative economic impacts on consumers and potential risks to public health and the environment. Therefore, the development of analytical techniques that can rapidly detect fraud and verify the authenticity of such products is of paramount importance. Traditionally, a wide variety of targeted approaches, such as chemical, chromatographic, molecular, and protein-based techniques, among others, have been frequently used to identify animal species, production methods, provenance, and processing of food products. Although these conventional methods are accurate and reliable, they are destructive, time-consuming, and can only be employed at the laboratory scale. On the contrary, alternative methods based mainly on spectroscopy have emerged in recent years as invaluable tools to overcome most of the limitations associated with traditional measurements. The number of scientific studies reporting on various authenticity issues investigated by vibrational spectroscopy, nuclear magnetic resonance, and fluorescence spectroscopy has increased substantially over the past few years, indicating the tremendous potential of these techniques in the fight against food fraud. It is the aim of the present manuscript to review the state-of-the-art research advances since 2015 regarding the use of analytical methods applied to detect fraud in food products of animal origin, with particular attention paid to spectroscopic measurements coupled with chemometric analysis. The opportunities and challenges surrounding the use of spectroscopic techniques and possible future directions will also be discussed.

Keywords: authentication; authenticity; chemometric; fish; origin; honey; meat; milk; spectroscopy; species

1. Introduction

In recent years, consumers have become more concerned about the quality and safety of food products and have become keenly interested in knowing more about food authenticity and food fraud. In other words, consumers demand more complete information about their food, including what they are really buying, where the food comes from, and when and how it was produced. Although fraud and adulteration have been practiced since ancient times, it is only in recent years that food authenticity issues have been more exposed, and public attention has been intensively paid to the magnitude of this problem and the serious consequences of food fraud [1,2]. Furthermore, during the current pandemic period with coronavirus raging around the world, affecting every aspect of life, including food choices and nutrition habits, consumers have become even more concerned about safety, accessibility, affordability, and the origin of food products than any time before. This increased interest in food authenticity may also be explained by the numerous food scandals over the last few years (e.g., horsemeat scandal in 2013 and rotten meat from Brazil in 2017) and the increased consumer awareness about the impacts of food fraud in terms of illegal economic gain, as well as negative effects on the public health and the environment. Nonetheless, several recent studies have indicated that fraud or mislabeling is still a widespread practice, especially in food products of animal origin, which are often considered among the most frequently adulterated foods [3–6]. Market globalization and increases in international trade, driven by fewer obstacles to the export and import of food, a complex food production chain, and the complex nature of food products of animal origin, the huge variety of these products, as well as the emergence of tricky and more sophisticated forms of fraud are some of the reasons that could explain this rise in food fraud and why detection and prevention are challenging tasks [7–10].

Fraud in animal origin products can take many forms, including mislabeling of the provenance (geographical or botanical origin), species substitution, discrepancies in the production method and farming or breading technique, addition of non-declared substances, as well as fraudulent treatments and non-declaration of processes, such as previous freezing, irradiation, and microwave heating (Figure 1). To support this review and obtain the research published in the last few years on the authenticity of food products of animal origin, Scopus database was queried in May 2020, using the keyword "authenticity" or "authentication" and the different categories of animal origin food products. It can be noticed that a huge amount of studies dealing with authenticity and detection of fraud in fish, meat, milk, honey, and eggs has been published in recent years; the number of published works increased from 530 between 2010 and 2014 to 1000 between 2015 and 2019 (Figure 2a).

Fraud in fish and other seafood is a widespread issue, and seafood products are often ranked among the top food product categories that are susceptible to fraud. Substitution of a high-value fish species with a cheaper alternative and mislabeling of the geographical origin are among the most common fraudulent activities practiced in the fish and seafood sector. Determining whether fish is wild or farmed, tracing farming systems, and differentiating between fresh and frozen–thawed seafoods are among the seafood authenticity topics that have been widely investigated [8]. According to our literature review, meat and meat products are the most studied animal origin foods with regards to authenticity (Figure 2b). Meat authenticity has similar issues to those of fish. To address authentication issues related to muscle foods (fish and meat), a wide range of protein- and DNA-based techniques, chromatography, elemental profiling, and isotopic analysis, among many other measurements, have been frequently applied to this problem [10–12]. Similar techniques have also been established in routine analysis for detecting fraud that occurs in other foods of animal origin (e.g., milk and dairy products, honey, and eggs).

Figure 1. Most reported authenticity issues in food products of animal origin.

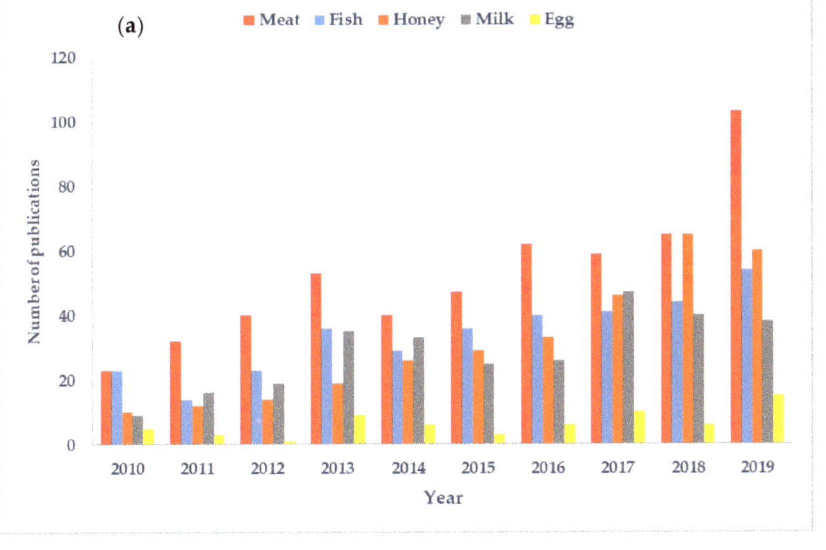

Figure 2. *Cont.*

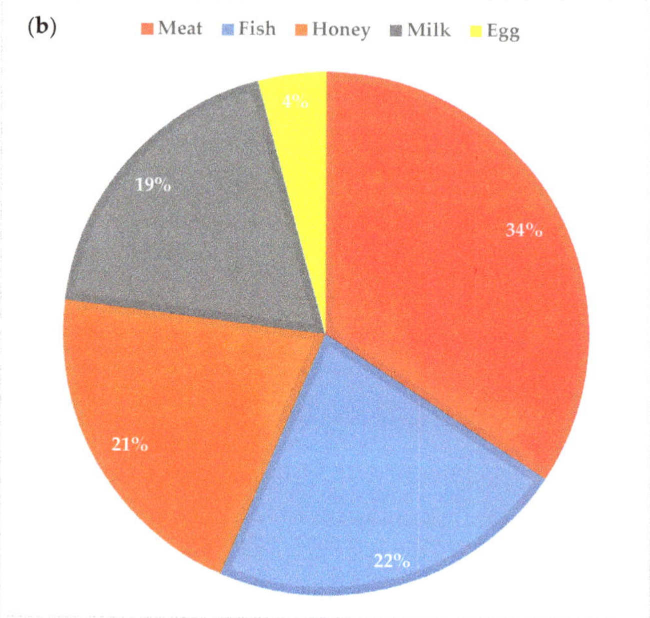

Figure 2. Temporal evolution of published work on the authenticity of different categories of food products of animal origin during the last decade (**a**) and publications distributed between the different food categories (**b**).

However, most of the aforementioned analytical methods are associated with several drawbacks, mostly related to the destructive nature of the measurements and the time required to perform the analysis. Therefore, there is still great interest in the development of non-destructive, rapid, accurate, robust, and high-throughput analytical methods for on-site and real-time food authentication. Spectroscopic techniques have gained much importance during the last few years, and spectroscopy has been a popular "buzz word" in the context of fighting fraud and verifying the authenticity of food products. The considerable interest in these non-targeted fingerprinting techniques may be due to the advancements in the analytical instruments and the increasing awareness in the food industry and research on the advantageous aspects of applying such techniques [13]. The number of scientific works regarding the use of spectroscopy for food authenticity increased from 134 papers during 2010–2014 to 369 papers during 2015–2019 (Figure 3a), while the number of total citations (Figure 3b) doubled during the last five years (20,784 citations between 2015 and 2019 versus 9666 citations between 2010 and 2014). Some examples of recent applications of spectroscopic techniques for authentication of food products of animal origin include detection of adulteration in meat [14,15] identification of milk species [16,17], detection of thawed muscle foods [18,19] identification of muscle foods species [20–22], and determination of the botanical origin of honey [23,24], among many others.

Over the last few years, several review papers have been published focusing on either one of the authenticity issues, such as the geographical origin [25,26] or species [27]; or one category of food products of animal origin, such as fish [7,8], meat [28,29], or honey [30]. Other papers have reviewed one specific type of analytical method, such as multielement and stable isotype techniques [11], volatilomics [31], DNA-based methods [32,33], or infrared spectroscopy [34]. The current review will cover the most recent studies that shed light on the various authenticity-related issues (i.e., geographical or botanical origin, species, production method, farming or breeding technique, and processing method) for all food products of animal origin (fish, meat, milk, honey, and egg), highlighting a wide range of both traditional and emerging techniques. This review will first introduce a brief description of the common multivariate data analysis and analytical techniques related to detecting fraud in food products of animal origin. Several examples of applications of conventional and spectroscopic techniques will be then presented, covering the most relevant works published during the last five years. Finally, some difficulties and challenges, as well as future trends in applications of these techniques, will be discussed. To the best of our knowledge, this review paper is the first to combine results from recent studies on a wide range of analytical methods applied to authenticate fish, meat, milk, honey, and egg, as well as their products.

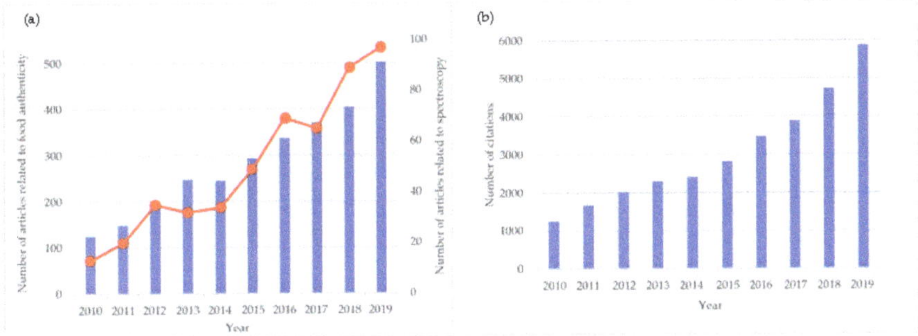

Figure 3. Numbers of published works related to food authenticity (blue bars) and use of spectroscopic techniques in relation to food authenticity (red line) (**a**). Numbers of citations including the words authenticity or authentication and spectroscopy (**b**) since 2010. Data obtained from Scopus database on 25 May 2020.

2. Multivariate Data Analysis

Traditional chemometric methods are based on linear projections onto a lower dimensional latent variable space, and these powerful and simple methods still dominate the field. However, more flexible and data-intensive machine learning methods have gained traction lately. These methods have the ability to model complex, non-linear relationships; however, the curve fitting procedures, interpretation, and validation are often more complicated. In general, the choice of data analysis strategy depends on the research question, as well as the type and size of the available data.

The data analysis pipeline consists of preprocessing, data exploration, modeling, and validation. The following sections give a brief description of each of these steps, with the main emphasis on recent trends and developments. For detailed overviews of data analysis in food authenticity, please refer to [35–38].

2.1. Data Preprocessing

The aim of preprocessing is to reduce non-relevant variations in the signal stemming from instrumental artifacts, surrounding effects, or sample preparation. The most used methods include standard normal variate (SNV), (extended) multiplicative signal correction ((E)MSC), derivatives, smoothing, baseline corrections, and peak alignments, which are often used in combination. The choice of preprocessing method is critical for the subsequent modeling and interpretation [39,40], and should be chosen based on knowledge of the samples and the measurement platform. Recent research suggests various strategies for making the modeling less sensitive to preprocessing, for instance by using a boosting approach [41], through Tikhonov regularization [42], or by using convolutional neural networks [43–45].

2.2. Data Exploration

Data exploration is an important step prior to the actual modeling. The aim is to gain an overview of the data, deal with outliers, evaluate the effects of preprocessing, and get a first impression of the potential for discriminating between samples. Principal component analysis (PCA) is the most used tool for data exploration, providing a linear transformation of the original data by maximizing the explained variance. Cluster analysis is another group of exploratory methods based on a certain distance or similarity measure between samples. These methods can be more flexible than PCA, depending on the chosen similarity metric, and may be useful for very large sample sizes.

2.3. Modeling

Authentication tasks mainly aim to determine which category a food item belongs to, i.e., classification. There are two main approaches to classification: class modeling and class discrimination [46–48]. While class modeling focuses on modeling the similarities among samples from the same category, class discrimination focuses on finding the differences between a set of predefined categories. The most used methods in the scientific literature are the soft independent modeling of class analogies (SIMCA) and partial least squares discriminant analysis (PLS-DA) classical chemometric methods for class modeling and discrimination, respectively; however, methods such as support vector machines (SVM), random forests (RF), k-nearest neighbor (k-NN), and different types of neural networks (NN) are also frequently applied. Quantitative prediction models are also relevant in some cases, for instance when the objective is to quantify the amounts of specific adulterants. An overview of alternative methods for class modeling, discriminant analysis, and quantitative prediction can be found in [35–38].

Data Fusion: Data or sensor fusion is an emerging topic within food authentication. A combination of several instrumental techniques can lead to more accurate results, either by providing complementary information or by reducing uncertainty [49–53]. Data fusion is also an active research area in fields other than authenticity, and new methods for explorative analysis, classification, and prediction are presented frequently. In principle, all multivariate methods can be used for data fusion by (1) combining all the measured variables directly, called low-level data fusion; (2) combining extracted features such as principal components, called mid- or feature-level data fusion; or (3) combining predictions or classifications from different techniques through voting, called high- or decision-level fusion. There are also several methods that are tailored for data fusion problems. Examples of newly developed explorative techniques include methods that separate common and distinctive variations in multiple data blocks [54,55], whereas sequentially orthogonalized PLS (SO-PLS) [56,57] is a common example of multiblock regression methods.

From Small to Big Data: In general, the traditional chemometric methods, such as PCA, SIMCA, and PLS-DA/PLSR, are suited for small feasibility studies, while larger studies allow for use of more data-intensive methods, such as SVM, RF, and NN. In industrial applications, however, databases with hundreds of thousands of samples are often available. Such huge data sets call for completely different data analysis strategies. There has so far been little focus on authentication models based on large databases in the scientific literature, mainly because these databases are not open. There are, however, a few exceptions showing that local modeling is a promising strategy [58,59]. In local modeling, a new model is fitted for each new sample to be predicted, using only a subset of spectrally similar samples as a calibration set. More research is needed on the use of local modeling for classification and on the analysis of large databases in general.

2.4. Validation

One of the main barriers for the successful implementation of fingerprinting techniques in food authenticity is the lack of proper validation schemes [2,60–62]. A full validation scheme consists of four phases: (1) optimization of the analytical procedure, (2) statistical model selection and parameter optimization, (3) testing of the model performance, and (4) stability testing by system challenges [60]. Most published feasibility studies stop at phase two or three, while phase four is essential for successful implementation.

Phase one is specific for the analytical technique and will not be covered here. The aim of phase two is to select an optimal modeling strategy and model parameters. This is usually done by resampling methods, such as cross-validation or bootstrapping. Phase three involves testing the model performance using an independent test set, while phase four tests the extrapolation of the model, e.g., overtime or for different instruments and locations. Thorough reviews of both numerical and conceptual aspects of validation are given in [63,64].

3. Overview of Fraud Detection Techniques

3.1. Spectroscopic Techniques

3.1.1. Vibrational Spectroscopy

Innovation pathways in vibrational spectroscopy during this past half decade are preludes to potential impacts and further practical achievements in the next half decade. Vibrational spectroscopy techniques, including infrared spectroscopy in the near (NIR)- and mid (MIR)-infrared spectral ranges, as well as Raman spectroscopy, enable a fingerprinting chemical analysis of an intact food sample in situ for adulteration in real time. The sample remains intact for confirmatory analysis using other techniques. Spectroscopic technologies require high levels of rigor in the evidence for authentication of both the food or food product and of the adulterant or contaminant.

Variance in the spectral signature of the food always can complicate the capacity to distinguish the amount and composition of an adulterant or contaminant. Recent state-of-the-art authentication of milk products has been reported [65,66]. The authentication of raw milk involves a different process—knowing its fingerprint identity enables detecting adulteration [67]. Products made from milk have also been authenticated. Desi ghee made from buffalo and from cow milk can be differentiated [68], while butter containing lard [69] and cream and yogurt [70] can be distinguished with chemometrics.

Authentication in meats is required for foods that are labeled as individual meats [71]. Horsemeat in minced beef [72], beef and mutton in pork [73], and rainbow trout in Atlantic salmon [74] each require sufficient data specific to substances to be labeled to assure the meat contaminant is properly characterized in order to identify markers characteristic of each additional component. Spectral data on the primary meat preferentially needs to be oversampled relative to that of a contaminant, or of minor or occasional components that could be misinterpreted as unrelated to the original meat. Factors such as diet can alter vibrational fingerprints. Eggs from poultry fed omega-3 fatty acids contain an intentional adulterant that can be detected in the spectral signature of the eggs [75]. Work involving fish fillet authentication using vibrational spectroscopy has also been published [21,76].

3.1.2. Nuclear Magnetic Resonance

Nuclear magnetic resonance (NMR) spectroscopy, despite being a very well-established methodology in food analysis, has had limited new publications over the last five years. The major difficulties are that foods are inherently mixtures of components and adulterants may or may not be mixtures. Thus, identifying NMR chemical shifts that do not belong in a particular food first requires authentication of the fact that a particular set of peaks may not arise occasionally (i.e., more rarely) on its own. This is an innately complicated process because one needs to ascertain which chemical shifts are correlated. A major advantage of NMR is that modern NMR techniques can trace the fingerprints from finger to finger and ascertain one part of a fingerprint belongs to another hand. Publishing the results of such an effort is often difficult because someone else may have found the same compound in this (or another) food product. Further, if the compound found is of little apparent biological or food property relevance, journal reviewers can deem such research as having a correspondingly low relevance.

The specificity of NMR complicates the authentication of the composition of an adulterant. A unique and specific NMR peak at best detects only a single component of an adulterant. If an adulterant happens to be a mixture of components, NMR is useful only for detecting chemical components one at a time. Thus, if in minced meat labelled beef porcine fat can be detected as an adulterant, NMR can only identify a chemical shift, which identifies a site on a specific unsaturated lipid as foreign to beef. It cannot likely identify which animal (or plant for that matter) was the source of the product contamination. Once markers have been authenticated properly to a specific chemical structure, this fingerprint is treated as a positive result awaiting verification by some other technique. Verifying the food commodity that has been used for adulteration requires significantly more spectroscopic data. Each and every spectral identification result that can be detected in a specific matrix can be a significant challenge and are important to know. Publishing such a finding is a more complicated endeavor.

Three recent NMR manuscripts involved detection of adulteration in milk, powdered milk, or butter [77–79]. Two publications involving edible lipids, including milk, additionally used more complicated NMR experiments (time-domain NMR and ^{13}C inept NMR) [80,81]. The more complicated NMR techniques enhance the resolution and quality of data collected. A similar enhancement using improved technologies and methodologies in milk was reported using Raman chemical imaging techniques [82]. One manuscript reported on the authentication of krill oil using NMR techniques [83].

The focus on NMR research in authenticating lipid compositions in foods is because specific lipids in mixtures of lipids appear to be characteristic of their origin. The high resolution of NMR enables deconvolution of the specificity of the lipid composition at the molecular level. Solvent effects, however, appear to complicate spectral assignments. Two publications verified that NMR techniques can fully distinguish omega-3 from omega-6 fatty acids in mixtures [84] and among three omega-3 fatty acid structural analogs, each in an intact lipid environment [85]. Authenticating the fingerprints of lipids is an essential component of and prerequisite for verifying adulteration correctly.

3.1.3. Fluorescence Spectroscopy

Fluorescence spectroscopy is based on measurement of the spectral distribution of the intensity of the light emitted by electronically excited molecules. Fluorescence coupled with chemometrics has been widely used in food studies, including for products of animal origin [86–91]. The main advantages of fluorescence as compared to other spectroscopic techniques are its higher sensitivity and selectivity. Due to these features, fluorescence is particularly useful for studying minor and trace components in complex food matrices [87,91]. Characterization of real multifluorophoric food samples requires more advanced measurement techniques than conventional emission or excitation spectra. The advanced fluorescence techniques have often been used in food studies, including excitation–emission matrix (EEM) fluorescence spectroscopy, synchronous fluorescence spectroscopy (SFS), and total synchronous fluorescence spectroscopy (TSFS) [87,92].

Fluorescence patterns of food products are usually complex. Fluorophores in food include natural food components, process-derived compounds, food additives, and contaminants [89]. Autofluorescence of meat and fish originates mainly from collagen, adipose tissues, proteins, and oxidation products [89]. Milk and dairy products contain several intrinsic fluorophores, including free aromatic amino acids, nucleic acids, aromatic amino acids in proteins, vitamins A and B_2, nicotinamide adenine dinucleotide (NAD), chlorophyll, and oxidation and Maillard reaction products [86,90]. Fluorescence in honey is ascribed to proteins, polyphenolic compounds, and Maillard reaction products [23,93,94]. The unique fluorescence patterns of food products have been successfully utilized in authentication studies of food of animal origin, including meat [95,96], fish [97,98], milk [16,17] dairy products [86,90], and honey [23,88,99–102].

3.1.4. Other Spectroscopic Techniques

The number of studies on the potential use of novel spectroscopic techniques to detect fraudulent practices encountered in the food chain has gradually increased in recent years. In this section, information is given on the applications of laser-induced breakdown spectroscopy (LIBS), terahertz (THz) spectroscopy, and hyperspectral imaging (I) in food adulteration analysis.

LIBS has been presented as a potential alternative to the existing analytical atomic spectrometry techniques used to determine the elemental composition of food. Most of the samples need a minimum or no sample preparation to be analyzed by using LIBS. The simultaneous analysis of multiple elements can be achieved. It is highly applicable to at-, on-, and in-line measurements and remote sensing, enhancing its potential as an analytical technology process [103,104]. LIBS coupled with several chemometric methods has been widely used for species discrimination [105], determination of adulteration [106], and spatial mapping of the sample surfaces in meat, milk, and other products [107,108]. Recently, some studies utilized LIBS for analysis of honey adulteration [109,110] and determination of its geographical origin [111,112]. Although there is a significant amount of research in the literature reporting the high potential of LIBS as an at-line monitoring tool for the industry, there is still a need for further improvements in system components and configurations. Besides, more research is required to recommend alternatives to reduce the matrix effect and minimize sample preparation procedures in order to improve the predictive accuracy. Peng et al. have described the significant challenges and possible solutions to these in order to speed up the use of LIBS as an in situ monitoring tool [113,114].

Terahertz spectroscopy (THz) is another technique that provides an excellent alternative to X-rays in order to obtain high-resolution images from the interiors of solid objects. Frequency-domain and time-domain measurements are performed for both imaging and spectroscopy with THz waves [115]. There are a limited number of studies on the use of THz spectroscopy for the determination of food adulteration, which were previously compiled by Afsah-Hejri [115] and He [116]. Adulteration of milk with a fat powder [117], discrimination of honey samples [118], and determination of honey adulteration [119] were some of the recent study topics.

Hyperspectral imaging (HSI) is another relatively new technology, which has explicit potential to satisfy the needs for remote and real-time monitoring techniques. Being rapid, non-invasive, and providing spectral and spatial features simultaneously are some of its significant advantages. Numerous articles describe the pros and cons of HSI-based methods for food authenticity and adulteration analyses [14,15]. Nowadays, low-cost, rapid, and simple multispectral imaging systems are being designed for the determination of particular adulterations [120]. Efforts are being made to offer alternative methods for the interpretation of HSI data. The transition from the use of linear classifiers to machine learning and deep learning solutions offers a great variety of opportunities [121]. Another trend is to employ miniature devices called single shot or snapshot hyperspectral sensors, which are ultra-portable and able to acquire data at video rates [122]. The enormous potential of the HSI technique to detect many aspects of food adulterations has been shown in the literature. Even so, enhancement of the spectral and spatial resolution and presentation of alternative technologies for advanced data analysis would be positive contributions to the accuracy and cost-effectiveness of the developed methods.

3.2. Other Analytical Methods

3.2.1. DNA-Based Techniques

To date, many DNA-based detection methods have been developed to determine animal species in food products. In particular, DNA-based methods have been used to detect target species in processed foods, because DNA is stable at high temperatures and pressures. Sequencing-based techniques (such as DNA barcoding and minibarcoding), polymerase chain reaction (PCR) coupled with restriction fragment length polymorphism (PCR-RFLP), real-time PCR, multiplex PCR, and species-specific PCR are among the most used techniques [32,123,124].

Identification of short DNA sequences, called DNA barcodes, has been widely exploited for species discrimination. DNA barcoding and minibarcoding were used to authenticate animal-derived food products sold in the Chinese market [125] and to identify selected brands of locally-produced canned and dried sardines in the Philippines [126].

In PCR-RFLP, the PCR products are cleaved with restriction enzymes, followed by gel electrophoresis and blotting [32,123]. The technique was successfully applied to differentiate four commercial shrimp types in India, and the developed PCR-RFLP protocol was validated by analyzing 52 commercial shrimp products [127].

Real-time PCR and multiplex PCR methods are the most common detection technologies in meat and meat products, fish and seafood, and other food categories that are known to have a high incidence of adulteration [124,128–130]. There are numerous reports in the literature demonstrating that real-time PCR is a powerful method that can be used as a reliable and sensitive technique for meat identification. For example, in one of the recent studies, a real-time PCR assay was developed for the detection of raw donkey meat and different processed meat mixtures [131]. Fang and Zhang used real-time PCR and TaqMan-specific probes for the detection of murine components in mutton meat products [129]. The results showed that the limit of detection was lower than 1 pg of DNA per reaction and 0.1% murine contamination in meat mixtures.

Many researchers have applied multiplex PCR methods for identification of meat species for simultaneous and rapid detection of multiple species in a single reaction. For example, two direct-triplex real-time PCR systems based on intercalating dyes were applied as a robust and precise quantitative PCR assay for meat species identification [124]. No DNA extraction was required and 92.5% of market samples of six commonly eaten meat species were successfully amplified. The multiplex PCR method was also applied to detect chicken, duck, and goose in beef, mutton, pork, or quail meat samples [132]. In a similar study, a multiplex PCR assay was used to identify lamb, beef, and duck in a meat mixture before and after heat treatment [133]. Similar approaches were developed to monitor commercial jerky products [134]; to detect chicken and pigeon in raw and heat-treated meats [135]; and to detect chicken, turkey, and duck in processed meat products [130]. Recently, a fast multiplex real-time PCR with TaqMan probes was performed to simultaneously detect pork, chicken, and beef in processed meat samples [136].

The species-specific PCR method has been used to a great extent for meat species identification in foods because of its high specificity and rapidity. For instance, El-Razik and co-authors used a species-specific PCR test to differentiate donkey and horse tissue in cooked beef meat products in Egypt [137]. In another study, a species-specific PCR was developed for the identification of beef in India [138].

In addition, more advanced high-throughput DNA sequencing methods, such as next-generation sequencing (NGS) [139,140], have emerged in recent years as valuable techniques for carrying out untargeted screening of complex samples.

3.2.2. Protein-Based Techniques and Related Methods

Chromatographic, electrophoretic, and immunological methods have been widely used for different authenticity issues for food products of animal origin [29,123,141,142]. Different mass spectrometry (MS) techniques have emerged in recent years, and along with chromatographic and NMR techniques have become some of the most commonly applied approaches for metabolomic fingerprinting [142,143]. Traditionally, MS methods are coupled with chromatographic separation techniques, such as liquid chromatography mass spectrometry (LC-MS) [142]. More recently, direct MS analysis approaches, such as matrix-assisted laser desorption/ionization-time of flight (MALDI-TOF), real-time techniques (e.g., direct analysis in real-time (DART) technique), and high-resolution mass spectrometry (HR-MS), among others, have been developed and applied to many authentication problems [123,144–147]. For example, a DART HR-MS method was developed to discriminate Canadian wild salmon from the farmed fish produced in Canada, Chile, and Norway [144]. The results showed that PCA applied to the 30 most abundant signals generated from fatty acids after the DART HR-MS analysis of fillet lipid extracts enabled a rapid discrimination between farmed and wild fish, whereas discriminant analysis (DA) gave a correct classification rate of 100%. In another study, the differences between rainbow trout and king and Atlantic salmons were studied using a lipidomical method based on hydrophilic interaction chromatography MS [147]. PCA was applied to recognize the variance among these fish species, which was attributed to the genetic origin, living environment, and feed ingredients, among others. A novel method based on quadrupole time-of-flight (Q-TOF) MS coupled with a surgical diathermy device was recently developed to distinguish cod from oilfish in real time [145]. PCA demonstrated that the clusters of oilfish were well separated from those of cod, while the application of discriminant analysis models showed that the fish tissue can be authenticated with 96–100% accuracy. Another recent study investigated the potential of ultra-performance liquid chromatography–triple time-of-flight–tandem mass spectrometry (UPLC−triple TOF−MS/MS) to determine lipid composition in the muscle tissue of four popularly consumed shrimp species [146]. About 600 lipid compounds from 14 classes were characterized and quantified, and PCA results of lipid profiles allowed the different species to be distinguished. In a similar investigation, the use of LC-TOF−MS allowed the detection of commercially available, highly processed mixed-meat products, including duck, goose, and chicken, along with pork and beef [148].

Besides the chromatographic and mass spectrometry techniques, enzyme-linked immunosorbent assay (ELISA) is one of the most widely used methods for meat identification, because it is cheap and easy to perform [141,149].

Although the aforementioned techniques have several advantages, such as stability during thermal processing and high sensitivity and selectivity, most of these measurements are time-consuming because several steps are required for sample preparation, protein extraction, and lipid extraction. In addition, the technical difficulty with MS and PCR in food adulteration is that they are useful mainly and sometimes only after the rest of the chemistry and spectroscopy work has been completed. Such techniques are especially valuable for verifying adulterations detected in situ by other technologies.

3.2.3. Isotopic Technique

As the isotopic compositions of the plants or animals reflect the condition of natural environment where they grew up, the light stable isotopes $^{13}C/^{12}C$, $^{18}O/^{16}O$, $^{2}H/^{1}H$, and $^{15}N/^{14}N$, $^{34}S/^{32}S$; and the heavy isotopes $^{11}B/^{10}B$ and $^{87}Sr/^{86}Sr$ are commonly used in food authentication [11]. Preliminary studies have demonstrated the usefulness of stable isotope analysis in determining the origins of animal origin products [150–154]. However, the animal origin products had more complicated life cycles than the plant origin products. The stable isotopes such as $\delta^{2}H$ and $\delta^{18}O$ were more likely to be affected by the ambient environment [151,155]. Camin and co-authors [151] reported the H/O ratios of Italian rainbow trout fillets were positively interrelated with the O ratio of tank water. However, the other stable isotopes $^{13}C/^{12}C$, $^{15}N/^{14}N$, and $^{34}S/^{32}S$ were reported to be affected by diet [11,156,157]. Taking shrimp as an example, the $\delta^{13}C$ and $\delta^{15}N$ values in shrimp are significantly related to the food sources [158]. During shrimp culture, the farmers may use several brands of commercial feeds with different ingredients and isotopic signatures. Li et al. [156] reported that the $\delta^{13}C$ and $\delta^{15}N$ values in 16 commercial feeds used in shrimp culture in China ranged from −23.03 to −24.75‰, and from 2.1 to 8.18‰, respectively. The dietary shifts could influence the stable isotope signature of shrimp. The effects of diet on the stable isotope signature of animal origin products should be considered when using traceability methods. Moreover, animals can only be sampled for traceability purposes when they are in isotopic equilibrium with their diet. In a recent study, Li and others suggested the sampling of shrimp that have been consistently fed with the same feed for more than twenty days [158].

The stable isotopes of animal origin products could also be affected by other environmental factors, such as culture seasons and salinity [159,160]. Compared with the marine ecosystem, the freshwater ecosystem generally has low $\delta^{13}C$ and $\delta^{15}N$ values [159,161]. Previous studies reported different $\delta^{13}C$ and $\delta^{15}N$ values in shrimp and fish cultured in freshwater and seawater [156,159]. Hence, all of these factors should be compared when using isotopic traceability methods to allow for deter animal origin product fraud.

3.2.4. Elemental Technique

The isotopic technique is usually combined with an elemental profiling technique to increase the accuracy of the traceability methods [157,159,162–165]. Elemental profiling techniques rely on digestion of samples into ions, then concentration of the ions is followed by spectroscopic analysis, including atomic absorption spectroscopy (AAS), inductively coupled plasma–optical emission spectroscopy (ICP-OES), and ICP–mass spectrometry (ICP–MS) [11]. The analyzed elements include K, Ca, Na, Mg, Cu, Fe, Mn, Al, Zn, As, Cd, Cr, Mn, Ni, Zn, Ba, Sr, Li, Se, Co, Ti, and V. Those elements include bulk structural elements (P, S, Si, etc.), macroelements (K, Ca, Na, Mg, etc.), trace elements (Cu, Fe, Zn etc.), and ultratrace elements (As, Cd, Cr, Mn, Ni, Co, V, etc.). Both non-metal elements (P, S, As, etc.) and metals (Mn, Fe, Cu, etc.) have been used in analysis [166]. In recent years, the rare earth elements (REEs), including Y, Ce, Nd, Pr, Sm, Er, and Eu, have also been used in traceability methods [159,167]. Databases generated by chemical analysis are subjected to multivariate analysis for data exploration.

Elemental profiling was used in geographic traceability testing of plant origin products, because element compositions of the specimen were believed to be a distorted reflection of the elemental profiling of the soil environment in which they grew [11]. This fact is more complicated for animals, who derive their elements not only from the environment but also the food they consume. Hence, feed is a factor that needs to be seriously considered in the traceability of animal origin food products. Mineral concentrations of feed, such as fish feed, vary greatly due to differences in raw ingredients, addition of specific macro or trace mineral premixes, or contamination [11]. The culture environment of animals is also more complicated than plants and the elemental profiling of animals can be affected by factors such as the culture season, size of the animal, species, and water quality [160,168,169]. For example, Han and others [168] reported that the element compositions of salmonid obtained from the reservoir were vulnerable to seasonal changes. Although studies have demonstrated the usefulness of elemental profiling in tracing the origin of animal origin products, all factors should be considered in future studies to strengthen the accuracy of the method.

4. Examples of Recent Use of Spectroscopic and Traditional Methods to Detect Fraud

4.1. Fish and Seafood Products

Identification of Geographical Origin: Provenance or geographical origin has become one of the most important authenticity issues for fish and seafood due to the increasing awareness among consumers of the impacts of their purchasing choice of seafood on the marine environment. Many consumers are becoming more worried about fraud, which occurs when fraudsters conceal the geographical origin or hide an illegally harvested protected species or a species from a protected area. Thus, reporting of the country of origin or place of provenance of seafood is essential in the fight to preserve sustainable fisheries, for better management of fish stocks, and to prevent unreported and unregulated fishing. This is why a requirement with respect to a clear indication of the geographical origin of seafood products has been implemented in many countries, such as the European Union and the USA [123,170].

Several analytical methods have been developed in order to identify the origin of seafood. Trace elements fingerprinting, stable isotope analysis, and DNA-based methods are among the most used approaches for this purpose. While these techniques show promise for definitively identifying the geographical origin of fish and other seafood [32,171–173], they have certain drawbacks, especially in terms of the required time and the destructive nature of measurements.

Recently, some studies have demonstrated the usefulness of spectroscopic techniques for monitoring the geographical origin of seafood [174–176] (Table 1). In one of these studies, NIR spectroscopy was applied to classify tilapia fillets according to their 4 geographical origins, namely Guangdong, Hainan, Guangxi, and Fujian in China [174]. SIMCA performed on the spectra showed a classification ability ranging from 75% for the Guangxi provenance to more than 80% for the other origins. In another study, a better classification efficiency of sea cucumber originating from nine Chinese locations was obtained using NIR spectroscopy combined with PCA [175]. More recently, a similar technique was used to trace the geographical origin of European sea bass collected from the Western, Central, or Eastern Mediterranean Sea [176]. Results showed correct classification rates of 100% 88%, and 85% for the fish originating from the Eastern, Central, and Western Mediterranean Seas, respectively, with lipid absorption bands being the major contributor to the discrimination ability of the spectra.

In the literature, there are few studies regarding the use of NMR or fluorescence spectroscopy for monitoring of the geographical origin of seafood. In one of these scarce studies [177], ^{1}H NMR spectroscopy combined with SIMCA and PLS-DA was successfully applied to discriminate caviar cans originating from producers in the Aquitaine region in France from other producers. Therefore, more spectroscopic studies should be conducted on this topic in order to draw valid conclusions about the potential of these techniques for determining the geographical origin of fish and other seafood.

Tracing Wild and Farmed Seafood and Farming Methods: During the last few years, there has been a rapid expansion of aquaculture as a result of overfishing and decreasing wild fish stocks. Consumers generally prefer wild fish over farmed fish, and when it comes to farming, organically farmed fish is usually believed to be healthier and of higher quality in terms of animal welfare and environmental perspectives compared to conventionally farmed fish. This is why labeling farmed fish as wild or conventionally raised fish as organic is considered a fraudulent practice.

Various approaches have been proposed over the years to trace production methods and farming systems. Elemental profiling, stable isotopes, fatty acid analysis, or combinations of these methods have been extensively applied [144,173,178,179]. For example, a technique based on stable isotope analysis allowed differentiation of organically farmed from conventionally farmed salmon and brown trout, independent of the type of processing, i.e., raw, smoked, or graved [180]. In another study, the combination of stable isotope ratio analysis with multielement analysis gave a correct classification of 100% of shrimp samples according to their geographical origin and production method (i.e., wild or farmed), while 93.5% of the samples were correctly classified according to species [163]. A more recent study confirmed the positive effects of combining the stable isotopes and elemental profiling techniques to determine the origin and production method of Asian sea bass collected from Australian and Asian sources [160].

Only a few studies regarding the use of spectroscopic techniques for distinguishing between wild and farmed fish or between different farming regimes have been published so far. Xu and co-authors studied the possibility of discriminating wild and farmed salmon with different geographical origins and farming systems using HSI operating in two spectral ranges (spectral set I: 400–1000 nm; spectral set II: 897–1753 nm) coupled with different chemometric tools [181]. The best results were obtained with SVM applied to spectral set I, giving a correct classification rate of 98.2%. In a more recent study, NIR spectroscopy in the range of 1100–2500 nm was applied to authenticate European sea bass [176]. Slight separation was observed between fish groups when PCA was applied. However, PLS-DA allowed a clear discrimination between wild and farmed fish with a correct classification rate of 100% being achieved. Moreover, the different farming systems, including extensive, semi-intensive, and intensive farming, were discriminated from each other with correct classification rates of 67%, 80%, and 100%, respectively. In this study, the absorption bands of proteins were reported to be the greatest contributors to the discrimination ability of the spectra.

Detection of Species Fraud: Substitution of valuable marine species with less desirable or cheaper ones is the most common type of fraud in fish and other seafood. Detection of this type of fraud is difficult, especially if the fish is in the form of a fillet without skin or if the seafood product has been processed [182,183]. Given the widespread practice of species fraud and the serious consequences it can have, it is no wonder that a wide variety of conventional methods and spectral fingerprinting techniques have been investigated in order to aid in addressing this issue. DNA analysis and MS methods are among the most commonly used techniques in this regard [126,145,146,184,185].

Table 1. Examples of applications of spectroscopic techniques with respect to various authenticity issues for fish and other seafood.

Fish or Other Seafood	Authenticity Issue	Analytical Technique	Modeling Method	Reference
Horse mackerel, European anchovy, red mullet, bluefish, Atlantic salmon, and flying gurnard	Species identification/detection of thawed fish	Raman	PCA	[186]
Pacific white shrimp	Origin authentication	NIR HSI (874–1734 nm)	PLS-DA, LS-SVM, ELM	[187]
Norwegian salmon	Species identification	FT-IR (4000–450 cm^{-1})	PLS-DA	[188]
Carotenoid, salmonid/freshwater, saltwater fishes	Species identification/origin	Raman	HCA	[189]
Seven freshwater fish species	Species identification	NIR (1000–1799 nm)	PCA, LDA	[190]
Fish surimi; white croaker, hairtail, red coat	Species identification	FT-IR (2500–25000 nm)	PCR	[191]
Fish surimi; white croaker, hairtail, red coat	Species identification	NIR (1000–2500 nm)	DA	[192]
Freshwater shrimps	Addition adulterant	LF-NMR, MRI	PCA, PLSR	[193]
Tilapia	Detection of thawed fish	NIR (1000–2500 nm)	PCA	[194]
Crucian carp	Detection of thawed fish	VIS/NIR HSI (400–1000 nm)	PLS-DA	[195]
Grass carp	Detection of thawed fish	VIS/NIR HIS (400–1000 nm)	SIMCA, PLS-DA, LS-SVM, and PNN	[196]
Shelled shrimp	Detection of thawed products	VIS/NIR HSI (400–1000 nm)	SIMCA, RF	[197]

PCA, Principal Component Analysis; PCR, Principle Component Regression; LDA, Linear Discriminant Analysis; DA, Discriminant Analysis; RF, Random Forest; SIMCA, Soft Independent Modeling of Class Analogy; PLS-DA, Partial Least Squares Discriminant Analysis; PLSR, Partial Least Squares Regression; LS-SVM, Least Squares Support Vector Machines; PNN, Probabilistic Neural Network; VIS/NIR; Visible–Near-Infrared Spectroscopy; his, Hyper Spectral Imaging; LF-NMR, Low-Field Nuclear Magnetic Resonance; MRI, Magnetic Resonance Imaging; FT-IR, Fourier-Transform Infrared Spectroscopy; ELM, Extreme Learning Machine; HCA, Hierarchical Cluster Analysis.

Several spectroscopic techniques in conjunction with chemometric tools have been used to identify fish species and detect fraud. Alamprese and Casiraghi used FT-NIR and FT-MIR data coupled with two classification techniques (i.e., SIMCA and linear discriminant analyses (LDA)) in order to discriminate valuable fish species (i.e., red mullet and plaice) substituted with cheaper ones, namely Atlantic mullet and flounder [76]. The best results were obtained by the LDA model, giving a 100% correct classification rate for red mullet and Atlantic mullet, regardless of the used spectroscopic techniques. Regarding discrimination between plaice and flounder species, the best results were obtained using FT-IR, with more than 83% prediction ability and 100% specificity being achieved.

The progress in miniaturization accompanied by software development has led to the emergence of several handheld and portable devices based on spectroscopy for many applications in the food industry [198,199]. In this respect, an investigation based on a handheld NIR device and FT-NIR benchtop spectrometer was carried out in order to discriminate Atlantic cod from haddock fillets and patties [200]. The results obtained by applying LDA and SIMCA models to the spectra using the portable device demonstrated an equivalent discrimination power to those obtained by the stationary benchtop instrument.

Besides NIR spectroscopy, other vibrational spectroscopic techniques have been widely employed to detect fraud in seafood species. For instance, MIR spectroscopy was applied to detect fraud involving substituting Atlantic salmon with rainbow trout in mini-burgers [201]. Using PCA, the authors succeeded in discriminating 11 formulations with different percentages of these two species, and the percentage of the fraud in the mixture was successfully predicted using PLSR. The same authenticity issue (i.e., species identification) was later studied in a similar investigation, but with a different vibrational spectroscopic technique, namely Raman spectroscopy [74].

Again, few or no studies have been found in the literature regarding the application of NMR or fluorescence spectroscopy. A recent study investigated the use of HSI in 4 different spectroscopic modes, including reflectance in the VIS/NIR region, fluorescence, reflectance in the short-wave infrared region, and Raman spectroscopy for discriminating between 6 fish species and differentiating between fresh and frozen–thawed fish [21]. By testing several machine learning classifiers, the authors obtained the best results when using the VIS/NIR and the short-wave infrared techniques for the identification of fish species and detection of thawed fish, respectively.

Checking of Processing Treatments: Fish and other seafood products are highly perishable foods that must be processed or preserved properly and rapidly after catch or harvest in order to extend their shelf life and maintain quality. Freezing has been widely applied as one of the most common ways of achieving this purpose. However, fresh products are often considered by consumers to be of superior quality and are usually sold at higher prices then frozen food. Therefore, discrimination between fresh and frozen products is one of the most important authenticity issues. Enzymatic, electrophoretic, and histological methods have been commonly used to detect thawed fish and seafood [202–205].

Vibrational spectroscopy, NMR, and fluorescence spectroscopy have shown considerable potential as interesting alternatives to traditional measurements used to differentiate fresh from frozen–thawed seafood. For example, differentiation of fresh and frozen–thawed Atlantic mullet fillets was successfully reported with the use of SIMCA applied to FT-IR, with values of more than 98%, 88%, and 95% being obtained for classification ability, prediction ability, and specificity, respectively [76]. Similar results were reported by using PLS-DA on VIS/NIR spectra obtained for fresh and frozen–thawed tuna, and high sensitivity, specificity, and accuracy of the model were achieved [206].

Unlike the other vibrational spectroscopy, very little work has been devoted to examining the potential of Raman spectroscopy to differentiate fresh and frozen–thawed fish. Fat extracted from six fish species, namely horse mackerel, European anchovy, red mullet, bluefish, Atlantic salmon, and flying gurnard, was analyzed using Raman spectroscopy in order to discriminate between fresh, once-frozen–thawed, and twice-frozen–thawed fish [186]. PCA models were developed and displayed a clear discrimination between the 3 states of each fish species, indicating a strong ability of this technique to rapidly detect changes in the lipid structures of fish species compared to gas chromatography, which is usually used in classical analysis.

Although NMR has been widely used to monitor changes in fish occurring during freezing and frozen storage [207], little work has been done regarding the use of this technique to differentiate between fresh and frozen–thawed fish. Recently, NMR was used to deal with freshness authentication of Atlantic salmon by analyzing metabolic changes that occur during the thawing process [19]. A PCA score plot showed distinct fresh and frozen–thawed groupings, while the discrimination ability was attributed to the formation of aspartate in the thawed salmon.

Few studies on fluorescence spectroscopy have been reported in the scientific literature, showing the possibility of the application of this technique to study different authenticity issues in seafood. For instance, the potential of front-face fluorescence spectroscopy was investigated to discriminate between fresh and frozen–thawed sea bass [208]. In this study, four fluorophores were examined, including NADH (excitation at 340 nm), tryptophan (excitation at 290 nm) riboflavin (excitation at 380 nm), and vitamin A (emission set at 410 nm). The results showed that this technique coupled with some appropriate chemometric tools was able to discriminate not only between fresh and frozen–thawed fish, but also between frozen fish of differing quality before freezing and storage.

Many studies have demonstrated the potential use of HSI for various authentication purposes [209]. Discrimination between fresh and frozen–thawed cod fillets was investigated by using VIS/NIR HSI adapted for online measurements of fish fillets moving on a conveyor belt at a speed of 40 cm/s, a rate that meets the industrial production requirements [210]. The results showed that the technique was able to differentiate between both fresh and frozen–thawed cod fillets and between the fillets according to different freezing and thawing protocols as a function of sample freeze–thaw history. In this study, the discrimination ability was attributed to variations in the visible region of the spectrum induced by oxidation of hemoglobin and myoglobin and to scattering changes caused by protein denaturation and other structural modifications during the freezing–thawing processes.

In light of the herein reviewed results, it can be noticed that various spectroscopic methods have tremendous potential for the detection of fraud and verification of several authentication issues in fish and other seafood. Our literature review revealed that the detection of species fraud and thawed fish are the most studied topics, while vibrational spectroscopic techniques, particularly NIR spectroscopy, are the most investigated techniques. Our literature review shows that few spectroscopic studies have been conducted with respect to the determination of geographical origins and detection of the modality of production (capture or aquaculture) of fish and other seafood. The low number of studies regarding authenticity issues, such as geographical origin, may be due to the difficulty associated with modeling variability in the spectra due to challenges related to many factors affecting measurements, such as biological variability, water temperature, and salinity [8,176]. Surprisingly, only a few applications of fluorescence spectroscopy have been reported, although the high sensitivity and specificity of this technique compared to the other spectroscopic techniques is well known. Therefore, fluorescence spectroscopic techniques should be investigated more extensively in future works.

4.2. Meat and Meat Products

Meat Species Adulteration: Meat and meat products can have a wide range of market values, depending on several factors. Among other factors, the biological origin is one of the most relevant. In fact, some animals are considered of greater value because of their renowned organoleptic characteristics; consequently, they have a higher selling price. One of the most common adulterations in meat products is the addition of the flesh of a different animal of a lower market value.

In recent years, a lot of effort has been put into developing non-destructive approaches for detecting meat adulterations. In this regard, the choice has often been spectroscopy, especially infrared spectroscopy, which limits or completely avoids any loss of sample material [27] (Table 2). Among the different flesh used as an adulterant, pork, which undesirable for several reasons [29], is probably one of the most investigated and reported materials in the literature. For instance, Kuswandi and collaborators [211] very successfully exploited FT-IR spectroscopy (equipped with attenuated total reflection cell) to detect porcine meat in beef jerky. In order to achieve this goal, the authors exploited three different classifiers, namely LDA, SIMCA, and SVM, and the best results were provided by LDA, giving a total classification rate of 100%. Beside FT-IR, NIR spectroscopy has also been widely exploited in this regard. For instance, Kuswandi et al. used NIR coupled with PLS-DA to detect pork adulteration in beef meatballs [212]. This approach provided extremely satisfying results, since the optimal classification model detected all the adulterated samples. In a similar study proposed by Rady and Adedeji [213], pork adulteration in minced beef was evaluated by NIR spectroscopy combined with PLS-DA. This research provided slightly lower but very promising results.

After pork, another common adulterant in beef meat is poultry. Several studies have used spectroscopy to detect this kind of adulteration. One example is the work from Deniz and collaborators [214], who demonstrated the possibility of using a fast and non-destructive spectroscopic technique to detect chicken or turkey in beef minced meat. In more detail, adulterated samples of different proportions (5%, 10%, 20%, 40%, and 100%) were prepared and analyzed by FT-IR combined with hierarchical cluster analysis (HCA) and PCA. The data obtained by HCA gave less information than those obtained by PCA, while different spectral bands, especially those of lipids, exhibited noticeable differences between the different meat products (beef, chicken, turkey). A similar study was proposed by Alamprese and collaborators in 2016 [215], who also investigated beef adulteration with turkey, however they inspected fresh, thawed, and cooked meat samples using NIR spectroscopy. Eventually, they used PLS-DA to identify the adulterant and were able to distinguish between samples presenting a low level of adulteration ($<20\%$) and highly adulterated ones ($\geq 20\%$).

HSI has been widely used and has shown promise in overcoming the challenges related to measurements of heterogeneous food matrices, such as muscle foods (meat, fish). For instance, Kamruzzaman et al. used this technique coupled with PCA to detect pork [216] and chicken [217] adulteration in beef. Similarly, HSI was applied to detect fraud in minced beef [218]. The data were preprocessed by MSC and SNV, and the performance of two classification models (SVM and RF) was compared. The best results were obtained using the optimized RF model developed on selected wavelengths, achieving an accuracy of 96.87%.

One of the main advantages of HSI is the possibility to generate a distribution map, allowing the visualization of adulteration levels [14,20]. On the other hand, the data generated from HSI are extremely vast, requiring complex data handling. Multispectral imaging (MSI), however, uses a lower number of spectral bands, thus the acquisition time and complexity of MSI are comparably lower than that of HSI. MSI was successfully used recently in order to detect minced beef adulteration with horsemeat [219]. In this study, the performance of three classification models, namely PLS-DA, RF, and SVM, was explored, and the best results were obtained by the SVM model, giving a correct classification rate of more than 95%.

Beside spectroscopic methods, the traditional ones (e.g., PCR) are still widely used in this field of quality control. For example, Hou et al. used a PCR method to detect different adulterants (duck, chicken, and goose) in pork, beef, and mutton [132]. Similarly, Kim et al. used it to detect undesired donkey meat in mixtures [131]. Several similar studies have been conducted recently for the same purpose [220–222]. Very recently, Yin and co-workers proposed a novel and highly sensitive molecular assay (PCR-based) for the fast revelation of pork components at a concentration of 0.01% in adulterated meat [223]. A relatively novel technique, which is widely used to detect adulterated meat, is DNA barcoding. As an example, Xing et al. successfully exploited DNA barcoding and DNA mini-barcoding to detect mislabeling of several products on the Chinese market [125]. In addition to the previously mentioned approaches, ELISA is another common tool used for species identification in food authentication. For example, it has been used to detect pork-adulterated beef by Mandli and collaborators [141], whereas Perestam et al. compared the performance of the ELISA and of PCR for detecting beef and pork—both approaches have advantages and disadvantages for this purpose [149].

Table 2. Examples of applications of spectroscopic techniques with respect to various authenticity issues in meat and meat products.

Meat and Meat Products	Authenticity Issue	Analytical Technique	Modeling Method	Reference
Bovine meat	Detection of non-meat ingredients	FT-IR (4000–525 cm^{-1})	PLS-DA, data fusion	[52]
Mutton, beef, pork	Species identification	FT-IR (4000–450 cm^{-1})	SVM, PLS-DA	[73]
Porcine, poultry, bovine, ovine	Species identification	FT-IR (4000–550 cm^{-1})	PCA, PLS-DA, and PLS	[224]
Pig	Identification of feeding regime	Portable NIR (900–1700 nm)	LDA, QDA, and non-parametric Bayes	[225]
Beef, lamb, pork	Species identification	FT-NIR (1100–1938 nm)	One-class classifier partial least squares (OC-PLS), SIMCA	[226]
Pig lard	Origin identification	FT-NIR (750–2500 nm)	PLS-DA	[227]
Lamb, beef, pork	Species identification	HSI VIS/NIR (548–1701 nm)	SVM, CNN	[228]
Beef, meat of rat	Species identification	FT-IR (4000–400 cm^{-1})	PCA, PLSR	[229]
Veal sausages, pork	Species identification	Various FT-NIR equipment	PCA, SVM	[230]
Fresh and rotten beef	Meat identification	VIS/NIR HSI (496–1000 nm)	SVM, LS-SVM, PLSR	[231]
Turkey cuts, processed products	Meat identification	VIS/NIR (400–2500 nm)	PCA, LDA	[232]
Lamb, beef	Species identification	NIR (1100–2300 nm)	PCA, PLS-DA	[233]
Duck, beef, pork	Species identification	NIR (12500–5400 cm^{-1})	DA, PLSR	[234]
Beef, pork, beef heart, beef tallow	Species identification	VIS/NIR (350–2500 nm)	SVM, RF, PLSR, DCNN	[235]
Tan mutton	Detection of thawed meat	NIR HSI (900–1700 nm)	PLS-DA	[236]

PCA, Principal Component Analysis; PCR, Principle Component Regression; LDA, Linear Discriminant Analysis; DA, Discriminant Analysis; QDA, Quadratic Discriminant Analysis; RF, Random Forest; SIMCA, Soft Independent Modeling of Class Analogy; PLS-DA, Partial Least Squares Discriminant Analysis; PLSR, Partial Least Squares Regression; LS-SVM, Least Squares Support Vector Machines; VIS/NIR, Visible–Near-Infrared Spectroscopy; HSI, Hyper Spectral Imaging; FT-IR; Fourier-Transform Infrared Spectroscopy; (D)CNN, (Deep) Convolution Neural Networks.

Distinction Between Fresh and Thawed Meat: Beside adulteration with undesired meats, scams concerning meat freshness are unfortunately common, and consequently in the literature it is possible to find different studies aiming to detect this kind of fraudulent action. It is not always easy to discern the freshness of meat by sight, and mislabeling can occur accidentally or intentionally to make illicit profits by selling thawed meat as fresh. Regardless of the reason, it is important to possess suitable tools for the authentication of fresh meat. Once again, in recent years, spectroscopy has played a key role in the detection of this kind of fraud.

One of the meats investigated the most in this context is chicken, mainly because of the few visual differences that differentiate fresh and thawed products. Nevertheless, Grunert and collaborators have suggested that discrimination can be achieved by FT-IR spectroscopy; in fact, in their study they showed the possibility of using this technique coupled with artificial neural networks (ANN) to discern fresh and thawed samples (frozen and stored for time periods from 2 up to 85 days) [237]. The results were extremely satisfying, since twenty samples (of the twenty-one investigated) were correctly classified. A similar study was proposed by Parastar and collaborators, where fresh and thawed chicken samples were analyzed using a portable NIR instrument and then classified by different methods (random subspace discriminant ensemble (RSDE), PLS-DA, ANN, and SVM); the best results were obtained by using RSDE, providing extremely satisfying results with a classification accuracy higher than 95% [18].

Detection of the Geographical Origin and Production Method: The traceability of meat and meat products is relevant from different standpoints; for this reason, several approaches have been proposed to assess the origins of meat samples [238]. Traditionally, meat and meat products are traced by means of protein- and DNA-based methods [239]. An example is a recently published paper by Muñoz and collaborators, who focused on Iberian pork meat, which is used to prepare a Spanish typical cured meat product [240]. The authors proposed a single nucleotide variant genotyping panel suitable for recognizing purebreds (Duroc and Iberian) or crossbreds. Interesting solutions for the origin assessment of edible meats were also provided by means of stable isotope ratio analysis. For instance, Erasmus and co-workers showed that $\delta^{15}N$ and $\delta^{13}C$ can be used to discriminate South-African lamb breeds in diverse regions [241]. These authors related the isotope abundancies to the pedo-climatic conditions of the different areas. A similar study on a diverse animal species was conducted by Monahan et al., who investigated the possibility of using stable isotope ratio analysis to recognize Irish chickens [242]. Further applications can be found in [243].

Despite the tools mentioned above providing noteworthy outcomes, they are time-consuming, destructive, relatively expensive, and require complex sample preparation. During the first decade of this century (2000–2010), a lot of effort has been put into developing fast and non-destructive spectroscopy-based approaches to achieve the same purpose. However, during the last five years, not many novel strategies have been proposed. For example, recently Zhang and co-authors demonstrated that FT-IR spectroscopy integrated with second derivative infrared spectroscopy (SD-IR) and two-dimensional correlation infrared spectroscopy (2DCOS-IR) coupled with computer vision methodologies represent suitable choices for discrimination of different hams produced in three different locations [244].

There are few studies on the potential of spectroscopic techniques for the determination of the production method (dietary background) of meat. One example is a study conducted by Huang and co-authors [245], who applied reflectance spectroscopy in two spectral ranges (400–700 nm and 400–2500 nm) coupled with PLS-DA to discriminate carcasses of lambs reared with 3 feeding regimes, involving perirenal fat from pasture-fed, concentrate-fed, and concentrate-finished after pasture feeding diets. The results demonstrated that the 3 feeding regimes could be distinguished with overall correct classification rates of 95.1% and 99% for the 400–700 nm and 400–2500 nm spectral ranges, respectively.

Other Common Adulterants or Contaminants in Meat: A number of foreign ingredients can be introduced (voluntarily or accidentally) in meat and meat products. Some contaminants can be unintentional, while others are conceived to alter the characteristics of the treated food in order to make it more palatable to the consumers. For instance, the addition of food dyes in meat products is allowed by law, but the types of colorants are strictly regulated; consequently, the possible presence of forbidden dyes has to be checked [243]. Other forms of fraud in meat may involve unwanted or forbidden physical pretreatments, as is the case with irradiation. This practice, which is generally used to extend the shelf-life of food products, is allowed for some foods (for instance dry aromatic herbs) but it is banned for meat. As a consequence, different research studies have been conducted with the aim of developing analytical approaches suitable for the detection of this illicit practice, as in the case discussed by Varrà and co-authors, where irradiated and non-irradiated sausages were discriminated by NIR spectroscopy coupled with orthogonal partial least square–discriminant analysis (OPLS-DA) [246].

One further illegal practice is fraudulent mislabeling, consisting of substituting a high-value cut meat with a cheaper alternative, as in the case reported by Sanz and his group [247]. In their study, the authors investigated four different types of lamb muscles using HSI and discriminated the four diverse categories using seven classifiers. The most accurate outcome was achieved using linear least mean squares, which led to a total correct classification rate of 96.67%.

Only limited research has been found in the literature about the use of fluorescence spectroscopy for studying authenticity issues in meat and meat products. In one of the scarce studies, FFFS combined with chemometric tools (PLS and PLS-DA) was successfully applied to classify three different beef muscles, namely *the semitendinosus*, *rectus abdominis*, and *infraspinatus* muscles [248]. These results were confirmed recently in a similar study [95]; in this study, FFFS achieved better accuracy in discrimination of beef muscles than synchronous fluorescence spectroscopy.

4.3. Milk and Dairy Products

Thanks to its enhanced nutritional value provided by the presence of high-quality protein and minerals, milk is an essential food for people of all ages, from infants to elderly people [249]. Adulteration of milk by the addition of undeclared substances is a widely encountered problem in the dairy industry. Whey, melamine, starch, water, chlorine, formalin, and hydrogen peroxide are the most frequently used adulterants for this type of practice. Mixing milk from different species, replacement of milk fat with non-milk fats or oils, labelling a conventional product as an organic farming product, and false declaration of the processing technology and geographical origin are the other primary fraudulent practices. Several physicochemical methods, liquid and gas chromatography, isotope ratio analysis, and DNA-based techniques have been used for these issues, which involve drawbacks such as having a high cost and being labor-intensive. Spectroscopic techniques (Table 3), being rapid, easy to operate, and applicable to on-line and at-line measurements, as well as providing a high amount of data, are alternatives that can be used to overcome the disadvantages of existing methods [250].

Addition of Non-Declared Substances: Urea, melamine, dicyandiamide, sodium bicarbonate, ammonium sulfate, and sucrose are the most frequently used adulteration agents for milk and dairy products [251,252]. Infrared spectroscopy, FT-MIR, and MIRS have been widely applied to determine raw milk and milk powder adulteration by using waste whey [253,254]. In a comprehensive study by Coitinho et al. [67], the FT-IR MilkoScan FT1 device was calibrated and validated using a large number of raw milk samples. Then, the sensitivity (80–90%) and specificity (80–100%) of the method were designated for adulteration of raw milk with different adulterants. Several NIR spectroscopic methods have been utilized to detect milk and milk powder adulteration [255]. In a recent study, a non-targeted method employing benchtop FT-NIR and portable NIR devices coupled with SIMCA was developed to determine eleven potential adulterants in milk powder. The portable device provided lower sensitivity and specificity due to its lower spectral resolution and narrower spectral range [256].

Table 3. Examples of applications of spectroscopic techniques with respect to various authenticity issues in milk and dairy products.

Milk or Dairy Products	Authenticity Issue	Analytical Technique	Modeling Method	Reference
Yogurt and cheese	Species identification	Front-face fluorescence	PLS-DA and PLSR	[17]
Raw milk	Detection of adulterants	Time Domain NMR	PCA, PLS, and SIMCA	[257]
Milk powder	Detection of adulterants	^{1}H NMR	PCA and Conformity Index	[78]
Ultra-heat-treated bovine milk	Detection of adulterants	^{1}H and 2D NMR	PLS-DA	[258]
Goat milk	Detection of adulterants	FT-NIR (10000–4000 cm^{-1})	PCA, Q-control, k-NN, SIMCA, and PLS-DA	[255]
Milk powder	Detection of adulterants	NIR (850–2499.5 nm)	PLSR	[259]
Dairy cream	Detection of adulterants	Raman spectroscopy	LDA	[260]
Milk	Species identification	2DCOS-SFS	Relative auto-peak intensity	[261]
Milk	Species identification	NIR (700–2500 nm)	PLS-DA	[262]
Raw and pasteurized milk	Species identification	Raman	PLS-DA	[263]
Milk	Identification of geographical origin	MIR (926–3050 cm^{-1})	GA-LDA	[264]
Cow and goat milk	Detection of adulterants	MIR and Raman	PLSR	[265]
Milk	Species identification	FT-IR (1700–600 cm^{-1})	PCA and HCA	[266]

PCA, Principal Component Analysis; LDA; Linear Discriminant Analysis; DA, Discriminant Analysis; SIMCA, Soft Independent Modeling of Class Analogy; PLS-DA, Partial Least Squares Discriminant Analysis; PLSR, Partial Least Squares Regression; ^{1}H NMR, High-Field Nuclear Magnetic Resonance; 2D-NMR, Two-Dimensional Nuclear Magnetic Resonance; FT-IR, Fourier-Transform Infrared Spectroscopy; HCA, Hierarchical Cluster Analysis; (D)CNN, (Deep) Convolution Neural Networks; k-NN, k-Nearest Neighbors; Q-control, Control Chart Q; GA-LDA, Genetic Algorithm Linear Discriminant Analysis; 2DCOS-SFS, Synchronous Fluorescence Spectroscopy coupled with Two-Dimensional Correlation Spectroscopy.

Raman spectroscopy is another vibrational spectroscopic technique that has been widely investigated for adulteration purposes. For example, a portable Raman spectrometer was employed to detect melamine, dicyandiamide, urea, ammonium sulfate, and sucrose adulteration of milk. The standard error of prediction and relative standard deviation values were 39 to 72 ppm and 8% for nitrogen-rich compounds, and 1400 ppm and 10% for sucrose, respectively. The selectivity and efficiency values were 100% for the PLS-DA model in discriminating pure milk samples from adulterated ones [267]. The obtained results were found to be comparable with those of a previous study of the same group, in which a Raman microprobe system was employed [268]. Considering the high-throughput Raman chemical-imaging-based method, it was possible to visualize the spatial distributions of melamine and urea in milk powder and quantify these at the 50 ppm level [82]. Moreover, vegetable oils that were fraudulently added to dairy cream and yogurt were detected by Raman spectroscopy [70,260]. Finding alternative sample preparation procedures is an essential point to be highlighted for efficient Raman spectroscopic analysis in milk and dairy products. Nedeljković et al. [269] performed a preheating process to butter and margarine samples before Raman measurements. In a recent study, the successful use of a portable Raman spectrometer to assess lard adulteration in butter was reported. Samples were melted and mixed thoroughly prior to the Raman measurements [69]. Lohumi et al. developed a line scan spatially offset Raman spectroscopy (SORS) technique that can collect data from packaged butter and margarine samples [270].

Detection of Species Fraud: Successful discrimination and quantification of milk from undeclared species have been carried out using infrared spectroscopy [271]. Equivalent promising results were reported with Raman spectroscopy [272]. Nonetheless, it is important to emphasize the fluorescence interference problem during Raman spectroscopy measurements, especially with 532 nm lasers. Studies employing lasers with different wavelengths (e.g., 785 and 1064 nm) have extended the use of this technique for milk and dairy product analyses.

There have been very few studies in the literature reporting the use of NMR for the determination of adulteration. Nonetheless, one study succeeded in discriminating soymilk, bovine milk, goat milk, and their adulterants after coupling chemometrics and metabolite analysis using 1D- and 2D-NMR, with limit of quantification values ranging between 2% and 5% [273]. Some other studies highlighted the changing sensitivity and specificity of the ^{1}H time-domain NMR (TD-NMR) method, depending on the used adulterant [81,257].

The identification of milk species by employing different measurement techniques involving fluorescence spectroscopy has been studied by several authors [16,274]. Boukria et al. [261] highlighted that cow milk adulteration in camel milk could be detected through the application of the two-dimensional correlation spectroscopy method on SFS spectra. Inclusion of a higher number of samples in the calibration model and scanning of a more comprehensive wavelength range were emphasized as determinant factors in obtaining satisfying discrimination results.

The successful use of several DNA-based analytical methods has been reported for milk authentication and traceability in the dairy sector [275]. In recent applications, entirely satisfactory limit of detection values were achieved [276,277]. Efforts have been made to develop low-cost and user-friendly PCR devices with accuracy and stability comparable to commercial alternatives [278]. Commercial PCR-based assays designed for the detection and quantitative authentication of animal species in a specific dairy product are also available in the market [279,280].

Identification of Geographical Origin and Production Method: Over the last five years, various studies have been reported regarding the authentication of Mozzarella di Bufala Campana Protected Designation of Origin (MBC-PDO) cheese. For example, to combat fraud, Bontempo et al. [281] have successfully proposed the use of the stable isotope method combined with elemental analysis to differentiate both milk and cheese products produced in the PDO area from other products produced outside the PDO area. In another study, Salzano et al. [282] demonstrated that it was possible to distinguish MBC-PDO milk and cheese from non-MBC-PDO products using an advanced GC-MS method and metabolite identification.

Concerning spectroscopic techniques, most of the reported studies were performed in the infrared wavelength range. In more detail, Caredda et al. [264] showed that MIR correctly identified 99% of the ewe's milk from different geographical regions. In another study, Liu et al. [283] conducted a study to assess the interest in a portable micro-NIR spectrometer to discriminate organic milk from pasture and conventional milk. It was shown that the micro-NIRS device could distinguish between organic and conventional milk as efficiently as the FT-NIRS device (i.e., laboratory device).

The abovementioned studies prove how frequently spectroscopic techniques are used to detect adulteration of milk and dairy products. Nonetheless, there is an imbalance in use between the different available spectroscopic techniques. Vibrational spectroscopy has been clearly the most preferable applied method used to detect and identify the most common adulterants in milk. However, more studies comparing the performance of NIR, MIR, and Raman spectroscopy for detecting adulteration of milk samples are necessary. Based on the existing literature, it can be noticed that Raman spectroscopy has particular potential for use for routine analysis of milk and dairy products. However, there is still a need for further studies investigating the simultaneous use of adulterants and extending the scope by developing novel untargeted approaches. Regarding the identification or authentication of milk and dairy products based on their geographical origin and processing treatments, surprisingly only a few studies were conducted during the last five years using spectroscopic techniques. This conclusion is similar to that discussed above for fish and meat products. Thus, the use of spectroscopic techniques for differentiation of fresh and frozen–thawed milk and dairy products and investigation of the effects of the applied processes (milk preparation, cheese processing, etc.) or storage conditions that are important for compliance with specifications (such as PDO, protected geographical indication, etc.) are some of the issues that need to be further studied.

4.4. Honey and Other Products of Animal Origin

Honey is a natural sweet product made by bees from the nectar of plants or plant excretions combined with bees' own specific substances and maturated in the honeycomb. The characteristic flavor, nutritional value, and health benefits of honey depend on its origin and production methods. As a high-quality food product with a high price, honey is often subjected to fraudulent practices, which include mislabeling and adulteration. Development of methods for assessing honey authenticity is of interest to consumers, the honey industry, and food law agencies. Several papers have reviewed the methods used for honey analysis [30,284–288].

Botanical Origin: The price of honey strictly depends on its botanical origin. According to botanical origin, honey is classified as unifloral, multifloral (polyfloral), and honeydew [30]. The monofloral honeys are often more expensive than multifloral honeys and are subject to mislabeling or adulteration with cheaper honeys [289].

The most used conventional method for determining honey quality related to its origin is melissopalynological analysis based on the identification and quantification of pollen grains in honey sediment [30]. The physicochemical (profiles) parameters, such as sugars, moisture, proline, and hydroxymethylfurfural (HMF) contents; acidity; electrical conductivity; diastase; and invertase activity are used to establish the origin of a honey. Analytical techniques including gas and liquid chromatography are often used to measure markers of honey origin, such as sugar, phenolic compounds, and flavor compounds. The profiling techniques, stable isotope ratio, and trace element analysis can provide an indication of the geographical origin of honey. The identification of plant species and varieties of honey by DNA fingerprinting is also utilized to assess honey origin.

Spectroscopic techniques have shown considerable potential as rapid and often non-destructive methods used to study the authenticity of honey. In recent years, several studies have demonstrated the potential use of various spectroscopic techniques for evaluation of the botanical origin of honeys (Table 4). For example, NIR spectroscopy and chemometrics were applied to palynological and mineral characteristics of honey collected from Northwestern Spain [290]. Prediction models using a modified PLSR for the main pollen types (Castanea, Eucalyptus, Rubus, and Erica) in honeys and their mineral compositions were established. The ratio of performance to deviation exhibited a good prediction capacity for Rubus pollen and for Castanea pollen, whereas these ratios were excellent for minerals, Eucalyptus pollen, and Erica pollen.

The benefit of data fusion obtained using different analytical techniques was demonstrated for classification tasks of honey according to the botanical origin. The honey samples from three different botanical origins were analyzed by attenuated total reflection IR spectroscopy (ATR/FT-IR) and headspace gas chromatography–ion mobility spectrometry (HS-GC-IMS) [291]. The obtained datasets were combined in a low-level data fusion approach with subsequent multivariate classification by principal component analysis–linear discriminant analysis (PCA-LDA) or PLS-DA. The results showed that data fusion is an effective strategy for improving the classification performance.

Raman spectroscopy techniques complement information obtained from infrared spectral data and can be used in honey authenticity assessment [287]. Raman spectroscopy, performed using fiber optics, was successfully used to distinguish the botanical origin of unifloral (chestnut, citrus, and acacia) honeys produced in the Italian region of Calabria [292]. Moreover, predictive models were built to quantify important marker indicators in nutraceuticals, such as the main sugars, potassium, and selected sensory properties.

A promising quick, automatic, and non-invasive approach for honey botanical origin classification was developed using a combination of VIS/NIR hyperspectral imaging and machine learning, namely SVM and k-NN [24]. The developed techniques include noisy band elimination, spectral normalization, and hierarchical classification. The proposed model showed promising results under several classification scenarios, achieving high classification performances.

The blending of expensive (pure and rare) honey with a cheaper (pure and plentiful) one is another form of honey adulteration. NMR spectroscopy allows the rapid detection of adulterants in honey, as well as the simultaneous quantification of various chemical compounds from a spectrum [287]. For example, ^{1}H NMR spectroscopy combined with chemometric techniques was applied to detect and quantify adulteration of acacia honey with cheaper rape honey [293]. The highest prediction accuracy for rape honey addition of −89.7% was obtained using canonical discriminant analysis (CDA), determined from compounds located in the spectral range corresponding to the aliphatic compounds and carbohydrates (3.00–6.00 ppm). Orthogonal projection to latent structure discriminant analysis (OPLS-DA) was used to further discriminate samples of pure acacia honey adulterated with different amounts of rape honey. A PLSR model established a linear fit between the actual and predicted adulterant concentrations, with an R^2 value of up to 0.9996.

The fluorescence of honey originates from several groups of compounds, such as amino acids, proteins, phenolic acids, vitamins, fluorescent Maillard reaction products, and other bioactive molecules [23,102,294]. Few studies have demonstrated the potential of fluorescence for authenticity assessment. Fluorescence spectroscopy in EEM mode coupled with parallel factor analysis (PARAFAC) and PLS-DA was applied for classification of honey samples of different botanical origin, including acacia, sunflower, linden, meadow, and fake honey [100]. The classes of honey of different botanical origin were differentiated mainly by emissions from phenolic compounds and Maillard reaction products. PLS-DA constructed from the PARAFAC model provided detection of fake honey samples with 100% sensitivity and specificity. Moreover, PLS-DA classification results gave errors of only 0.5% for linden, 10% for acacia, and about 20% for both sunflower and meadow mixes.

Table 4. Examples of applications of spectroscopic techniques with respect to various authenticity issues of honey.

Honey	Authenticity Issue	Analytical Technique	Modeling Method	Reference
Acacias, lindens, sunflowers, and meadow mixes	Identification of fake honey produced by feeding of bee colonies with a sucrose solution	Fluorescence	LDA	[99]
Honey of various botanical origins, collected from different parts of Ethiopia	Identification of botanical origin	Fluorescence	SIMCA	[101]
Commercial honey from two different provinces of Ecuador	Adulteration	Raman	SIMCA	[295]
Acacia honey	Adulteration of acacia honey with cheaper rape honey	^{1}H NMR	CDA, OPLS-DA	[294]
Honey samples (Vitex, Jujube, and Acacia)	Identification of botanical origin	Electronic nose, electronic tongue, NIR, and MIR	PLS-DA, SVM, iPLS	[296]
South African honey	Differntiation between authentic South African and imported or adulterated honey	NIR	PLS-DA	[297]
Honey samples from the Granada Protected Designation of Origin (Spain)	Quantification of the level of adulteration	VIS/NIR	HCA, PCA, LDA, PLS	[298]
High-quality honey (Granada Protected Designation of Origin, Spain)	Identification and quantification of different types of adulterants (inverted sugar, rice syrup, brown cane sugar, and fructose syrup)	VIS/NIR	HCA, PCA, LDA, PLS	[299]
Honey samples belonging to seven different varieties	Identification of botanical origin	FT-NIR HPLC-DAD	PLS-DA	[300]

PCA, Principal Component Analysis; LDA, Linear Discriminant Analysis; SIMCA, Soft Independent Modeling of Class Analogy; PLS-DA, Partial Least Squares Discriminant Analysis; PLSR, Partial Least Squares Regression; SVM, Support Vector Machines; VIS/NIR, Visible–Near-Infrared Spectroscopy; NMR, Nuclear Magnetic Resonance; FT-IR, Fourier-Transform Infrared Spectroscopy; HCA, Hierarchical Cluster Analysis; CDA, Canonical Discriminant Analysis; OPLS-DA, Orthogonal Projection to Latent Structure Discriminant Analysis; iPLS, Interval Partial Least Squares; HPLC-DAD, High-Performance Liquid Chromatography with Diode Array Detection.

Adulteration Detection: Honey is a natural product for which the addition of any other substance is prohibited by international regulations. However, due to its high economic value, it is often subject to adulteration. The most common adulterants in honey are sugars from high-fructose corn syrup, corn sugar syrup, inverted sugar syrup, and cane sugar syrup [287]. Adulteration of honey is not limited to direct addition of sugars into natural honey. A common fraudulent practice is overfeeding of bees with concentrated sugar solutions during the main nectar flowing season [30]. Among analytical methods, spectroscopic techniques have become popular for detecting the adulterants in honey [287].

FT-IR and PLSR were utilized for the determination of sucrose syrup adulteration of Turkish honeys [301]. The results indicated that the predicted sucrose concentration of honey samples by the spectroscopic method ranged between 4.52 and 15.16%, and that the obtained results were confirmed by chromatography. Several studies reported successful applications of NIR or VIS/NIR spectroscopy for evaluation of honey adulteration. For example, NIR spectra (1300–1800 nm) recorded with a fiber optic immersion probe were used for the detection of high-fructose corn syrup in four artisanal Robinia honeys [302]. The PLSR models developed using the spectral region containing absorption bands related to both water and carbohydrates allowed accurate (root mean squared error of cross-validation; RMSECV = 1.48; R^2_{CV} = 0.987) detection of the adulterant concentration. Recently, NIR and MIR spectroscopy coupled with SVM and data fusion were utilized to detect adulteration of 20 common honey types from 10 provinces in China [303]. Both pure honey and adulterated samples with different percentages of syrup were analyzed. Compared to low-level data fusion, intermediate-level data fusion significantly improved the detection model, achieving 100% accuracy, sensitivity, and specificity.

Fluorescence excitation–emission spectroscopy was effectively used for the non-destructive and fast detection of fake honey samples obtained during winter feeding of bee colonies with a sucrose solution [99]. Natural honey samples (acacias, lindens, sunflowers, and meadow mixes) were perfectly discriminated from fake honey samples using the developed LDA model. Natural and adulterated honey samples differed significantly in five spectral regions corresponding to aromatic amino acids, phenolic compounds, furosine, and Maillard reaction products.

Eggs are consumed worldwide and are well known as a source of vitamins, minerals, phospholipids, and high-quality proteins. EU regulation classifies egg production into four hen housing systems, including 0 for organic production, 1 for free range, 2 for barns, and 3 for cages. Consumers are willing to pay higher prices for eggs produced in a way that considers animal welfare [304], and chicken eggs are often a subject of food fraud. Therefore, there is a need for analytical methods that are suitable for classifying eggs and for detecting the fraudulent mislabeling of eggs obtained from different production systems.

Various procedures are used to discriminate eggs, including carotenoid profiling, fatty acid composition, and mineral content procedures. Eggs from various systems (1-, 2-, and 3-coded eggs) may be discriminated through fluorescent patterns on egg surfaces or stable nitrogen isotope compositions. Stable isotopes methods were used to develop authentication criteria of eggs laid under cage, barn, free range, and organic farming regimens [305]. Recently, discrimination of selected chicken eggs in China's retail market based on multielement and lipidomic analyses was reported [306].

UV-VIS/NIR spectroscopy and chemometrics were utilized for a complete detection of the housing systems declared on the eggs' label [307]. Eggs were perfectly classified into the four housing systems by applying quadratic discriminant analysis for UV-VIS/NIR spectra of the yolk lipid extracts. NMR spectroscopy was successfully utilized as a tool to screen eggs according to the different systems of husbandry [304]. In this study, ^{1}H NMR of freeze-dried egg yolk samples were analyzed using PCA followed by a linear discriminant analysis (PCA-LDA). The prediction model allowed for the correct classification of about 93% of the organic eggs, barn eggs, and free range eggs.

5. Challenges and Future Trends

Even though extensive research regarding the authenticity and detection of fraud by on-site and real-time approaches has been carried out in recent years, several key challenges still remain concerning both technique-related issues and the model validation framework.

Regardless of the non-destructive approach considered, the correct sampling procedure is pivotal to provide valuable information, and thus to embrace the complexity of modern food authentication [308]. Indeed, non-destructive approaches include non-targeted methods (i.e., fingerprint techniques) with the ability to detect multiple small modifications in the considered food product and to extract these modifications as relevant information using the proper multivariate statistic approach. However, the database used to address the authentication issue should consider the main sampling-related criteria, such as the definition of the sample unit, number of samples, sample variability, handling procedure, representativeness, and so on. The most important considerations that must be addressed when creating a food authenticity database are discussed in the position paper by Donarski and co-authors [309]. These issues are highly relevant, as the database is used to define an "authentication rule", which is applied to compare the unknown sample fingerprint with those of authentic reference samples [308]. Even though the creation of the foodstuff-specific database was done considering the perfect sampling procedure and can quickly cover the variability expected from test samples, continuous maintenance of the database is needed to ensure long-term ability to return reproducible results, and most of the scientific publications do not meet this requirement.

Once the authentication issue has been defined and the database creation has been designed accordingly, consideration needs to be given to the definition of a standard operating procedure (SOP) from the sample preparation to the analytical protocol.

DNA-based methods, protein-based methods, and isotopic techniques require specific consideration when defining SOP. Indeed, in these cases, the required analytical steps for sample preparation highly influence the results and their interpretation [2]. As for any analytical technique, different experimental factors can influence the obtained results, introducing an analytical deviation that is not related to the authentication issue under study. These deviations should be reduced to the lowest terms and controlled to ensure that they do not introduce confounding results in the analysis [309]. The influence of experimental factors cannot be avoided, even in spectroscopic technologies (e.g., vibrational spectroscopy, NMR, and fluorescence spectroscopy), despite being reproducible and barely influenced by changes in sensitivity over time. Indeed, they do not generally require any sample preparation, guaranteeing long-term stability and online or in-line application along the production chain. This is particularly true for liquid "homogeneous" samples, whereas solid heterogeneous products, such as meat, fish, and dairy products, may require moderate sample preparation or multiple point measurements. Moreover, the choice of the proper acquisition mode is fundamental to obtain reliable spectroscopic results according to the nature of the food product, including the type of radiation (NIR, IR, NMR, or fluorescence spectroscopy), sample presentation (transmission, absorbance, reflectance, excitation or emission fluorescence, synchronous fluorescence, EEM), type of sample holder (cuvette, fiber probe, attenuated reflectance holder, integration sphere), and working temperature, among others.

Actually, HSI technologies are a valid alternative to point spectral scanning, whereby the spatial distribution of components in heterogeneous products can be distinguished using site-to-site spectroscopic fingerprint specificity. Food quality and authenticity, especially referring to meat products, have been widely investigated by HSI technologies associated with NIR radiation. However, most of the reported works are feasibility studies at the laboratory scale, whereas there is a lack of studies proving the model's robustness at the processing plant level. Furthermore, the huge disadvantages of HSI technology are related to the large amount of produced data for each single measure and the relatively long processing times for these data. However, simplified instruments (multispectral imaging systems) developed for specific applications could reduce the spectral range to be scanned to a few selected wavelengths, thus minimizing both the acquisition time and generated data, which could be managed quickly with the proper ad hoc chemometric method [310]. Simplified, miniaturized, and portable instruments have been developed for the whole spectroscopic field, which are oriented toward food authentication [311]. Certainly, the performance of these instruments in terms of the electromagnetic range covered, resolution, signal-to-noise ratio, specificity, and sensitivity is lower if compared to the results obtained by benchtop instruments [198]. However, their use for ad hoc authentication purposes and their combination with robust chemometric algorithms for classification applications are expected to be major trends in the coming years.

As described in Section 2, multivariate data analysis is the fundamental step taken to produce a model able to classify samples as authentic or non-authentic from any emerging detection method result. No matter the algorithm used to solve an authentication issue, robust validation of the model is mandatory to guarantee reliable and reproducible results and to favor the acceptance of these methodologies in legislation. This theme is quite contentious, and it is one of the major reasons for the refusal of emerging detection methods, along with the standardization procedures [2,60]. Although several attempts have been made to meet the need for common and reliable validation protocols, there is still a lack of validation programs for method developers, which is also reflected in the scientific literature. In the paper by Oliveri [46], a detailed analysis of the key aspects of model evaluation is discussed. This paper could be a landmark when defining a global workflow to solve an authentication issue using spectroscopic techniques.

Thus, it is undeniable that spectroscopic techniques have enormous advantages over the targeted approaches when addressing a food authentication issue; however, their wide application outside of laboratories remains challenging. Meeting these challenges will align emerging spectroscopic methods with the needs of food fraud risk management systems, paving the way for their use for food integrity assurance, such as with the EU-wide Rapid Alert System for Food and Feed (RASFF).

6. Concluding Remarks

This paper has reviewed and discussed papers published in the last 5 years on the use of different analytical methods used to target issues related to fraud in both food and products of animal origin. The available literature in the field has shown an increase in the number of applications combining rapid analytical methods (e.g., DNA analysis, vibrational spectroscopy) with modern data analytics (e.g., multivariate data analysis). The body of research as a whole presents indisputable evidence that these methods and techniques have enormous advantages over other approaches when addressing food authentication. However, several challenges still exist related to the wide application and implementation of these technologies in both research and commercial laboratories. This calls for the need for a continuous exchange between the food authentication stakeholders, together with the growth of a new generation of scientists able to work in both academic and industrial environments and who are skilled in facing all aspects of food authentication using non-targeted techniques.

Author Contributions: Conceptualization, methodology, writing—original draft preparation, A.H.; writing—original draft preparation, I.M.; writing—original draft preparation, revision, W.F.S.; writing—original draft preparation, H.T.T.; writing—original draft preparation, L.L.; writing—original draft preparation, H.-Y.K.; project administration, supervision, manuscript revision, H.N.; writing—original draft preparation, A.B.; writing—original draft preparation, A.A.-K.; writing—original draft preparation, M.S.; writing—original draft preparation, E.S.; writing—original draft preparation, S.G.; writing—original draft preparation, D.C. All authors have read and agreed to the published version of the manuscript.

Funding: This research received no external funding.

Acknowledgments: This work was supported by the Norwegian Institute of Food, Fisheries, and Aquaculture Research (Nofima) through a Strategic Research Initiative (Spectec Project): Rapid and Non-Destructive Measurements to Enable Process Optimization.

Conflicts of Interest: The authors declare no conflict of interest.

References

1. Danezis, G.P.; Tsagkaris, A.S.; Camin, F.; Brusic, V.; Georgiou, C.A. Food authentication: Techniques, trends & emerging approaches. *TrAC-Trends Anal. Chem.* **2016**, *85*, 123–132. [CrossRef]
2. McGrath, T.F.; Haughey, S.A.; Patterson, J.; Fauhl-Hassek, C.; Donarski, J.; Alewijn, M.; van Ruth, S.; Elliott, C.T. What are the scientific challenges in moving from targeted to non-targeted methods for food fraud testing and how can they be addressed?—Spectroscopy case study. *Trends Food Sci. Technol.* **2018**, *76*, 38–55. [CrossRef]
3. Delpiani, G.; Delpiani, S.M.; Deli Antoni, M.Y.; Covatti Ale, M.; Fischer, L.; Lucifora, L.O.; Díaz de Astarloa, J.M. Are we sure we eat what we buy? Fish mislabelling in Buenos Aires province, the largest sea food market in Argentina. *Fish. Res.* **2020**, *221*, 105373. [CrossRef]
4. Sotelo, C.G.; Velasco, A.; Perez-Martin, R.I.; Kappel, K.; Schröder, U.; Verrez-Bagnis, V.; Jérôme, M.; Mendes, R.; Silva, H.; Mariani, S.; et al. Tuna labels matter in Europe: Mislabelling rates in different tuna products. *PLoS ONE* **2018**, *13*, 1–12. [CrossRef] [PubMed]

5. Cawthorn, D.M.; Baillie, C.; Mariani, S. Generic names and mislabeling conceal high species diversity in global fisheries markets. *Conserv. Lett.* **2018**, *11*, 1–12. [CrossRef]
6. Robson, K.; Dean, M.; Brooks, S.; Haughey, S.; Elliott, C. A 20-year analysis of reported food fraud in the global beef supply chain. *Food Control* **2020**, *116*, 107310. [CrossRef]
7. Fiorino, G.M.; Garino, C.; Arlorio, M.; Logrieco, A.F.; Losito, I.; Monaci, L. Overview on Untargeted Methods to Combat Food Frauds: A Focus on Fishery Products. *J. Food Qual.* **2018**, *2018*. [CrossRef]
8. Ghidini, S.; Varrà, M.O.; Zanardi, E. Approaching Authenticity Issues in Fish and Seafood Products by Qualitative Spectroscopy and Approaching Authenticity Issues in Fish and Seafood Products by Qualitative Spectroscopy and Chemometrics Products by Qualitative Spectroscopy and Chemomet. *Molecules* **2019**, *24*, 1812. [CrossRef]
9. Leal, M.C.; Pimentel, T.; Ricardo, F.; Rosa, R.; Calado, R. Seafood traceability: Current needs, available tools, and biotechnological challenges for origin certification. *Trends Biotechnol.* **2015**, *33*, 331–336. [CrossRef]
10. Creydt, M.; Fischer, M. Food authentication in real life: How to link nontargeted approaches with routine analytics? *Electrophoresis* **2020**, 1–15. [CrossRef]
11. Li, L.; Boyd, C.E.; Sun, Z. Authentication of fishery and aquaculture products by multi-element and stable isotope analysis. *Food Chem.* **2016**, *194*, 1238–1244. [CrossRef] [PubMed]
12. Medina, S.; Pereira, J.A.; Silva, P.; Perestrelo, R.; Câmara, J.S. Food fingerprints—A valuable tool to monitor food authenticity and safety. *Food Chem.* **2019**, *278*, 144–162. [CrossRef] [PubMed]
13. Hassoun, A.; Heia, K.; Lindberg, S.; Nilsen, H. Spectroscopic Techniques for Monitoring Thermal Treatments in Fish and Other Seafood: A Review of Recent Developments and Applications. *Foods* **2020**, *6*, 767. [CrossRef] [PubMed]
14. Jiang, H.; Cheng, F.; Shi, M. Rapid identification and visualization of jowl meat adulteration in pork using hyperspectral imaging. *Foods* **2020**, *9*, 154. [CrossRef] [PubMed]
15. Rady, A.; Adedeji, A.A. Application of Hyperspectral Imaging and Machine Learning Methods to Detect and Quantify Adulterants in Minced Meats. *Food Anal. Methods* **2020**, *13*, 970–981. [CrossRef]
16. Genis, D.O.; Sezer, B.; Bilge, G.; Durna, S.; Boyaci, I.H. Development of synchronous fluorescence method for identification of cow, goat, ewe and buffalo milk species. *Food Control* **2020**, *108*. [CrossRef]
17. Genis, D.O.; Bilge, G.; Sezer, B.; Durna, S.; Boyaci, I.H. Identification of cow, buffalo, goat and ewe milk species in fermented dairy products using synchronous fluorescence spectroscopy. *Food Chem.* **2019**, *284*, 60–66. [CrossRef]
18. Parastar, H.; van Kollenburg, G.; Weesepoel, Y.; van den Doel, A.; Buydens, L.; Jansen, J. Integration of handheld NIR and machine learning to "Measure & Monitor" chicken meat authenticity. *Food Control* **2020**, *112*, 107149. [CrossRef]
19. Shumilina, E.; Møller, I.A.; Dikiy, A. Differentiation of fresh and thawed Atlantic salmon using NMR metabolomics. *Food Chem.* **2020**, *314*. [CrossRef]
20. Jiang, H.; Wang, W.; Zhuang, H.; Yoon, S.C.; Yang, Y.; Zhao, X. Hyperspectral imaging for a rapid detection and visualization of duck meat adulteration in beef. *Food Anal. Methods* **2019**, *12*, 2205–2215. [CrossRef]
21. Qin, J.; Vasefi, F.; Hellberg, R.S.; Akhbardeh, A.; Isaacs, R.B.; Yilmaz, A.G.; Hwang, C.; Baek, I.; Schmidt, W.F.; Kim, M.S. Detection of fish fillet substitution and mislabeling using multimode hyperspectral imaging techniques. *Food Control* **2020**, *114*. [CrossRef]
22. Edwards, K.; Manley, M.; Hoffman, L.C.; Beganovic, A.; Kirchler, C.G.; Huck, C.W.; Williams, P.J. Differentiation of South African game meat using near-infrared (NIR) spectroscopy and hierarchical modelling. *Molecules* **2020**, *25*, 1845. [CrossRef] [PubMed]
23. Parri, E.; Santinami, G.; Domenici, V. Front-face fluorescence of honey of different botanic origin: A case study from Tuscany (Italy). *Appl. Sci.* **2020**, *10*, 1776. [CrossRef]
24. Noviyanto, A.; Abdulla, W.H. Honey botanical origin classification using hyperspectral imaging and machine learning. *J. Food Eng.* **2020**, *265*, 109684. [CrossRef]
25. Katerinopoulou, K.; Kontogeorgos, A.; Salmas, C.E.; Patakas, A.; Ladavos, A. Geographical origin authentication of agri-food products: A review. *Foods* **2020**, *9*, 489. [CrossRef]
26. Krajnc, B.; Bontempo, L.; Luis Araus, J.; Giovanetti, M.; Alegria, C.; Lauteri, M.; Augusti, A.; Atti, N.; Smeti, S.; Taous, F.; et al. Selective Methods to Investigate Authenticity and Geographical Origin of Mediterranean Food Products. *Food Rev. Int.* **2020**, *00*, 1–27. [CrossRef]

27. Kumar, Y.; Chandrakant Karne, S. Spectral analysis: A rapid tool for species detection in meat products. *Trends Food Sci. Technol.* **2017**, *62*, 59–67. [CrossRef]
28. Vlachos, A.; Arvanitoyannis, I.S.; Tserkezou, P. An Updated Review of Meat Authenticity Methods and Applications. *Crit. Rev. Food Sci. Nutr.* **2016**, *56*, 1061–1096. [CrossRef]
29. Zia, Q.; Alawami, M.; Mokhtar, N.F.K.; Nhari, R.M.H.R.; Hanish, I. Current Analytical Methods for Porcine Identification in Meat and Meat Products. *Food Chem.* **2020**, *324*, 126664. [CrossRef]
30. Mădaş, M.N.; Mărghitaş, L.A.; Dezmirean, D.S.; Bobiş, O.; Abbas, O.; Danthine, S.; Francis, F.; Haubruge, E.; Nguyen, B.K. Labeling Regulations and Quality Control of Honey Origin: A Review. *Food Rev. Int.* **2020**, *36*, 215–240. [CrossRef]
31. Lytou, A.E.; Panagou, E.Z.; Nychas, G.J.E. Volatilomics for food quality and authentication. *Curr. Opin. Food Sci.* **2019**, *28*, 88–95. [CrossRef]
32. Lo, Y.-T.; Shaw, P.-C. DNA-based techniques for authentication of processed food and food supplements. *Food Chem.* **2018**, *240*, 767–774. [CrossRef] [PubMed]
33. El Sheikha, A.F.; Montet, D. How to Determine the Geographical Origin of Seafood? *Crit. Rev. Food Sci. Nutr.* **2016**, *56*, 306–317. [CrossRef] [PubMed]
34. Valand, R.; Tanna, S.; Lawson, G.; Bengtström, L. A review of Fourier Transform Infrared (FTIR) spectroscopy used in food adulteration and authenticity investigations. *Food Addit. Contam.-Part A Chem. Anal. Control. Expo. Risk Assess.* **2020**, *37*, 19–38. [CrossRef] [PubMed]
35. Cocchi, M. Chemometrics for Food Quality Control and Authentication. In *Encyclopedia of Analytical Chemistry*; John Wiley & Sons, Ltd.: Chichester, UK, 2017; pp. 1–29.
36. Ropodi, A.I.; Panagou, E.Z.; Nychas, G.J.E. Data mining derived from food analyses using non-invasive/non-destructive analytical techniques; determination of food authenticity, quality & safety in tandem with computer science disciplines. *Trends Food Sci. Technol.* **2016**, *50*, 11–25. [CrossRef]
37. Oliveri, P.; Simonetti, R. Chemometrics for Food Authenticity Applications. *Adv. Food Authent. Test.* **2016**, 702–728. [CrossRef]
38. Jiménez-Carvelo, A.M.; González-Casado, A.; Bagur-González, M.G.; Cuadros-Rodríguez, L. Alternative data mining/machine learning methods for the analytical evaluation of food quality and authenticity—A review. *Food Res. Int.* **2019**, *122*, 25–39. [CrossRef]
39. Oliveri, P.; Malegori, C.; Simonetti, R.; Casale, M. The impact of signal pre-processing on the final interpretation of analytical outcomes—A tutorial. *Anal. Chim. Acta* **2019**, *1058*, 9–17. [CrossRef]
40. Engel, J.; Gerretzen, J.; Szymańska, E.; Jansen, J.J.; Downey, G.; Blanchet, L.; Buydens, L.M.C. Breaking with trends in pre-processing? *TrAC Trends Anal. Chem.* **2013**, *50*, 96–106. [CrossRef]
41. Roger, J.; Biancolillo, A.; Marini, F. Sequential preprocessing through ORThogonalization (SPORT) and its application to near infrared spectroscopy. *Chemom. Intell. Lab. Syst.* **2020**, *199*, 103975. [CrossRef]
42. Skogholt, J.; Liland, K.H.; Indahl, U.G. Baseline and interferent correction by the Tikhonov regularization framework for linear least squares modeling. *J. Chem* **2018**, 1–18. [CrossRef]
43. Wahl, J.; Sjödahl, M.; Ramser, K. Single-Step Preprocessing of Raman Spectra Using Convolutional Neural Networks. *Appl. Spectrosc.* **2020**, *74*, 427–438. [CrossRef] [PubMed]
44. Cui, C.; Fearn, T. Modern practical convolutional neural networks for multivariate regression: Applications to NIR calibration. *Chemom. Intell. Lab. Syst.* **2018**, *182*, 9–20. [CrossRef]
45. Acquarelli, J.; Van Laarhoven, T.; Gerretzen, J.; Tran, T.N. Convolutional neural networks for vibrational spectroscopic data analysis. *Anal. Chim. Acta* **2017**, *954*, 22–31. [CrossRef] [PubMed]
46. Oliveri, P. Class-modelling in food analytical chemistry: Development, sampling, optimisation and validation issues—A tutorial. *Anal. Chim. Acta* **2017**, *982*, 9–19. [CrossRef] [PubMed]
47. Oliveri, P. Class-Modelling Approaches: Advantages over Discriminant Methods for Compliance Verifications. *NIR news* **2016**, *27*, 29–30. [CrossRef]
48. Rodionova, O.Y.; Oliveri, P.; Pomerantsev, A.L. Rigorous and compliant approaches to one-class classification. *Chemom. Intell. Lab. Syst.* **2016**, *159*, 89–96. [CrossRef]
49. Borràs, E.; Ferré, J.; Boqué, R.; Mestres, M.; Aceña, L.; Busto, O. Data fusion methodologies for food and beverage authentication and quality assessment—A review. *Anal. Chim. Acta* **2015**, *891*, 1–14. [CrossRef]
50. Zhou, L.; Zhang, C.; Qiu, Z.; He, Y. Information fusion of emerging non-destructive analytical techniques for food quality authentication: A survey. *TrAC Trends Anal. Chem.* **2020**, *127*, 115901. [CrossRef]

51. Biancolillo, A.; Boqué, R.; Cocchi, M.; Marini, F. Data Fusion Strategies in Food Analysis. *Data Handl. Sci. Technol.* **2019**, *31*, 271–310. [CrossRef]
52. Nunes, K.M.; Andrade, M.V.O.; Santos Filho, A.M.P.; Lasmar, M.C.; Sena, M.M. Detection and characterisation of frauds in bovine meat in natura by non-meat ingredient additions using data fusion of chemical parameters and ATR-FTIR spectroscopy. *Food Chem.* **2016**, *205*, 14–22. [CrossRef] [PubMed]
53. Callao, M.P.; Ruisánchez, I. An overview of multivariate qualitative methods for food fraud detection. *Food Control* **2018**, *86*, 283–293. [CrossRef]
54. Smilde, A.K.; Måge, I.; Næs, T.; Hankemeier, T.; Lips, M.A.; Kiers, H.A.L.; Acar, E.; Bro, R. Common and Distinct Components in Data Fusion. *J. Chemom.* **2017**, *31*. [CrossRef]
55. Måge, I.; Smilde, A.K.; van der Kloet, F.M. Performance of methods that separate common and distinct variation in multiple data blocks. *J. Chemom.* **2019**, *33*. [CrossRef]
56. Biancolillo, A.; Måge, I.; Næs, T. Combining SO-PLS and linear discriminant analysis for multi-block classification. *Chemom. Intell. Lab. Syst.* **2015**, *141*, 58–67. [CrossRef]
57. Næs, T.; Tomic, O.; Afseth, N.K.; Segtnan, V.; Måge, I. Multi-block regression based on combinations of orthogonalisation, PLS-regression and canonical correlation analysis. *Chemom. Intell. Lab. Syst.* **2013**, *124*, 32–42. [CrossRef]
58. Shen, G.; Lesnoff, M.; Baeten, V.; Dardenne, P.; Davrieux, F.; Ceballos, H.; Belalcazar, J.; Dufour, D.; Yang, Z.; Han, L.; et al. Local partial least squares based on global PLS scores. *J. Chemom.* **2019**, *33*, e3117. [CrossRef]
59. Minet, O.; Baeten, V.; Lecler, B.; Dardenne, P.; Fernández Pierna, J. Local vs global methods applied to large near infrared databases covering high variability. In Proceedings of the 18th International Conference on Near Infrared Spectroscopy, Copenhagen, Denmark, 11–15 June 2017; pp. 45–49. [CrossRef]
60. Riedl, J.; Esslinger, S.; Fauhl-Hassek, C. Review of validation and reporting of non-targeted fingerprinting approaches for food authentication. *Anal. Chim. Acta* **2015**, *885*, 17–32. [CrossRef]
61. Ulberth, F. Tools to combat food fraud—A gap analysis. *Food Chem.* **2020**, *330*, 127044. [CrossRef]
62. Alewijn, M.; van der Voet, H.; van Ruth, S. Validation of multivariate classification methods using analytical fingerprints—Concept and case study on organic feed for laying hens. *J. Food Compos. Anal.* **2016**, *51*, 15–23. [CrossRef]
63. Westad, F.; Marini, F. Validation of chemometric models—A tutorial. *Anal. Chim. Acta* **2015**, *893*, 14–24. [CrossRef] [PubMed]
64. Esbensen, K.H.; Geladi, P. Principles of proper validation: Use and abuse of re-sampling for validation. *J. Chemom.* **2010**, *24*, 168–187. [CrossRef]
65. Chen, H.; Tan, C.; Lin, Z.; Wu, T. Detection of melamine adulteration in milk by near-infrared spectroscopy and one-class partial least squares. *Spectrochim. Acta-Part A Mol. Biomol. Spectrosc.* **2017**. [CrossRef] [PubMed]
66. Hansen, P.W.; Holroyd, S.E. Development and application of Fourier transform infrared spectroscopy for detection of milk adulteration in practice. *Int. J. Dairy Technol.* **2019**, *72*, 321–331. [CrossRef]
67. Coitinho, T.B.; Cassoli, L.D.; Cerqueira, P.H.R.; da Silva, H.K.; Coitinho, J.B.; Machado, P.F. Adulteration identification in raw milk using Fourier transform infrared spectroscopy. *J. Food Sci. Technol.* **2017**, *54*, 2394–2402. [CrossRef] [PubMed]
68. Ahmad, N.; Saleem, M. Raman spectroscopy based characterization of desi ghee obtained from buffalo and cow milk. *Int. Dairy J.* **2019**, *89*, 119–128. [CrossRef]
69. Taylan, O.; Cebi, N.; Tahsin Yilmaz, M.; Sagdic, O.; Bakhsh, A. Detection of lard in butter using Raman spectroscopy combined with chemometrics. *Food Chem.* **2020**, 127344. [CrossRef]
70. Yazgan Karacaglar, N.N.; Bulat, T.; Boyaci, I.H.; Topcu, A. Raman spectroscopy coupled with chemometric methods for the discrimination of foreign fats and oils in cream and yogurt. *J. Food Drug Anal.* **2019**. [CrossRef]
71. Zhao, M.; Downey, G.; Odonnell, C.P. Dispersive RAMAN spectroscopy and multivariate data analysis to detect offal adulteration of thawed beefburgers. *J. Agric. Food Chem.* **2015**, *63*, 1433–1441. [CrossRef] [PubMed]
72. Kamruzzaman, M.; Makino, Y.; Oshita, S.; Liu, S. Assessment of Visible Near-Infrared Hyperspectral Imaging as a Tool for Detection of Horsemeat Adulteration in Minced Beef. *Food Bioprocess Technol.* **2015**, *8*, 1054–1062. [CrossRef]

73. Yang, L.; Wu, T.; Liu, Y.; Zou, J.; Huang, Y.; Babu, S.V.; Lin, L. Rapid Identification of Pork Adulterated in the Beef and Mutton by Infrared Spectroscopy. *J. Spectrosc.* **2018**, *2018*. [CrossRef]
74. Chen, Z.; Wu, T.; Xiang, C.; Xu, X.; Tian, X. Rapid identification of rainbow trout adulteration in Atlantic salmon by Raman spectroscopy combined with machine learning. *Molecules* **2019**, *24*, 2851. [CrossRef] [PubMed]
75. De Oliveira Mendes, T.; Porto, B.L.S.; Almeida, M.R.; Fantini, C.; Sena, M.M. Discrimination between conventional and omega-3 fatty acids enriched eggs by FT-Raman spectroscopy and chemometric tools. *Food Chem.* **2019**, *273*, 144–150. [CrossRef] [PubMed]
76. Alamprese, C.; Casiraghi, E. Application of FT-NIR and FT-IR spectroscopy to fish fillet authentication. *LWT-Food Sci. Technol.* **2015**, *63*, 720–725. [CrossRef]
77. Fadzillah, N.A.; Man, Y.B.C.; Rohman, A.; Rosman, A.S.; Ismail, A.; Mustafa, S.; Khatib, A. Detection of butter adulteration with lard by employing 1H-NMR spectroscopy and multivariate data analysis. *J. Oleo Sci.* **2015**, *64*, 697–703. [CrossRef] [PubMed]
78. Bergana, M.M.; Adams, K.M.; Harnly, J.; Moore, J.C.; Xie, Z. Non-targeted detection of milk powder adulteration by 1H NMR spectroscopy and conformity index analysis. *J. Food Compos. Anal.* **2019**, *78*, 49–58. [CrossRef]
79. Tociu, M.; Todasca, M.C.; Bratu, A.; Mihalache, M.; Manolache, F. Fast approach for fatty acid profiling of dairy products fats using 1H-NMR spectroscopy. *Int. Dairy J.* **2018**, *83*, 52–57. [CrossRef]
80. Guyader, S.; Thomas, F.; Portaluri, V.; Jamin, E.; Akoka, S.; Silvestre, V.; Remaud, G. Authentication of edible fats and oils by non-targeted 13C INEPT NMR spectroscopy. *Food Control* **2018**, *91*, 216–224. [CrossRef]
81. Santos, P.M.; Pereira-Filho, E.R.; Colnago, L.A. Detection and quantification of milk adulteration using time domain nuclear magnetic resonance (TD-NMR). *Microchem. J.* **2016**. [CrossRef]
82. Qin, J.; Kim, M.S.; Chao, K.; Dhakal, S.; Lee, H.; Cho, B.K.; Mo, C. Detection and quantification of adulterants in milk powder using a high-throughput Raman chemical imaging technique. *Food Addit. Contam.-Part A Chem. Anal. Control. Expo. Risk Assess.* **2017**. [CrossRef]
83. Akanbi, T.O.; Barrow, C.J. Compositional Information Useful for Authentication of Krill Oil and the Detection of Adulterants. *Food Anal. Methods* **2018**, *11*, 178–187. [CrossRef]
84. Schmidt, W.F.; Chen, F.; Broadhurst, C.L.; Nguyen, J.K.; Qin, J.; Chao, K.; Kim, M.S. GTRS and 2D-NMR studies of alpha and gamma linolenic acids each containing the same H2C14-(H–C[dbnd]C–H)–C11H2–(H–C[dbnd]C–H)–C8H2 moiety. *J. Mol. Struct.* **2019**, *1196*, 258–270. [CrossRef]
85. Schmidt, W.F.; Chen, F.; Broadhurst, C.L.; Crawford, M.A. Liquid molecular model explains discontinuity between site uniformity among three N−3 fatty acids and their 13C and 1H NMR spectra. *J. Mol. Liq.* **2020**, *314*, 113376. [CrossRef]
86. Shaikh, S.; O'Donnell, C. Applications of fluorescence spectroscopy in dairy processing: A review. *Curr. Opin. Food Sci.* **2017**, *17*, 16–24. [CrossRef]
87. Sikorska, E.; Khmelinskii, I.; Sikorski, M. Fluorescence spectroscopy and imaging instruments for food quality evaluation. *Eval. Technol. Food Qual.* **2019**, 491–533. [CrossRef]
88. Dankowska, A. *Advances in Fluorescence Emission Spectroscopy for Food Authenticity Testing*; Woodhead Publishing: Sawston, UK, 2016; ISBN 9780081002209.
89. Christensen, J.; Nørgaard, L.; Bro, R.; Engelsen, S.B. Multivariate autofluorescence of intact food systems. *Chem. Rev.* **2006**, *106*, 1979–1994. [CrossRef]
90. Andersen, C.M.; Mortensen, G. Fluorescence Spectroscopy: A Rapid Tool for Analyzing Dairy Products. *J. Agric. Food Chem.* **2008**, *56*, 720–729. [CrossRef]
91. Karoui, R.; Blecker, C. Fluorescence Spectroscopy Measurement for Quality Assessment of Food Systems-a Review. *Food Bioprocess Technol.* **2011**, *4*, 364–386. [CrossRef]
92. Kumar, K.; Tarai, M.; Mishra, A.K. Unconventional steady-state fluorescence spectroscopy as an analytical technique for analyses of complex-multifluorophoric mixtures. *TrAC Trends Anal. Chem.* **2017**, *97*, 216–243. [CrossRef]
93. Bong, J.; Loomes, K.M.; Lin, B.; Stephens, J.M. New approach: Chemical and fluorescence profiling of NZ honeys. *Food Chem.* **2018**, *267*, 355–367. [CrossRef]
94. Bong, J.; Loomes, K.M.; Schlothauer, R.C.; Stephens, J.M. Fluorescence markers in some New Zealand honeys. *Food Chem.* **2016**, *192*, 1006–1014. [CrossRef] [PubMed]

95. Aït-Kaddour, A.; Loudiyi, M.; Ferlay, A.; Gruffat, D. Performance of fluorescence spectroscopy for beef meat authentication: Effect of excitation mode and discriminant algorithms. *Meat Sci.* **2018**, *137*, 58–66. [CrossRef] [PubMed]
96. Mita Mala, D.; Yoshimura, M.; Kawasaki, S.; Tsuta, M.; Kokawa, M.; Trivittayasil, V.; Sugiyama, J.; Kitamura, Y. Fiber optics fluorescence fingerprint measurement for aerobic plate count prediction on sliced beef surface. *LWT-Food Sci. Technol.* **2016**, *68*, 14–20. [CrossRef]
97. ElMasry, G.; Nakazawa, N.; Okazaki, E.; Nakauchi, S. Non-invasive sensing of freshness indices of frozen fish and fillets using pretreated excitation–emission matrices. *Sensors Actuators B Chem.* **2016**, *228*, 237–250. [CrossRef]
98. Hassoun, A.; Karoui, R. Monitoring changes in whiting (Merlangius merlangus) fillets stored under modified atmosphere packaging by front face fluorescence spectroscopy and instrumental techniques. *Food Chem.* **2016**, *200*, 343–353. [CrossRef] [PubMed]
99. Dramićanin, T.; Lenhardt Acković, L.; Zeković, I.; Dramićanin, M.D. Detection of Adulterated Honey by Fluorescence Excitation-Emission Matrices. *J. Spectrosc.* **2018**, *2018*. [CrossRef]
100. Lenhardt, L.; Bro, R.; Zeković, I.; Dramićanin, T.; Dramićanin, M.D. Fluorescence spectroscopy coupled with PARAFAC and PLS DA for characterization and classification of honey. *Food Chem.* **2015**, *175*, 284–291. [CrossRef]
101. Mehretie, S.; Al Riza, D.F.; Yoshito, S.; Kondo, N. Classification of raw Ethiopian honeys using front face fluorescence spectra with multivariate analysis. *Food Control* **2018**, *84*, 83–88. [CrossRef]
102. Wilczyńska, A.; Żak, N. The use of fluorescence spectrometry to determine the botanical origin of filtered honeys. *Molecules* **2020**, *25*, 1350. [CrossRef]
103. Bhatt, C.R.; Ghany, C.T.; Yueh, F.Y.; Singh, J.P.; McIntyre, D.L. Laser-Induced Breakdown Spectroscopy: Advanced Analytical Technique. In *Molecular and Laser Spectroscopy: Advances and Applications*; Elsevier: Amsterdam, The Netherlands, 2018; ISBN 9780128498828.
104. Cullen, P.; Bakalis, S.; Sullivan, C. Advances in control of food mixing operations. *Curr. Opin. Food Sci.* **2017**. [CrossRef]
105. Velioglu, H.M.; Sezer, B.; Bilge, G.; Baytur, S.E.; Boyaci, I.H. Identification of offal adulteration in beef by laser induced breakdown spectroscopy (LIBS). *Meat Sci.* **2018**. [CrossRef]
106. Dixit, Y.; Casado-Gavalda, M.P.; Cama-Moncunill, R.; Cama-Moncunill, X.; Markiewicz-Keszycka, M.; Cullen, P.J.; Sullivan, C. Laser induced breakdown spectroscopy for quantification of sodium and potassium in minced beef: A potential technique for detecting beef kidney adulteration. *Anal. Methods* **2017**. [CrossRef]
107. Dixit, Y.; Casado-Gavalda, M.P.; Cama-Moncunill, R.; Cama-Moncunill, X.; Markiewicz-Keszycka, M.; Jacoby, F.; Cullen, P.J.; Sullivan, C. Introduction to laser induced breakdown spectroscopy imaging in food: Salt diffusion in meat. *J. Food Eng.* **2018**. [CrossRef]
108. De Oliveira, A.P.; de Oliveira Leme, F.; Nomura, C.S.; Naozuka, J. Elemental imaging by Laser-Induced Breakdown Spectroscopy to evaluate selenium enrichment effects in edible mushrooms. *Sci. Rep.* **2019**. [CrossRef]
109. Nespeca, M.G.; Vieira, A.L.; Júnior, D.S.; Neto, J.A.G.; Ferreira, E.C. Detection and quantification of adulterants in honey by LIBS. *Food Chem.* **2020**. [CrossRef]
110. Peng, J.; Xie, W.; Jiang, J.; Zhao, Z.; Zhou, F.; Liu, F. Fast quantification of honey adulteration with laser-induced breakdown spectroscopy and chemometric methods. *Foods* **2020**, *9*, 341. [CrossRef]
111. Zhao, Z.; Chen, L.; Liu, F.; Zhou, F.; Peng, J.; Sun, M. Fast classification of geographical origins of honey based on laser-induced breakdown spectroscopy and multivariate analysis. *Sensors* **2020**, *20*, 1878. [CrossRef]
112. Lastra-Mejías, M.; Izquierdo, M.; González-Flores, E.; Cancilla, J.C.; Izquierdo, J.G.; Torrecilla, J.S. Honey exposed to laser-induced breakdown spectroscopy for chaos-based botanical classification and fraud assessment. *Chemom. Intell. Lab. Syst.* **2020**. [CrossRef]
113. Peng, J.; Liu, F.; Zhou, F.; Song, K.; Zhang, C.; Ye, L.; He, Y. Challenging applications for multi-element analysis by laser-induced breakdown spectroscopy in agriculture: A review. *TrAC-Trends Anal. Chem.* **2016**. [CrossRef]
114. Legnaioli, S.; Campanella, B.; Poggialini, F.; Pagnotta, S.; Harith, M.A.; Abdel-Salam, Z.A.; Palleschi, V. Industrial applications of laser-induced breakdown spectroscopy: A review. *Anal. Methods* **2020**. [CrossRef]
115. Afsah-Hejri, L.; Hajeb, P.; Ara, P.; Ehsani, R.J. A Comprehensive Review on Food Applications of Terahertz Spectroscopy and Imaging. *Compr. Rev. Food Sci. Food Saf.* **2019**. [CrossRef]

116. He, Y.; Bai, X.; Xiao, Q.; Liu, F.; Zhou, L.; Zhang, C. Detection of adulteration in food based on nondestructive analysis techniques: A review. *Crit. Rev. Food Sci. Nutr.* **2020**, 1–21. [CrossRef]
117. Liu, J. Terahertz spectroscopy and chemometric tools for rapid identification of adulterated dairy product. *Opt. Quantum Electron.* **2017**. [CrossRef]
118. Liu, W.; Zhang, Y.; Yang, S.; Han, D. Terahertz time-domain attenuated total reflection spectroscopy applied to the rapid discrimination of the botanical origin of honeys. *Spectrochim. Acta-Part A Mol. Biomol. Spectrosc.* **2018**. [CrossRef]
119. Liu, W.; Zhang, Y.; Li, M.; Han, D.; Liu, W. Determination of invert syrup adulterated in acacia honey by terahertz spectroscopy with different spectral features. *J. Sci. Food Agric.* **2020**. [CrossRef]
120. Feng, C.-H.; Makino, Y.; Oshita, S.; García Martín, J.F. Hyperspectral imaging and multispectral imaging as the novel techniques for detecting defects in raw and processed meat products: Current state-of-the-art research advances. *Food Control* **2018**, *84*, 165–176. [CrossRef]
121. Signoroni, A.; Savardi, M.; Baronio, A.; Benini, S. Deep learning meets hyperspectral image analysis: A multidisciplinary review. *J. Imaging* **2019**, *5*, 52. [CrossRef]
122. Al-Sarayreh, M.; Reis, M.M.; Yan, W.Q.; Klette, R. Potential of deep learning and snapshot hyperspectral imaging for classification of species in meat. *Food Control* **2020**. [CrossRef]
123. Verrez-Bagnis, V.; Sotelo, C.G.; Mendes, R.; Silva, H.; Kappel, K.; Schröder, U. Methods for Seafood Authenticity Testing in Europe. In *Bioactive Molecules in Food;* Spring: Cham, Switzerland; New York, NY, USA, 2018; Volume 1, pp. 1–55.
124. Thanakiatkrai, P.; Kitpipit, T. Meat species identification by two direct-triplex real-time PCR assays using low resolution melting. *Food Chem.* **2017**, *233*, 144–150. [CrossRef]
125. Xing, R.R.; Hu, R.R.; Han, J.X.; Deng, T.T.; Chen, Y. DNA barcoding and mini-barcoding in authenticating processed animal-derived food: A case study involving the Chinese market. *Food Chem.* **2020**, *309*, 125653. [CrossRef]
126. Labrador, K.; Agmata, A.; Palermo, J.D.; Follante, J.; Pante, M.J. Authentication of processed Philippine sardine products using Hotshot DNA extraction and minibarcode amplification. *Food Control* **2019**, *98*, 150–155. [CrossRef]
127. Wilwet, L.; Jeyasekaran, G.; Jeya, R.; Sivaraman, B. A single enzyme PCR-RFLP protocol targeting 16S rRNA/tRNA val region to authenticate four commercially important shrimp species in India. *Food Chem.* **2018**, *239*, 369–376. [CrossRef]
128. Kang, T.S. Development of four PCR-based methods to differentiate tilefish species (Branchiostegus japonicus and B. albus). *Food Chem.* **2019**, *271*, 1–8. [CrossRef]
129. Fang, X.; Zhang, C. Detection of adulterated murine components in meat products by TaqMan© real-time PCR. *Food Chem.* **2016**, *192*, 485–490. [CrossRef]
130. Kim, M.J.; Yoo, I.; Yang, S.M.; Suh, S.M.; Kim, H.Y. Development and validation of a multiplex PCR assay for simultaneous detection of chicken, turkey and duck in processed meat products. *Int. J. Food Sci. Technol.* **2018**, *53*, 2673–2679. [CrossRef]
131. Kim, M.J.; Suh, S.M.; Kim, S.Y.; Qin, P.; Kim, H.R.; Kim, H.Y. Development of a real-time PCR assay for the detection of donkey (Equus asinus) meat in meat mixtures treated under different processing conditions. *Foods* **2020**, *9*, 130. [CrossRef]
132. Bo, H.; Xianrong, M.; Liyuan, Z.; Jinyue, G.; Shaowen, L.; Hui, J. Development of a sensitive and speci fi c multiplex PCR method for the simultaneous detection of chicken, duck and goose DNA in meat products. *Meat Sci.* **2015**, *101*, 90–94. [CrossRef]
133. Qin, P.; Hong, Y.; Kim, H.Y. Multiplex-PCR Assay for Simultaneous Identification of Lamb, Beef and Duck in Raw and Heat-Treated Meat Mixtures. *J. Food Saf.* **2016**, *36*, 367–374. [CrossRef]
134. Kim, M.J.; Kim, H.Y. Species identification of commercial jerky products in food and feed using direct pentaplex PCR assay. *Food Control* **2017**, *78*, 1–6. [CrossRef]
135. Kim, M.J.; Kim, H.Y. Development of a fast duplex real-time PCR assay for simultaneous detection of chicken and pigeon in raw and heat-treated meats. *Food Control* **2018**, *85*, 1–5. [CrossRef]
136. Kim, M.J.; Kim, H.Y. A fast multiplex real-time PCR assay for simultaneous detection of pork, chicken, and beef in commercial processed meat products. *LWT* **2019**, *114*, 108390. [CrossRef]

137. El-razik, K.A.E.A.B.D.; Sayed, A.; Abuelnaga, M.; Younes, A.M.; Atta, N.S.; Arafa, A.A.; Kandil, M.M. Species–specific PCR test for the quick recognition of equine tissue in raw and processed beef meat mixtures. *Food Sci. Technol.* **2019**, *2061*, 166–172. [CrossRef]
138. Karabasanavar, N.; Girish, P.S.; Kumar, D.; Singh, S.P. Detection of beef adulteration by mitochondrial D-loop based species-specific polymerase chain reaction. *Int. J. Food Prop.* **2017**, *20*, 2264–2271. [CrossRef]
139. Cardoso, D.; Melo, R.; Gonçalves, M. Food metagenomics: Next generation sequencing identifies species mixtures and mislabeling within highly processed cod products. *Food Control* **2017**, *80*, 2014–2017. [CrossRef]
140. Xing, R.R.; Wang, N.; Hu, R.R.; Zhang, J.K.; Han, J.X.; Chen, Y. Application of next generation sequencing for species identification in meat and poultry products: A DNA metabarcoding approach. *Food Control* **2019**, *101*, 173–179. [CrossRef]
141. Mandli, J.; EL Fatimi, I.; Seddaoui, N.; Amine, A. Enzyme immunoassay (ELISA/immunosensor) for a sensitive detection of pork adulteration in meat. *Food Chem.* **2018**, *255*, 380–389. [CrossRef]
142. Stachniuk, A.; Sumara, A.; Montowska, M.; Fornal, E. Liquid chromatography–mass spectrometry bottom-up proteomic methods in animal species analysis of processed meat for food authentication and the detection of adulterations. *Mass Spectrom. Rev.* **2019**, 1–28. [CrossRef]
143. Danezis, G.P.; Tsagkaris, A.S.; Brusic, V.; Georgiou, C.A. Food authentication: State of the art and prospects. *Curr. Opin. Food Sci.* **2016**, *10*, 22–31. [CrossRef]
144. Fiorino, G.M.; Losito, I.; De Angelis, E.; Arlorio, M.; Logrieco, A.F.; Monaci, L. Assessing fish authenticity by direct analysis in real time-high resolution mass spectrometry and multivariate analysis: Discrimination between wild-type and farmed salmon. *Food Res. Int.* **2019**, *116*, 1258–1265. [CrossRef]
145. Shen, Q.; Li, L.; Song, G.; Feng, J.; Li, S.; Wang, Y.; Ma, J.; Wang, H. Development of an intelligent surgical knife rapid evaporative ionization mass spectrometry based method for real-time differentiation of cod from oilfish. *J. Food Compos. Anal.* **2020**, *86*, 103355. [CrossRef]
146. Sun, H.; Song, Y.; Zhang, H.; Zhang, X.; Liu, Y.; Wang, X.; Cong, P.; Xu, J.; Xue, C. Characterization of lipid composition in the muscle tissue of four shrimp species commonly consumed in China by UPLC−Triple TOF−MS/MS. *LWT* **2020**, *128*, 109469. [CrossRef]
147. Yu, X.; Li, L.; Wang, H.; Song, G.; Wang, J.; Li, S.; Wang, Y. Lipidomics study of rainbow trout (Oncorhynchus mykiss) and salmons (Oncorhynchus tshawytscha and Salmo salar) using hydrophilic interaction chromatography and mass spectrometry. *LWT-Food Sci. Technol.* **2020**, *121*, 108988. [CrossRef]
148. Fornal, E.; Montowska, M. Species-specific peptide-based liquid chromatography–mass spectrometry monitoring of three poultry species in processed meat products. *Food Chem.* **2019**, *283*, 489–498. [CrossRef] [PubMed]
149. Perestam; Fujisaki, K.K.; Nava, O.; Hellberg, R.S. Comparison of real-time PCR and ELISA-based methods for the detection of beef and pork in processed meat products. *Food Control* **2017**, *71*, 346–352. [CrossRef]
150. Camin, F.; Bontempo, L.; Perini, M.; Piasentier, E. Stable Isotope Ratio Analysis for Assessing the Authenticity of Food of Animal Origin. *Compr. Rev. Food Sci. Food Saf.* **2016**, *15*, 868–877. [CrossRef]
151. Camin, F.; Perini, M.; Bontempo, L.; Galeotti, M.; Tibaldi, E.; Piasentier, E. Stable isotope ratios of H, C, O, N and S for the geographical traceability of Italian rainbow trout (Oncorhynchus mykiss). *Food Chem.* **2018**, *267*, 288–295. [CrossRef]
152. Mekki, I.; Camin, F.; Perini, M.; Smeti, S.; Hajji, H.; Mahouachi, M.; Piasentier, E.; Atti, N. Differentiating the geographical origin of Tunisian indigenous lamb using stable isotope ratio and fatty acid content. *J. Food Compos. Anal.* **2016**, *53*, 40–48. [CrossRef]
153. Zhang, X.; Cheng, J.; Han, D.; Zhao, X.; Chen, X.; Liu, Y. Geographical origin traceability and species identification of three scallops (Patinopecten yessoensis, Chlamys farreri, and Argopecten irradians) using stable isotope analysis. *Food Chem.* **2019**, *299*, 125107. [CrossRef]
154. Yin, H.M.; Huang, F.; Shen, J.; Yu, H.M. Using Sr isotopes to trace the geographic origins of Chinese mitten crabs. *Acta Geochim.* **2020**, *39*, 326–336. [CrossRef]
155. Coulter, D.P.; Bowen, G.J.; Höök, T.O. Influence of diet and ambient water on hydrogen and oxygen stable isotope ratios in fish tissue: Patterns within and among tissues and relationships with growth rates. *Hydrobiologia* **2017**, *799*, 111–121. [CrossRef]
156. Li, L.; Ren, W.; Dong, S.; Feng, J. Investigation of geographic origin, salinity and feed on stable isotope profile of Pacific white shrimp (Litopenaeus vannamei). *Aquac. Res.* **2018**, *49*, 1029–1036. [CrossRef]

157. Gopi, K.; Mazumder, D.; Sammut, J.; Saintilan, N. Determining the provenance and authenticity of seafood: A review of current methodologies. *Trends Food Sci. Technol.* **2019**, *91*, 294–304. [CrossRef]
158. Li, L.; Kokkuar, N.; Han, C.; Ren, W.; Dong, S. Effects of dietary shifts on the stable isotope signature of Pacific white shrimp Litopenaeus vannamei and implications for traceability. *Mar. Freshw. Res.* **2020**. [CrossRef]
159. Li, L.; Han, C.; Dong, S.; Boyd, C.E. Use of elemental profiling and isotopic signatures to differentiate Pacific white shrimp (Litopenaeus vannamei) from freshwater and seawater culture areas. *Food Control* **2019**, *95*, 249–256. [CrossRef]
160. Gopi, K.; Mazumder, D.; Sammut, J.; Saintilan, N.; Crawford, J.; Gadd, P. Isotopic and elemental profiling to trace the geographic origins of farmed and wild-caught Asian seabass (Lates calcarifer). *Aquaculture* **2019**, *502*, 56–62. [CrossRef]
161. Kim, H.; Suresh Kumar, K.; Shin, K.H. Applicability of stable C and N isotope analysis in inferring the geographical origin and authentication of commercial fish (Mackerel, Yellow Croaker and Pollock). *Food Chem.* **2015**, *172*, 523–527. [CrossRef]
162. Carter, J.F.; Tinggi, U.; Yang, X.; Fry, B. Stable isotope and trace metal compositions of Australian prawns as a guide to authenticity and wholesomeness. *Food Chem.* **2015**, *170*, 241–248. [CrossRef]
163. Ortea, I.; Gallardo, J.M. Investigation of production method, geographical origin and species authentication in commercially relevant shrimps using stable isotope ratio and/or multi-element analyses combined with chemometrics: An exploratory analysis. *Food Chem.* **2015**, *170*, 145–153. [CrossRef]
164. Gopi, K.; Mazumder, D.; Sammut, J.; Saintilan, N.; Crawford, J.; Gadd, P. Combined use of stable isotope analysis and elemental profiling to determine provenance of black tiger prawns (Penaeus monodon). *Food Control* **2019**, *95*, 242–248. [CrossRef]
165. Liu, H.; Zhao, Q.; Guo, X.; Tang, C.; Yu, X.; Zhan, T.; Qin, Y.; Zhang, J. Application of isotopic and elemental fingerprints in identifying the geographical origin of goat milk in China. *Food Chem.* **2019**, *277*, 448–454. [CrossRef]
166. Boyd, C. *Water Quality An Introduction Second Edition*; Springer: New York, NY, USA, 2015; ISBN 9783319174457.
167. Danezis, G.P.; Pappas, A.C.; Zoidis, E.; Papadomichelakis, G.; Hadjigeorgiou, I.; Zhang, P.; Brusic, V.; Georgiou, C.A. Game meat authentication through rare earth elements fingerprinting. *Anal. Chim. Acta* **2017**, *991*, 46–57. [CrossRef] [PubMed]
168. Han, C.; Dong, S.; Li, L.; Wei, F.; Zhou, Y.; Gao, Q. The effect of the seasons on geographical traceability of salmonid based on multi-element analysis. *Food Control* **2020**, *109*. [CrossRef]
169. Jia, Y.; Wang, L.; Qu, Z.; Wang, C.; Yang, Z. Effects on heavy metal accumulation in freshwater fishes: Species, tissues, and sizes. *Environ. Sci. Pollut. Res.* **2017**, *24*, 9379–9386. [CrossRef] [PubMed]
170. Abbas, O.; Zadravec, M.; Baeten, V.; Mikuš, T.; Lešić, T.; Vulić, A.; Prpić, J.; Jemeršić, L.; Pleadin, J. Analytical methods used for the authentication of food of animal origin. *Food Chem.* **2018**, *246*, 6–17. [CrossRef]
171. Bennion, M.; Morrison, L.; Brophy, D.; Carlsson, J.; Abrahantes, J.C.; Graham, C.T. Trace element fingerprinting of blue mussel (Mytilus edulis)shells and soft tissues successfully reveals harvesting locations. *Sci. Total Environ.* **2019**, *685*, 50–58. [CrossRef]
172. Chaguri, M.P.; Maulvault, A.L.; Nunes, M.L.; Santiago, D.A.; Denadai, J.C.; Fogaça, F.H.; Sant'Ana, L.S.; Ducatti, C.; Bandarra, N.; Carvalho, M.L.; et al. Different tools to trace geographic origin and seasonality of croaker (Micropogonias furnieri). *LWT-Food Sci. Technol.* **2015**, *61*, 194–200. [CrossRef]
173. Farabegoli, F.; Pirini, M.; Rotolo, M.; Silvi, M.; Testi, S.; Ghidini, S.; Zanardi, E.; Remondini, D.; Bonaldo, A.; Parma, L.; et al. Toward the Authentication of European Sea Bass Origin through a Combination of Biometric Measurements and Multiple Analytical Techniques. *J. Agric. Food Chem.* **2018**, *66*, 6822–6831. [CrossRef]
174. Liu, Y.; Ma, D.H.; Wang, X.C.; Liu, L.P.; Fan, Y.X.; Cao, J.X. Prediction of chemical composition and geographical origin traceability of Chinese export tilapia fillets products by near infrared reflectance spectroscopy. *LWT-Food Sci. Technol.* **2015**, *60*, 1214–1218. [CrossRef]
175. Guo, X.; Cai, R.; Wang, S.; Tang, B.; Li, Y.; Zhao, W. Non-destructive geographical traceability of sea cucumber (Apostichopus japonicus) using near infrared spectroscopy combined with chemometric methods. *R. Soc. Open Sci.* **2018**, *5*. [CrossRef]
176. Ghidini, S.; Varrà, M.O.; Dall'Asta, C.; Badiani, A.; Ianieri, A.; Zanardi, E. Rapid authentication of European sea bass (Dicentrarchus labrax L.) according to production method, farming system, and geographical origin by near infrared spectroscopy coupled with chemometrics. *Food Chem.* **2019**, *280*, 321–327. [CrossRef]

177. Heude, C.; Elbayed, K.; Jezequel, T.; Fanuel, M.; Lugan, R.; Heintz, D.; Benoit, P.; Piotto, M. Metabolic Characterization of Caviar Specimens by 1H NMR Spectroscopy: Towards Caviar Authenticity and Integrity. *Food Anal. Methods* **2016**, *9*, 3428–3438. [CrossRef]
178. Arechavala-Lopez, P.; Milošević-González, M.; Sanchez-Jerez, P. Using trace elements in otoliths to discriminate between wild and farmed European sea bass (Dicentrarchus labrax L.) and Gilthead sea bream (Sparus aurata L.). *Int. Aquat. Res.* **2016**, *8*, 263–273. [CrossRef]
179. Wang, Y.V.; Wan, A.H.L.; Lock, E.-J.; Andersen, N.; Winter-Schuh, C.; Larsen, T. Know your fish: A novel compound-specific isotope approach for tracing wild and farmed salmon. *Food Chem.* **2018**, *256*, 380–389. [CrossRef] [PubMed]
180. Molkentin, J.; Lehmann, I.; Ostermeyer, U.; Rehbein, H. Traceability of organic fish—Authenticating the production origin of salmonids by chemical and isotopic analyses. *Food Control* **2015**, *53*, 55–66. [CrossRef]
181. Xu, J.L.; Riccioli, C.; Sun, D.W. Comparison of hyperspectral imaging and computer vision for automatic differentiation of organically and conventionally farmed salmon. *J. Food Eng.* **2017**, *196*, 170–182. [CrossRef]
182. FAO. *Overview of Food Fraud in the Fisheries Sector*; FAO: Rome, Italy, 2018; Volume 1165, ISBN 9789251304020.
183. Hong, E.; Lee, S.Y.; Jeong, J.Y.; Park, J.M.; Kim, B.H.; Kwon, K.; Chun, H.S. Modern analytical methods for the detection of food fraud and adulteration by food category. *J. Sci. Food Agric.* **2017**, *97*, 3877–3896. [CrossRef] [PubMed]
184. Kappel, K.; Eschbach, E.; Fischer, M.; Fritsche, J. Design of a user-friendly and rapid DNA microarray assay for the authentication of ten important food fish species. *Food Chem.* **2020**, *311*, 125884. [CrossRef] [PubMed]
185. Chatterjee, N.S.; Chevallier, O.P.; Wielogorska, E.; Black, C.; Elliott, C.T. Simultaneous authentication of species identity and geographical origin of shrimps: Untargeted metabolomics to recurrent biomarker ions. *J. Chromatogr. A* **2019**, *1599*, 75–84. [CrossRef]
186. Velioğlu, H.M.; Temiz, H.T.; Boyaci, I.H. Differentiation of fresh and frozen-thawed fish samples using Raman spectroscopy coupled with chemometric analysis. *Food Chem.* **2015**, *172*, 283–290. [CrossRef]
187. Sun, D.; Weng, H.; He, X.; Li, L.; He, Y.; Cen, H. Combining near-infrared hyperspectral imaging with elemental and isotopic analysis to discriminate farm-raised pacific white shrimp from high-salinity and low-salinity environments. *Food Chem.* **2019**, *299*. [CrossRef]
188. Wu, T.; Zhong, N.; Yang, L. Identification of Adulterated and Non-adulterated Norwegian Salmon Using FTIR and an Improved PLS-DA Method. *Food Anal. Methods* **2018**, *11*, 1501–1509. [CrossRef]
189. Rašković, B.; Heinke, R.; Rösch, P.; Popp, J. The Potential of Raman Spectroscopy for the Classification of Fish Fillets. *Food Anal. Methods* **2016**, *9*, 1301–1306. [CrossRef]
190. Lv, H.; Xu, W.; You, J.; Xiong, S. Classification of freshwater fish species by linear discriminant analysis based on near infrared reflectance spectroscopy. *J. Near Infrared Spectrosc.* **2017**, *25*, 54–62. [CrossRef]
191. Liu, Y.; Hu, W.; Guo, X.-X.; Wang, X.-C.; Sun, S.-Q.; Xu, C.-H. Rapid discrimination of three marine fish surimi by Tri-step infrared spectroscopy combined with Principle Component Regression. *Spectrochim. Acta Part A Mol. Biomol. Spectrosc.* **2015**, *149*, 516–522. [CrossRef] [PubMed]
192. Zhang, X.Y.; Hu, W.; Teng, J.; Peng, H.H.; Gan, J.H.; Wang, X.C.; Sun, S.Q.; Xu, C.H.; Liu, Y. Rapid recognition of marine fish surimi by one-step discriminant analysis based on near-infrared diffuse reflectance spectroscopy. *Int. J. Food Prop.* **2017**, *20*, 2932–2943. [CrossRef]
193. Wang, H.; Wang, R.; Song, Y.; Kamal, T.; Lv, Y.; Zhu, B.; Tao, X.; Tan, M. A fast and non-destructive LF-NMR and MRI method to discriminate adulterated shrimp. *J. Food Meas. Charact.* **2018**, *12*, 1340–1349. [CrossRef]
194. Wang, W.L.; Chen, W.H.; Tian, H.Y.; Liu, Y. Detection of Frozen-Thawed Cycles for Frozen Tilapia (Oreochromis) Fillets Using Near Infrared Spectroscopy. *J. Aquat. Food Prod. Technol.* **2018**, *27*, 609–618. [CrossRef]
195. Shan, J.; Wang, X.; Russel, M.; Zhao, J.; Zhang, Y. Comparisons of Fish Morphology for Fresh and Frozen-Thawed Crucian Carp Quality Assessment by Hyperspectral Imaging Technology. *Food Anal. Methods* **2018**, *11*, 1701–1710. [CrossRef]
196. Cheng, J.-H.; Sun, D.-W.; Pu, H.-B.; Chen, X.; Liu, Y.; Zhang, H.; Li, J.-L. Integration of classifiers analysis and hyperspectral imaging for rapid discrimination of fresh from cold-stored and frozen-thawed fish fillets. *J. Food Eng.* **2015**, *161*, 33–39. [CrossRef]
197. Qu, J.-H.; Cheng, J.-H.; Sun, D.-W.; Pu, H.; Wang, Q.-J.; Ma, J. Discrimination of shelled shrimp (Metapenaeus ensis) among fresh, frozen-thawed and cold-stored by hyperspectral imaging technique. *LWT-Food Sci. Technol.* **2015**, *62*, 202–209. [CrossRef]

198. Beć, K.B.; Grabska, J.; Siesler, H.W.; Huck, C.W. Handheld near-infrared spectrometers: Where are we heading? *NIR News* **2020**, *31*, 28–35. [CrossRef]
199. Beganovic, A.; Hawthorne, L.M.; Bach, K.; Huck, C.W. Critical review on the utilization of handheld and portable Raman spectrometry in meat science. *Foods* **2019**, *8*, 49. [CrossRef] [PubMed]
200. Grassi, S.; Casiraghi, E.; Alamprese, C. Handheld NIR device: A non-targeted approach to assess authenticity of fish fillets and patties. *Food Chem.* **2018**, *243*, 382–388. [CrossRef] [PubMed]
201. Sousa, N.; Moreira, M.; Saraiva, C.; de Almeida, J. Applying Fourier Transform Mid Infrared Spectroscopy to Detect the Adulteration of Salmo salar with Oncorhynchus mykiss. *Foods* **2018**, *7*, 55. [CrossRef] [PubMed]
202. Ethuin, P.; Marlard, S.; Delosière, M.; Carapito, C.; Delalande, F.; Van Dorsselaer, A.; Dehaut, A.; Lencel, V.; Duflos, G.; Grard, T. Differentiation between fresh and frozen-thawed sea bass (Dicentrarchus labrax) fillets using two-dimensional gel electrophoresis. *Food Chem.* **2015**, *176*, 294–301. [CrossRef]
203. Diop, M.; Watier, D.; Masson, P.-Y.; Diouf, A.; Amara, R.; Grard, T.; Lencel, P. Assessment of freshness and freeze-thawing of sea bream fillets (Sparus aurata) by a cytosolic enzyme: Lactate dehydrogenase. *Food Chem.* **2016**, *210*, 428–434. [CrossRef]
204. Marlard, S.; Doyen, P.; Grard, T. Rapid Multiparameters Approach to Differentiate Fresh Skinless Sea Bass (Dicentrarchus labrax) Fillets from Frozen-Thawed Ones. *J. Aquat. Food Prod. Technol.* **2019**, *28*, 253–262. [CrossRef]
205. Pezzolato, M.; Baioni, E.; Maurella, C.; Varello, K.; Meistro, S.; Balsano, A.; Bozzetta, E. Distinguishing between fresh and frozen-thawed smoked salmon: Histology to detect food adulteration in high-value products. *J. Food Prot.* **2020**, *83*, 52–55. [CrossRef]
206. Reis, M.M.; Martínez, E.; Saitua, E.; Rodríguez, R.; Pérez, I.; Olabarrieta, I. Non-invasive differentiation between fresh and frozen/thawed tuna fillets using near infrared spectroscopy (Vis-NIRS). *LWT-Food Sci. Technol.* **2017**, *78*, 129–137. [CrossRef]
207. Gudjónsdóttir, M.; Romotowska, P.E.; Karlsdóttir, M.G.; Arason, S. Low field nuclear magnetic resonance and multivariate analysis for prediction of physicochemical characteristics of Atlantic mackerel as affected by season of catch, freezing method, and frozen storage duration. *Food Res. Int.* **2019**, *116*, 471–482. [CrossRef]
208. Karoui, R.; Hassoun, A.; Ethuin, P. Front face fluorescence spectroscopy enables rapid differentiation of fresh and frozen-thawed sea bass (Dicentrarchus labrax) fillets. *J. Food Eng.* **2017**, *202*, 89–98. [CrossRef]
209. Kamruzzaman, M.; Makino, Y.; Oshita, S. Non-invasive analytical technology for the detection of contamination, adulteration, and authenticity of meat, poultry, and fish: A review. *Anal. Chim. Acta* **2015**, *853*, 19–29. [CrossRef] [PubMed]
210. Washburn, K.E.; Stormo, S.K.; Skjelvareid, M.H.; Heia, K. Non-invasive assessment of packaged cod freeze-thaw history by hyperspectral imaging. *J. Food Eng.* **2017**, *205*, 64–73. [CrossRef]
211. Kuswandi, B.; Putri, F.K.; Gani, A.A.; Ahmad, M. Application of class-modelling techniques to infrared spectra for analysis of pork adulteration in beef jerkys. *J. Food Sci. Technol.* **2015**, *52*, 7655–7668. [CrossRef] [PubMed]
212. Kuswandi, B.; Cendekiawan, K.A.; Kristiningrum, N.; Ahmad, M. Pork adulteration in commercial meatballs determined by chemometric analysis of NIR Spectra. *J. Food Meas. Charact.* **2015**, *9*, 313–323. [CrossRef]
213. Rady, A.; Adedeji, A. Assessing different processed meats for adulterants using visible-near-infrared spectroscopy. *Meat Sci.* **2018**, *136*, 59–67. [CrossRef] [PubMed]
214. Deniz, E.; Güneş Altuntaş, E.; Ayhan, B.; İğci, N.; Özel Demiralp, D.; Candoğan, K. Differentiation of beef mixtures adulterated with chicken or turkey meat using FTIR spectroscopy. *J. Food Process. Preserv.* **2018**, *42*, 1–12. [CrossRef]
215. Alamprese, C.; Amigo, J.M.; Casiraghi, E.; Engelsen, S.B. Identification and quantification of turkey meat adulteration in fresh, frozen-thawed and cooked minced beef by FT-NIR spectroscopy and chemometrics. *Meat Sci.* **2016**, *121*, 175–181. [CrossRef]
216. Kamruzzaman, M.; Makino, Y.; Oshita, S. Hyperspectral imaging in tandem with multivariate analysis and image processing for non-invasive detection and visualization of pork adulteration in minced beef. *Anal. Methods* **2015**, *7*, 7496–7502. [CrossRef]
217. Kamruzzaman, M.; Makino, Y.; Oshita, S. Rapid and non-destructive detection of chicken adulteration in minced beef using visible near-infrared hyperspectral imaging and machine learning. *J. Food Eng.* **2016**, *170*, 8–15. [CrossRef]

218. Guo, B.; Zhao, J.; Weng, S.; Yin, X.; Tang, P. Rapid determination of minced beef adulteration using hyperspectral reflectance spectroscopy and multivariate methods. *IOP Conf. Ser. Earth Environ. Sci.* **2020**, *428*. [CrossRef]
219. Ropodi, A.I.; Panagou, E.Z.; Nychas, G.-J.E. Multispectral imaging (MSI): A promising method for the detection of minced beef adulteration with horsemeat. *Food Control* **2017**, *73*, 57–63. [CrossRef]
220. Dalsecco, L.S.; Palhares, R.M.; Oliveira, P.C.; Teixeira, L.V.; Drummond, M.G.; de Oliveira, D.A.A. A Fast and Reliable Real-Time PCR Method for Detection of Ten Animal Species in Meat Products. *J. Food Sci.* **2018**, *83*, 258–265. [CrossRef]
221. Iwobi, A.; Sebah, D.; Kraemer, I.; Losher, C.; Fischer, G.; Busch, U.; Huber, I. A multiplex real-time PCR method for the quantification of beef and pork fractions in minced meat. *Food Chem.* **2015**, *169*, 305–313. [CrossRef] [PubMed]
222. Kim, M.; Yoo, I.; Lee, S.Y.; Hong, Y.; Kim, H.Y. Quantitative detection of pork in commercial meat products by TaqMan®real-time PCR assay targeting the mitochondrial D-loop region. *Food Chem.* **2016**, *210*, 102–106. [CrossRef]
223. Yin, R.; Sun, Y.; Wang, K.; Feng, N.; Zhang, H.; Xiao, M. Development of a PCR-based lateral flow strip assay for the simple, rapid, and accurate detection of pork in meat and meat products. *Food Chem.* **2020**, *318*, 126541. [CrossRef]
224. Gao, F.; Zhou, S.; Han, L.; Yang, Z.; Liu, X. A novel FT-IR spectroscopic method based on lipid characteristics for qualitative and quantitative analysis of animal-derived feedstuff adulterated with ruminant ingredients. *Food Chem.* **2017**, *237*, 342–349. [CrossRef]
225. Piotrowski, C.; Garcia, R.; Garrido-Varo, A.; Pérez-Marín, D.; Riccioli, C.; Fearn, T. Short Communication: The potential of portable near infrared spectroscopy for assuring quality and authenticity in the food chain, using Iberian hams as an example. *Animal* **2019**, *13*, 3018–3021. [CrossRef]
226. Pieszczek, L.; Czarnik-Matusewicz, H.; Daszykowski, M. Identification of ground meat species using near-infrared spectroscopy and class modeling techniques—Aspects of optimization and validation using a one-class classification model. *Meat Sci.* **2018**, *139*, 15–24. [CrossRef]
227. Chiesa, L.; Panseri, S.; Bonacci, S.; Procopio, A.; Zecconi, A.; Arioli, F.; Cuevas, F.J.; Moreno-Rojas, J.M. Authentication of Italian PDO lard using NIR spectroscopy, volatile profile and fatty acid composition combined with chemometrics. *Food Chem.* **2016**, *212*, 296–304. [CrossRef]
228. Al-sarayreh, M.; Reis, M.M.; Yan, W.Q. Detection of Red-Meat Adulteration by Deep Spectral–Spatial Features in Hyperspectral Images. *J. Imaging* **2018**, *4*, 63. [CrossRef]
229. Rahmania, H.; Sudjadi; Rohman, A. The employment of FTIR spectroscopy in combination with chemometrics for analysis of rat meat in meatball formulation. *Meat Sci.* **2015**, *100*, 301–305. [CrossRef] [PubMed]
230. Schmutzler, M.; Beganovic, A.; Böhler, G.; Huck, C.W. Methods for detection of pork adulteration in veal product based on FT-NIR spectroscopy for laboratory, industrial and on-site analysis. *Food Control* **2015**, *57*, 258–267. [CrossRef]
231. Zhao, H.T.; Feng, Y.Z.; Chen, W.; Jia, G.F. Application of invasive weed optimization and least square support vector machine for prediction of beef adulteration with spoiled beef based on visible near-infrared (Vis-NIR) hyperspectral imaging. *Meat Sci.* **2019**, *151*, 75–81. [CrossRef] [PubMed]
232. Barbin, D.F.; Badaró, A.T.; Honorato, D.C.B.; Ida, E.Y.; Shimokomaki, M. Identification of Turkey meat and processed products using near infrared spectroscopy. *Food Control* **2020**, *107*, 106816. [CrossRef]
233. López-Maestresalas, A.; Insausti, K.; Jarén, C.; Pérez-Roncal, C.; Urrutia, O.; Beriain, M.J.; Arazuri, S. Detection of minced lamb and beef fraud using NIR spectroscopy. *Food Control* **2019**, *98*, 465–473. [CrossRef]
234. Leng, T.; Li, F.; Xiong, L.; Xiong, Q.; Zhu, M.; Chen, Y. Quantitative detection of binary and ternary adulteration of minced beef meat with pork and duck meat by NIR combined with chemometrics. *Food Control* **2020**, *113*, 107203. [CrossRef]
235. Weng, S.; Guo, B.; Tang, P.; Yin, X.; Pan, F.; Zhao, J.; Huang, L.; Zhang, D. Rapid detection of adulteration of minced beef using Vis/NIR reflectance spectroscopy with multivariate methods. *Spectrochim. Acta-Part A Mol. Biomol. Spectrosc.* **2020**, *230*, 118005. [CrossRef]
236. Li, D.; Peng, Y.; Zhang, H.; Delgado-Pando, G. Investigation on Texture Changes and Classification between Cold-Fresh and Freeze-Thawed Tan Mutton. *J. Food Qual.* **2019**, *2019*. [CrossRef]
237. Grunert, T.; Stephan, R.; Ehling-Schulz, M.; Johler, S. Fourier Transform Infrared Spectroscopy enables rapid differentiation of fresh and frozen/thawed chicken. *Food Control* **2016**, *60*, 361–364. [CrossRef]

238. Arvanitoyannis, I.S. (Ed.) *Authenticity of Foods of Animal Origin*; CRC Press, Taylor & Francis Group: Boca Raton, FL, USA, 2016; ISBN 9781482201338.
239. Downey, G. (Ed.) *Advances in Food Authenticity Testing*; Elsevier: Amsterdam, The Nederlands, 2016; ISBN 9780081002209.
240. Muñoz, M.; García-Casco, J.M.; Alves, E.; Benítez, R.; Barragán, C.; Caraballo, C.; Fernández, A.I.; García, F.; Núñez, Y.; Óvilo, C.; et al. Development of a 64 SNV panel for breed authentication in Iberian pigs and their derived meat products. *Meat Sci.* **2020**, *167*, 108152. [CrossRef] [PubMed]
241. Erasmus, S.W.; Muller, M.; Butler, M.; Hoffman, L.C. The truth is in the isotopes: Authenticating regionally unique South African lamb. *Food Chem.* **2018**, *239*, 926–934. [CrossRef] [PubMed]
242. Monahan, F.J.; Schmidt, O.; Moloney, A.P. Meat provenance: Authentication of geographical origin and dietary background of meat. *Meat Sci.* **2018**, *144*, 2–14. [CrossRef] [PubMed]
243. Iammarino, M.; Marino, R.; Albenzio, M. How meaty? Detection and quantification of adulterants, foreign proteins and food additives in meat products. *Int. J. Food Sci. Technol.* **2017**, *52*, 851–863. [CrossRef]
244. Zhang, D.; Feng, X.; Xu, C.; Xia, D.; Liu, S.; Gao, S.; Zheng, F.; Liu, Y. Rapid discrimination of Chinese dry-cured hams based on Tri-step infrared spectroscopy and computer vision technology. *Spectrochim. Acta-Part A Mol. Biomol. Spectrosc.* **2020**, *228*, 117842. [CrossRef]
245. Huang, Y.; Andueza, D.; De Oliveira, L.; Zawadzki, F.; Prache, S. Comparison of visible and near infrared reflectance spectroscopy on fat to authenticate dietary history of lambs. *Animal* **2015**, *9*, 1912–1920. [CrossRef]
246. Varrà, M.O.; Fasolato, L.; Serva, L.; Ghidini, S.; Novelli, E.; Zanardi, E. Use of near infrared spectroscopy coupled with chemometrics for fast detection of irradiated dry fermented sausages. *Food Control* **2020**, *110*, 107009. [CrossRef]
247. Sanz, J.A.; Fernandes, A.M.; Barrenechea, E.; Silva, S.; Santos, V.; Gonçalves, N.; Paternain, D.; Jurio, A.; Melo-Pinto, P. Lamb muscle discrimination using hyperspectral imaging: Comparison of various machine learning algorithms. *J. Food Eng.* **2016**, *174*, 92–100. [CrossRef]
248. Sahar, A.; Dufour, É. Classification and characterization of beef muscles using front-face fluorescence spectroscopy. *Meat Sci.* **2015**, *100*, 69–72. [CrossRef]
249. Pereira, P.C. Milk nutritional composition and its role in human health. *Nutrition* **2014**. [CrossRef]
250. Kamal, M.; Karoui, R. Analytical methods coupled with chemometric tools for determining the authenticity and detecting the adulteration of dairy products: A review. *Trends Food Sci. Technol.* **2015**. [CrossRef]
251. Schmidt, W.F.; Broadhurst, C.L.; Qin, J.; Lee, H.; Nguyen, J.K.; Chao, K.; Hapeman, C.J.; Shelton, D.R.; Kim, M.S. Continuous temperature-dependent Raman spectroscopy of melamine and structural analog detection in milk powder. *Appl. Spectrosc.* **2015**. [CrossRef] [PubMed]
252. Karunathilaka, S.R.; Farris, S.; Mossoba, M.M.; Moore, J.C.; Yakes, B.J. Non-targeted detection of milk powder adulteration using Raman spectroscopy and chemometrics: Melamine case study. *Food Addit. Contam.-Part A Chem. Anal. Control. Expo. Risk Assess.* **2017**. [CrossRef] [PubMed]
253. Vieira, S.M.; de Souza, L.M.; França, A.S.; Oliveira, L.S.; Neto, W.B. FTMIR-PLS as a promising method for rapid detection of adulteration by waste whey in raw milk. *Dairy Sci. Technol.* **2016**. [CrossRef]
254. De Carvalho, B.M.A.; De Carvalho, L.M.; Dos Reis Coimbra, J.S.; Minim, L.A.; De Souza Barcellos, E.; Da Silva Júnior, W.F.; Detmann, E.; De Carvalho, G.G.P. Rapid detection of whey in milk powder samples by spectrophotometric and multivariate calibration. *Food Chem.* **2015**, *174*, 1–7. [CrossRef] [PubMed]
255. Da Paixão Teixeira, J.L.; dos Santos Caramês, E.T.; Baptista, D.P.; Gigante, M.L.; Pallone, J.A.L. Vibrational spectroscopy and chemometrics tools for authenticity and improvement the safety control in goat milk. *Food Control* **2020**, *112*, 107105. [CrossRef]
256. Karunathilaka, S.R.; Yakes, B.J.; He, K.; Chung, J.K.; Mossoba, M. Non-targeted NIR spectroscopy and SIMCA classification for commercial milk powder authentication: A study using eleven potential adulterants. *Heliyon* **2018**. [CrossRef]
257. Coimbra, P.T.; Bathazar, C.F.; Guimarães, J.T.; Coutinho, N.M.; Pimentel, T.C.; Neto, R.P.C.; Esmerino, E.A.; Freitas, M.Q.; Silva, M.C.; Tavares, M.I.B.; et al. Detection of formaldehyde in raw milk by time domain nuclear magnetic resonance and chemometrics. *Food Control* **2020**. [CrossRef]
258. Cui, J.; Zhu, D.; Su, M.; Tan, D.; Zhang, X.; Jia, M.; Chen, G. The combined use of 1H and 2D NMR-based metabolomics and chemometrics for non-targeted screening of biomarkers and identification of reconstituted milk. *J. Sci. Food Agric.* **2019**. [CrossRef]

259. Kene Ejeahalaka, K.; On, S.L.W. Effective detection and quantification of chemical adulterants in model fat-filled milk powders using NIRS and hierarchical modelling strategies. *Food Chem.* **2020**. [CrossRef]
260. Nedeljkovic, A.; Tomasevic, I.; Miocinovic, J.; Pudja, P. Feasibility of discrimination of dairy creams and cream-like analogues using Raman spectroscopy and chemometric analysis. *Food Chem.* **2017**. [CrossRef] [PubMed]
261. Boukria, O.; El Hadrami, E.M.; Sultanova, S.; Safarov, J.; Leriche, F.; Aït-Kaddour, A. 2D-Cross Correlation Spectroscopy Coupled with Molecular Fluorescence Spectroscopy for Analysis of Molecular Structure Modification of Camel Milk and Cow Milk Mixtures during Coagulation. *Foods* **2020**, *9*, 724. [CrossRef] [PubMed]
262. Mabood, F.; Jabeen, F.; Ahmed, M.; Hussain, J.; Al Mashaykhi, S.A.A.; Al Rubaiey, Z.M.A.; Farooq, S.; Boqué, R.; Ali, L.; Hussain, Z.; et al. Development of new NIR-spectroscopy method combined with multivariate analysis for detection of adulteration in camel milk with goat milk. *Food Chem.* **2017**. [CrossRef] [PubMed]
263. Yazgan Karacaglar, N.N.; Genis, H.E.; Bulat, T.; Topcu, A.; Durna, S.; Yetisemiyen, A.; Boyaci, I.H. Discrimination of milk species using Raman spectroscopy coupled with PLS-DA in raw and pasteurized milk. *J. Sci. Food Agric.* **2020**. [CrossRef]
264. Caredda, M.; Addis, M.; Ibba, I.; Leardi, R.; Scintu, M.F.; Piredda, G.; Sanna, G. Building of prediction models by using Mid-Infrared spectroscopy and fatty acid profile to discriminate the geographical origin of sheep milk. *LWT-Food Sci. Technol.* **2017**. [CrossRef]
265. Keller, M.D.; Lee, C.; Wang, W.; Wilson, B.; Connett, M. *Detecting Adulterants in Milk with Lower Cost Mid-Infrared and Raman Spectroscopy*; International Society for Optics and Photonics: Bellingham, WA, USA, 2018.
266. Cirak, O.; Icyer, N.C.; Durak, M.Z. Rapid detection of adulteration of milks from different species using Fourier Transform Infrared Spectroscopy (FTIR). *J. Dairy Res.* **2018**. [CrossRef]
267. Nieuwoudt, M.K.; Holroyd, S.E.; McGoverin, C.M.; Simpson, M.C.; Williams, D.E. Rapid, sensitive, and reproducible screening of liquid milk for adulterants using a portable Raman spectrometer and a simple, optimized sample well. *J. Dairy Sci.* **2016**. [CrossRef]
268. Nieuwoudt, M.K.; Holroyd, S.E.; McGoverin, C.M.; Simpson, M.C.; Williams, D.E. Raman spectroscopy as an effective screening method for detecting adulteration of milk with small nitrogen-rich molecules and sucrose. *J. Dairy Sci.* **2016**. [CrossRef]
269. Nedeljković, A.; Rösch, P.; Popp, J.; Miočinović, J.; Radovanović, M.; Pudja, P. Raman Spectroscopy as a Rapid Tool for Quantitative Analysis of Butter Adulterated with Margarine. *Food Anal. Methods* **2016**. [CrossRef]
270. Lohumi, S.; Lee, H.; Kim, M.S.; Qin, J.; Cho, B. Through-packaging analysis of butter adulteration using line-scan spatially offset Raman spectroscopy. *Anal. Bioanal. Chem.* **2018**, *410*, 5663–5673. [CrossRef]
271. Jaiswal, P.; Jha, S.N.; Borah, A.; Gautam, A.; Grewal, M.K.; Jindal, G. Detection and quantification of soymilk in cow-buffalo milk using Attenuated Total Reflectance Fourier Transform Infrared spectroscopy (ATR-FTIR). *Food Chem.* **2015**. [CrossRef] [PubMed]
272. Amjad, A.; Ullah, R.; Khan, S.; Bilal, M.; Khan, A. Raman spectroscopy based analysis of milk using random forest classification. *Vib. Spectrosc.* **2018**. [CrossRef]
273. Li, Q.; Yu, Z.; Zhu, D.; Meng, X.; Pang, X.; Liu, Y.; Frew, R.; Chen, H.; Chen, G. The application of NMR-based milk metabolite analysis in milk authenticity identification. *J. Sci. Food Agric.* **2017**, *97*, 2875–2882. [CrossRef] [PubMed]
274. Brandao, M.P.; Neto, M.G.; de Carvalho dos Anjos, V.; Bell, M.J.V. Detection of adulteration of goat milk powder with bovine milk powder by front-face and time resolved fluorescence. *Food Control* **2017**. [CrossRef]
275. Kalogianni, D.P. DNA-based analytical methods for milk authentication. *Eur. Food Res. Technol.* **2018**. [CrossRef]
276. Tsirigoti, E.; Katsirma, Z.; Papadopoulos, A.I.; Samouris, G.; Ekateriniadou, L.V.; Boukouvala, E. Application of triplex-PCR with an innovative combination of 3 pairs of primers for the detection of milk's animal origin in cheese and yoghurt. *J. Dairy Res.* **2020**. [CrossRef]
277. Cosenza, G.; Iannaccone, M.; Gallo, D.; Pauciullo, A. A fast and reliable polymerase chain reaction method based on short interspersed nuclear elements detection for the discrimination of buffalo, cattle, goat, and sheep species in dairy products. *Asian-Australasian J. Anim. Sci.* **2019**. [CrossRef]

278. An, J.; Jiang, Y.; Shi, B.; Wu, D.; Wu, W. Low-cost battery-powered and user-friendly real-time quantitative PCR system for the detection of multigene. *Micromachines* **2020**, *11*, 435. [CrossRef]
279. Šnirc, M.; Fekete, T.; Belej, L.; Židek, R.; Golian, J.; Hašcík, P.; Zajác, P.; Capla, J. Detection of ovine milk adulteration using taqman real-time PCR assay. *Potravin. Slovak J. Food Sci.* **2017**. [CrossRef]
280. Fekete, T.; Šnirc, M.; Belej, L.; Židek, R.; Golian, J.; Haščík, P.; Zeleňáková, L.; Zajác, P. Authentication of caprine milk and cheese by commercial qPCR assay. *Potravin. Slovak J. Food Sci.* **2017**. [CrossRef]
281. Bontempo, L.; Barbero, A.; Bertoldi, D.; Camin, F.; Larcher, R.; Perini, M.; Sepulcri, A.; Zicarelli, L.; Piasentier, E. Isotopic and elemental profiles of Mediterranean buffalo milk and cheese and authentication of Mozzarella di Bufala Campana PDO: An initial exploratory study. *Food Chem.* **2019**, *285*, 316–323. [CrossRef] [PubMed]
282. Salzano, A.; Manganiello, G.; Neglia, G.; Vinale, F.; De Nicola, D.; D'Occhio, M.; Campanile, G. A preliminary study on metabolome profiles of buffalo milk and corresponding mozzarella cheese: Safeguarding the authenticity and traceability of protected status buffalo dairy products. *Molecules* **2020**, *25*, 304. [CrossRef] [PubMed]
283. Liu, N.; Parra, H.A.; Pustjens, A.; Hettinga, K.; Mongondry, P.; van Ruth, S.M. Evaluation of portable near-infrared spectroscopy for organic milk authentication. *Talanta* **2018**, *184*, 128–135. [CrossRef] [PubMed]
284. Soares, S.; Amaral, J.S.; Oliveira, M.B.P.P.; Mafra, I. A Comprehensive Review on the Main Honey Authentication Issues: Production and Origin. *Compr. Rev. Food Sci. Food Saf.* **2017**, *16*, 1072–1100. [CrossRef]
285. Pita-Calvo, C.; Vázquez, M. Honeydew Honeys: A Review on the Characterization and Authentication of Botanical and Geographical Origins. *J. Agric. Food Chem.* **2018**, *66*, 2523–2537. [CrossRef]
286. Pascual-Maté, A.; Osés, S.M.; Fernández-Muiño, M.A.; Sancho, M.T. Métodos analíticos en mieles. *J. Apic. Res.* **2018**, *57*, 38–74. [CrossRef]
287. Se, K.W.; Wahab, R.A.; Syed Yaacob, S.N.; Ghoshal, S.K. Detection techniques for adulterants in honey: Challenges and recent trends. *J. Food Compos. Anal.* **2019**, *80*, 16–32. [CrossRef]
288. Maione, C.; Barbosa, F.; Barbosa, R.M. Predicting the botanical and geographical origin of honey with multivariate data analysis and machine learning techniques: A review. *Comput. Electron. Agric.* **2019**, *157*, 436–446. [CrossRef]
289. Stanek, N.; Teper, D.; Kafarski, P.; Jasicka-Misiak, I. Authentication of phacelia honeys (Phacelia tanacetifolia) based on a combination of HPLC and HPTLC analyses as well as spectrophotometric measurements. *LWT* **2019**, *107*, 199–207. [CrossRef]
290. Escuredo, O.; González-Martín, M.I.; Rodríguez-Flores, M.S.; Seijo, M.C. Near infrared spectroscopy applied to the rapid prediction of the floral origin and mineral content of honeys. *Food Chem.* **2015**, *170*, 47–54. [CrossRef]
291. Schwolow, S.; Gerhardt, N.; Rohn, S.; Weller, P. Data fusion of GC-IMS data and FT-MIR spectra for the authentication of olive oils and honeys—is it worth to go the extra mile? *Anal. Bioanal. Chem.* **2019**, *411*, 6005–6019. [CrossRef] [PubMed]
292. Grazia Mignani, A.; Ciaccheri, L.; Mencaglia, A.A.; Di Sanzo, R.; Carabetta, S.; Russo, M. Dispersive raman spectroscopy for the nondestructive and rapid assessment of the quality of southern Italian honey types. *J. Light. Technol.* **2016**, *34*, 4479–4485. [CrossRef]
293. Song, X.; She, S.; Xin, M.; Chen, L.; Li, Y.; Heyden, Y.V.; Rogers, K.M.; Chen, L. Detection of adulteration in Chinese monofloral honey using 1H nuclear magnetic resonance and chemometrics. *J. Food Compos. Anal.* **2020**, *86*. [CrossRef]
294. Cebrero, G.; Sanhueza, O.; Pezoa, M.; Báez, M.E.; Martínez, J.; Báez, M.; Fuentes, E. Relationship among the minor constituents, antibacterial activity and geographical origin of honey: A multifactor perspective. *Food Chem.* **2020**, *315*, 126296. [CrossRef] [PubMed]
295. Aykas, D.P.; Shotts, M.-L.; Rodriguez-Saona, L.E. Authentication of commercial honeys based on Raman fingerprinting and pattern recognition analysis. *Food Control* **2020**, *117*, 107346. [CrossRef]
296. Gan, Z.; Yang, Y.; Li, J.; Wen, X.; Zhu, M.; Jiang, Y.; Ni, Y. Using sensor and spectral analysis to classify botanical origin and determine adulteration of raw honey. *J. Food Eng.* **2016**, *178*, 151–158. [CrossRef]
297. Guelpa, A.; Marini, F.; du Plessis, A.; Slabbert, R.; Manley, M. Verification of authenticity and fraud detection in South African honey using NIR spectroscopy. *Food Control* **2017**, *73*, 1388–1396. [CrossRef]

298. Ferreiro-González, M.; Espada-Bellido, E.; Guillén-Cueto, L.; Palma, M.; Barroso, C.G.; Barbero, G.F. Rapid quantification of honey adulteration by visible-near infrared spectroscopy combined with chemometrics. *Talanta* **2018**, *188*, 288–292. [CrossRef] [PubMed]
299. Aliaño-González, M.J.; Ferreiro-González, M.; Espada-Bellido, E.; Palma, M.; Barbero, G.F. A screening method based on Visible-NIR spectroscopy for the identification and quantification of different adulterants in high-quality honey. *Talanta* **2019**, *203*, 235–241. [CrossRef]
300. Ghanavati Nasab, S.; Javaheran Yazd, M.; Marini, F.; Nescatelli, R.; Biancolillo, A. Classification of honey applying high performance liquid chromatography, near-infrared spectroscopy and chemometrics. *Chemom. Intell. Lab. Syst.* **2020**, *202*, 104037. [CrossRef]
301. Cengiz, M.F.; Durak, M.Z. Rapid detection of sucrose adulteration in honey using Fourier transform infrared spectroscopy. *Spectrosc. Lett.* **2019**, *52*, 267–273. [CrossRef]
302. Bázár, G.; Romvári, R.; Szabó, A.; Somogyi, T.; Éles, V.; Tsenkova, R. NIR detection of honey adulteration reveals differences in water spectral pattern. *Food Chem.* **2016**, *194*, 873–880. [CrossRef] [PubMed]
303. Huang, F.; Song, H.; Guo, L.; Guang, P.; Yang, X.; Li, L.; Zhao, H.; Yang, M. Detection of adulteration in Chinese honey using NIR and ATR-FTIR spectral data fusion. *Spectrochim. Acta-Part A Mol. Biomol. Spectrosc.* **2020**, *235*, 118297. [CrossRef] [PubMed]
304. Ackermann, S.M.; Lachenmeier, D.W.; Kuballa, T.; Schütz, B.; Spraul, M.; Bunzel, M. NMR-based differentiation of conventionally from organically produced chicken eggs in Germany. *Magn. Reson. Chem.* **2019**, *57*, 579–588. [CrossRef] [PubMed]
305. Rogers, K.M.; Van Ruth, S.; Alewijn, M.; Philips, A.; Rogers, P. Verification of Egg Farming Systems from the Netherlands and New Zealand Using Stable Isotopes. *J. Agric. Food Chem.* **2015**, *63*, 8372–8380. [CrossRef] [PubMed]
306. Mi, S.; Shang, K.; Zhang, C.H.; Fan, Y.Q. Characterization and discrimination of selected chicken eggs in China's retail market based on multi-element and lipidomics analysis. *Food Res. Int.* **2019**, *126*. [CrossRef] [PubMed]
307. Puertas, G.; Vázquez, M. Fraud detection in hen housing system declared on the eggs' label: An accuracy method based on UV-VIS-NIR spectroscopy and chemometrics. *Food Chem.* **2019**, *288*, 8–14. [CrossRef]
308. Ballin, N.Z.; Laursen, K.H. To target or not to target? Definitions and nomenclature for targeted versus non-targeted analytical food authentication. *Trends Food Sci. Technol.* **2019**, *86*, 537–543. [CrossRef]
309. Donarski, J.; Camin, F.; Fauhl-Hassek, C.; Posey, R.; Sudnik, M. Sampling guidelines for building and curating food authenticity databases. *Trends Food Sci. Technol.* **2019**, *90*, 187–193. [CrossRef]
310. Amigo, J.M.; Grassi, S. Configuration of hyperspectral and multispectral imaging systems. *Data Handl. Sci. Technol.* **2020**, *32*, 17–34. [CrossRef]
311. Rodriguez-Saona, L.; Aykas, D.P.; Borba, K.R.; Urbina, A.U. Miniaturization of Optical Sensors and their Potential for High-Throughput Screening of Foods. *Curr. Opin. Food Sci.* **2020**. [CrossRef]

foods

MDPI

Article

Flash Gas Chromatography in Tandem with Chemometrics: A Rapid Screening Tool for Quality Grades of Virgin Olive Oils

Sara Barbieri [1], Chiara Cevoli [1], Alessandra Bendini [1,*], Beatriz Quintanilla-Casas [2,3], Diego Luis García-González [4] and Tullia Gallina Toschi [1]

1 Department of Agricultural and Food Science, Alma Mater Studiorum-Università di Bologna, 47521 Cesena, Italy; sara.barbieri@unibo.it (S.B.); chiara.cevoli3@unibo.it (C.C.); tullia.gallinatoschi@unibo.it (T.G.T.)

2 Department de Nutrició, Ciències de l'Alimentació i Gastronomia, Campus De l'Alimentació Torribera, Facultat de Farmàcia i Ciències de l'Alimentació, Universitat de Barcelona, 08921 Santa Coloma de Gramenet, Spain; beatrizquintanilla@ub.edu

3 Institut de Recerca en Nutrició i Seguretat Alimentària (INSA-UB), Universitat de Barcelona (UB), 08921 Santa Coloma de Gramenet, Spain

4 Instituto de la Grasa (CSIC), 41013 Sevilla, Spain; dlgarcia@ig.csic.es

* Correspondence: alessandra.bendini@unibo.it; Tel.: +39-0547-338121

Received: 11 June 2020; Accepted: 29 June 2020; Published: 2 July 2020

Abstract: This research aims to develop a classification model based on untargeted elaboration of volatile fraction fingerprints of virgin olive oils (n = 331) analyzed by flash gas chromatography to predict the commercial category of samples (extra virgin olive oil, EVOO; virgin olive oil, VOO; lampante olive oil, LOO). The raw data related to volatile profiles were considered as independent variables, while the quality grades provided by sensory assessment were defined as a reference parameter. This data matrix was elaborated using the linear technique partial least squares-discriminant analysis (PLS-DA), applying, in sequence, two sequential classification models with two categories (EVOO vs. no-EVOO followed by VOO vs. LOO and LOO vs. no-LOO followed by VOO vs. EVOO). The results from this large set of samples provide satisfactory percentages of correctly classified samples, ranging from 72% to 85%, in external validation. This confirms the reliability of this approach in rapid screening of quality grades and that it represents a valid solution for supporting sensory panels, increasing the efficiency of the controls, and also applicable to the industrial sector.

Keywords: virgin olive oil; quality; volatile compounds; sensory analysis; chemometrics

---

1. Introduction

The official methodology for sensory evaluation of virgin olive oils (VOOs), known as a panel test, is a fundamental tool to assess the quality of products that cannot be replaced by instrumental methods, considering that the overall and complex perceptual attributes (e.g., fruity and defects) are the indicators of the quality of VOOs. Despite its proven effectiveness in evaluating the quality grades of samples, tested in EU countries since 1991 [1,2], the scientific community has highlighted some drawbacks on its application that are mainly related to the following: (i) the reproducibility of results among different panels; (ii) critical attribution of the category when, e.g., a defect is borderline; (iii) costs, assessor fatigue and other limitations associated with a method working with humans.

Specifically, according to decisions taken at International Olive Council (IOC) level, the Reg. (EU) 1348/2013 [3] recommends the number of oils to be assessed by the sensory panels, fixing a maximum number of four samples at each session. Moreover, a maximum of three sessions per day is specified, to leave enough time between a session and another, thus avoiding the contrast effect that could be produced by immediately tasting sequences of samples. These specifications strongly limit the number of samples that can be assessed by one panel per day. On the other hand, to enhance panel skills in recognizing, identifying, and quantifying sensory attributes, the introduction of new artificial reference materials (obtained by chemical or biotechnological approaches), could improve the proficiency of the individual panels and their global alignment by overcoming some limitations associated with a natural matrix (e.g., limited amounts available, difficultly obtaining, low homogeneity year by year) and offering advantages such as preparation in each laboratory, reproducibility over time, possibility of purchase, and therefore their availability for the market.

In this context, the development of an instrumental method for rapid screening of quality grades of samples (extra virgin olive oil, EVOO; virgin olive oil, VOO; lampante olive oil, LOO) could represent a solution to support sensory panels (particularly for large private industries), decreasing their daily work by reducing the samples that need to be assessed (e.g., by excluding those definitely compliant), with a consequent increase in the efficiency of quality controls and reducing the number of samples that need to be controlled.

In this way, improvement of the activity of sensory panels, whose work remains central to ensuring the quality of the product, would be achieved by focusing sensory analysis only on uncertain samples (i.e., borderline oils between two product categories that can be the object of disagreement among panels).

It is well known that volatile compounds are crucial to determine VOO quality and that they are responsible for the different VOO sensory profiles [4–6]; their determination in a rapid way (e.g., screening method) could support sensory analysis and represents one of the current challenges in the olive oil sector where fast, accurate, and easy-to-use approaches providing real-time results are required.

Recently, different analytical techniques combined with chemometric statistical approaches have been proposed to predict sensory information [7–9].

Alongside the traditional techniques (targeted) in which specific and selected molecular markers are monitored during the analysis to assess the presence or absence of compounds and their quantification, untargeted analyses, based on a holistic approach and able to provide information such as a spectral fingerprint, giving a simplified and overall picture of the food under analysis, have gained an increasing relevance over the last years [10].

Among the latter, different analytical methods for determination of volatile compounds combined with multivariate chemometric techniques for VOOs quality testing have been described in the literature and proposed to the industrial sector as fast and high throughput screening techniques [9,11–18].

In particular, as an alternative to headspace gas chromatography-mass spectrometry (HS-GC/MS), which is the most widely used technique to quantify and characterize the profiles of volatile compounds of VOOs thanks to its high sensitivity and selectivity, the application of the HS-GC ion mobility spectrometry (HS-GC-IMS) has been proposed. This technique combines high selectivity and sensitivity with high robustness and cost-efficiency, and has given promising results in discriminating VOOs according to quality grades [9,11,12,14,18] or geographical origin [13,15].

The need to support organoleptic analysis was also reported in a specific call of the Horizon 2020 EU program (H2020-SFS-14a-2014) and is one of the main objectives of the OLEUM project (Horizon 2020, Grant Agreement No. 635690). In the framework of this project, two analytical instrumental techniques, headspace-solid phase micro extraction–gas-chromatography/mass spectrometry (HS-SPME-GC/MS) and flash gas chromatography (FGC) based on the determination of volatile compounds, have been proposed as the most promising rapid screening methods that can support sensory panels in the determination of quality grades.

In a recent work by Quintanilla-Casas and co-authors (2020) [17], the results obtained with HS-SPME-GC/MS with a fingerprinting approach to classify VOO categories has been demonstrated. Herein, a classification model based on minor fraction fingerprints that is able to predict the commercial category of olive oil samples (EVOOs, VOOs, LOOs) obtained by FGC is presented. The FGC is an innovative analytical approach for analysis of volatile compounds of VOOs based on the FGC separation: the headspace of VOOs, previously conditioned, is sampled by a syringe, the volatile organic compounds are adsorbed on a Tenax trap and subsequently desorbed by rapid heating, and, finally, transferred to a FGC step. The elution of analytes runs in parallel using two metal capillary columns with different polarity of the stationary phase. This gives rise to slight differences in the separation capability of molecules that are detected by a flame ionization detector (FID) located at the end of each column.

The main advantage of the FGC technique is its short analysis time (total separation time is 100 s); moreover, its application associated with sensory analysis for calibration and chemometric tools is promising to support the work of panel tests in discriminating samples of different product categories. A classification model, once built, could be easily applied in any laboratory or industry.

The effectiveness of this technique is already demonstrated by previous works aimed to differentiate VOOs according to their geographical origin declared by labels such as "100% Italian" and "non-100% Italian" oils [19] or "EU" and "extra-EU" [16].

The aim of this study was to classify VOOs according to quality grade, combining FGC data with the multivariate classification technique partial least squares discriminant analysis (PLS-DA). To provide robustness to our model, a set of 331 oils belonging to the three different commercial categories (EVOO, VOO, LOO) involving two harvesting/production years was analyzed. The adopted validation protocol (repeatability and reproducibility tests) and related performance are also shown.

2. Materials and Methods

2.1. Olive Oil Samples

An initial set of 334 EVOOs, VOOs, and LOOs oils representative of the most common olive cultivars, geographical origin, sensory positive attributes, and sensory defects were sampled. Specifically, in addition to a first set of 180 oils collected during the first year of the OLEUM project (2016–2017 olive season), another set of 154 samples (2017–2018 olive season) was collected and analyzed during the second year (Tables S1–S4 in the Supplementary Materials).

The panel test method was carried out by six panels involved in the OLEUM project as described by Barbieri et al. 2020 [20] and sensory data were expressed as mean of medians. The procedure deals with possible disagreement between panels with a decision tree in order to have definitive classification of samples in which definitive agreement is reached. In agreement with the sensory results reported in Tables S1 and S2 (Supplementary Materials), in the first year of the project 178 of 180 samples were immediately classified by panels (54 EVOO, 78 VOO, and 48 LOO). Classification was not possible for only two samples (UN_10, UP_14), as agreement among panels was not reached on the category (V/L). The sensory evaluation of oils from the second sampling allowed classification of 153 oils (69 EVOO, 51 VOO and 33 LOO); 1 sample was not classified due to an anomalous lemon smell (ZRS_1) and was therefore excluded from the set [20]. For these reasons, the classification model was built on 331 samples.

The oils collected were representative of possible commercial samples and borderline samples that can be the object of disagreement between panels in terms of sensory characteristics. Different aliquots of the samples, stored in the lab at 10–12 °C (for sensory analysis) and at −18 °C (for instrumental analysis), were reconditioned at room temperature before analysis.

2.2. Analytical Conditions

The FGC system (FGC-E-nose Heracles II, AlphaMos, Toulouse, France) is based on the technology of ultra-fast gas-chromatography.

The FGC is equipped with two columns working in parallel: a non-polar column (MXT5: 5% diphenyl, 95% methylpolysiloxane, 10 m length and 180 μm diameter) and a polar column (MXT-1701:

14% cyanopropylphenyl/86% dimethyl polysiloxane, 10 m length, 180 μm diameter). At the end of each column, a FID detector is placed and the acquired signal is digitalized every 0.01 s.

The analytical conditions applied were the same described by Melucci et al. 2016 [19]. The only difference was related to the temperature of the conditioning step of the samples before injection: the vial is placed in the auto-sampler (HS 100, CTC Analytics), which moves it in a shaker oven where it remains for 20 min at 40 °C, shaken at 500 rpm.

2.3. Validation Protocol

To confirm that the analytical procedure employed has performance capabilities consistent with the required application, a validation strategy for non-targeted approaches was performed.

A QC (quality control) sample, representative of the qualitative and quantitative VOO volatile composition (presence of volatile compounds along the entire interval of the chromatogram), was used. In this study, the QC sample was obtained by pooling the same volume of three case-control samples (1 EVOO, 1 VOO with median of 1.9 for fusty-muddy defect, and 1 VOO with a median of 2.5 for rancid defect) and seven replicates were taken into consideration.

The quality of the instrumental performance intended for fingerprinting analysis was checked by the calculation of the relative standard deviation (RSD) as proposed by the Food and Drug Administration [21]. Specifically, the repeatability (intra-day repeatability and inter-day repeatability performed according to EC 657/2002) [22] of the chromatographic signal evaluated in terms of RSD% of each chromatogram data point, with intensities above noise signal of the replicates of the same QC samples, was considered [23,24].

Prior to RSD calculation, data were aligned using the COW algorithm (correlation optimized warping) [25] and autoscaled (mean-centering followed by division of variable by the standard deviation of that column) to correct shifts in retention time and possible differences in the signal amplification of the instrument. All elaborations were made using PLS Toolbox for Matlab (MatlabR2018a®) (Natick, MA, USA).). For calculation of RSD% for each chromatogram data point, the evaluation and exclusion of noise signal is carried out to avoid considering non-relevant RSD%.

For precision, the FDA recommends a RSD not higher than 15% regarding the analytical variability for target analysis, except for concentrations close to the detection limit where a RSD of 20% is acceptable (FDA Bioanalytical Method Validation-Guidance for Industry, 2018). This, in agreement with the trend described by the Horwitz equation for targeted methods [26], demonstrates that the repeatability is strongly correlated with the intensity of the variables.

Although fingerprinting represents a different analytical approach and more variation is expected when doing untargeted analysis, these guidelines are used as a benchmark towards repeatability evaluation. Specifically, for intra-day repeatability, the acceptance criteria were as follows: more than 90% of signals with RSD < 15%; more than 95% of signals with RSD < 20% and distribution of RSD% vs. signal intensity in accordance with the Horwitz equation. For inter-day repeatability or within-lab reproducibility, the acceptance criteria were as follows: more than 85% of signals with RSD < 15%, more than 90% of signals with RSD < 20% and distribution of RSD% vs. signal intensity in accordance with Horwitz's equation.

In addition, the examination of system performance by checking the signal to noise ratio in standard solutions (instead of the evaluation of representative VOO profiles) to facilitate the assessment and comparison of method sensitivity for other laboratories was proposed. The sensitivity of the analytical system was evaluated by analyzing 2 g of each standard solution in refined olive oil (ethanol 0.05 mg·kg^{-1}, CAS Number 64-17-5; assay ≤ 97.2%; density 0.789 g/mL at 25 °C; hexanal, 0.1 mg·kg^{-1} CAS Number 66-25-1; assay ≥ 95% (GC); density 0.815 g/mL at 25 °C; (*E*)-2-hexenal, 0.75 mg·kg^{-1} CAS Number 6728-26-3; assay ≥ 97.0% (GC); density 0.846 g/mL at 25 °C). The S/N (S = intensity of the peak of the compound; N = mean intensity of the noise measured considering the baseline of the chromatographic zone between 43 and 50 s) for the selected analytes in the chromatograms should be >3 (acceptance criteria).

2.4. Classification Models

In order to predict the assignment of samples to a specific quality grade, full chromatograms were used to develop classification models. The raw data of each chromatogram, for a total of 19,900 points, were aligned by the COW algorithm and autoscaled using PLS Toolbox for Matlab (MatlabR2018a®). Subsequently, the noise was excluded and 8401 points were consecutively selected from first to last peak observed in the chromatogram.

Subsequently, PLS-DA (partial least square discriminant analysis) models [27] were built by using the intensity values of the points as variables X (matrix X), while the commercial categories (EVOO, VOO, LOO) were considered as variable Y. In particular, classification models with 2 categories were developed in sequence: EVOO vs. no-EVOO followed by VOO vs. LOO and LOO vs. no-LOO followed by EVOO vs. VOO, as proposed by Quintanilla-Casas et al. 2020 [17].

The sample dataset was split in calibration (venetian blinds cross validation, including 75% of the samples) and external validation set (25% of the samples) by using the Kennard–Stone method [28]. The dataset was deposited for possible consultation in an on-line repository [29].

The threshold value able to identify the belonging category of each sample into one of the groups was defined by using a probabilistic approach based on Bayes's rule [30]. Finally, to assess the goodness of the method, the receiver operating characteristic (ROC) curves were evaluated.

3. Results and Discussion

3.1. Performance of FGC

Most of the procedures proposed in the literature for validation of non-targeted methods focus on post-analytical data treatment and validation of statistical models. Nevertheless, a few studies have investigated control procedures as well as performance criteria and requirements to ensure the consistence of the analytical signal (fingerprint) [24,31].

Conventional performance criteria adopted for targeted methods are not applicable as such to fingerprinting methods. Fingerprinting methods intended for sample classification are not aimed at identification and quantification of analytes, but on finding distinctive patterns that are specific for a given food category (i.e., VOO commercial category) in raw analytical signals (i.e., chromatograms). Therefore, the main constraint of the fingerprinting analytical method is to provide a repeatable and reproducible signal with sufficient sensitivity to collect the information from samples for the final purpose of the method, i.e., quality classification.

For evaluation of intra-day repeatability, the pooled QC sample was analyzed by the same operator with the same equipment and in the same instrument operative conditions within the same day. For each variable (data points), mean value, SD, and RSD% were calculated considering the seven replicates. More than 97.5% of signals presented RSD < 10%, while it achieves 99.8% in correspondence of RSD < 20% (Table 1). To analyze the variability as related to the magnitude of the variables, RSD% was plotted versus signal intensity (data not shown). As expected, data points with RSD > 10% are characterized by low values of intensity. This is in agreement with the trend described by the Horwitz equation for targeted methods [26].

In the case of the inter-day repeatability (within-lab reproducibility), seven replicates of the pooled QC sample were analyzed by the same operator with the same equipment but on different days, consequently involving different environmental conditions, and the mean value, SD, and RSD% were calculated. More than 91% and 99.4% of the signals presented RSD < 10% and RSD < 20%, respectively (Table 1). A relation between intensity and RDS% was also observed in this study, similarly to that previously observed in the intra-day repeatability test.

As the fingerprinting approach intended for sample classification is not aimed in determining the concentration of single analytes, limits of detection or quantification cannot be calculated for the analytical outcome. However, the analytical method needs to be sufficiently sensitive to allow detection of minor constituents to avoid missing any valuable information.

Table 1. Frequency of each relative standard deviation percentage (RSD%) class obtained for intra-day and inter-day repeatability evaluated on the quality control (QC) sample.

Intra-Day Repeatability			Inter-Day Repeatability		
Class (RSD%)	**Frequency**	**%**	**Class (RSD%)**	**Frequency**	**%**
10	8185	97.5	10	7690	91.5
15	166	1.5	15	552	6.6
20	33	0.8	20	106	1.3
30	17	0.2	30	16	0.2
40	0	0	40	37	0.4
50	0	0	50	0	0
60	0	0	60	0	0
70	0	0	70	0	0
80	0	0	80	0	0
90	0	0	90	0	0
100	0	0	100	0	0

On this basis, the method's sensitivity needs to be set as a reference parameter to be evaluated in the validation process. A target-type strategy applied to standard solutions was proposed.

Standard solution compounds were chosen as most representative of the qualitative and quantitative volatile composition of VOOs, especially regarding the presence of volatile compounds over the entire interval of the chromatogram considered in fingerprinting analysis. Differences between the concentrations used for each compound are related to their different amounts generally present in a VOO sample. Results of the S/N are reported in Table 2.

Table 2. Concentration (mg·kg^{-1}) of each compound included in the standard solution used for method's sensitivity evaluation and related S/N. The standard mix were prepared by spiking refined olive oil with each compound and analysed by flash gas chromatography (FGC). S = intensity of the peak of the compound; N = mean intensity of the noise measured considering the baseline of the chromatographic zone between 43 and 50 s.

Compound	Retention Time	Concentration (mg·kg^{-1})	S/N
Ethanol	21.8	0.05	3.84 ± 0.99
Hexanal	55.6	0.1	5.55 ± 0.96
(*E*)-2-Hexenal	62.0	0.75	4.42 ± 1.82

3.2. Classification Models

A fingerprinting approach involving chemometric elaboration of the entire profiles in volatile molecules without identification and quantification was applied.

Two different classification strategies were taken into account: (i) a classification model able to discriminate EVOO and no-EVOO samples, followed by a model to classify VOO vs. LOO samples; (ii) a classification model able to discriminate LOO and no-LOO samples, followed by a model to classify VOO vs. EVOO samples.

The results, in terms of percentage and number of correctly classified samples, are reported in Table 3 for cross and external validation, respectively. Regarding the first classification strategy, the percentages of correctly classified samples ranged from 72 to 89% and from 72 to 85%, for cross and external validation, respectively. In particular, the best results were obtained during the second step useful to discriminate VOO vs. LOO. For the second strategy, conceptually more correct in terms of sequence because it first discriminates LOO which are not edible if not refined, the percentage ranged from 78 to 92% and from 73 to 85%, for cross and external validation, respectively. In this case, the highest percentages were reached using the first PLS-DA model (LOO vs. no-LOO). Furthermore, this latter model was the best of all PLS-DA models developed.

In general, the percentages are in the same range as those obtained by other authors who proposed chemometric models to discriminate VOO quality grades according to their volatile profile analyzed by different instrumental techniques [9,17].

The ROC curves (Figure 1) evaluated the sensitivity (number of samples predicted as in the class divided by number actually in the class) and the specificity (number of samples predicted as not in the class divided by actual number not in the class) of all PLS-DA models (external validation) [16]. In particular, the area under the curve (AUC) identifies the degree of discrimination (ranged 0.8148 to 0.8899) and suggests that all the models are characterized by a good degree of discrimination.

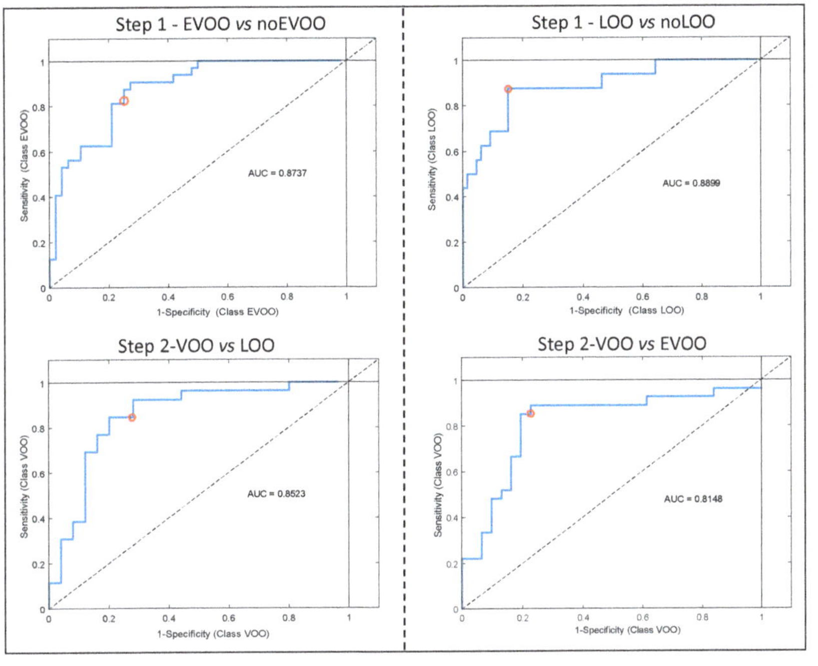

Figure 1. Receiver operating characteristic (ROC) curves of all developed PLS-DA models used to discriminate samples according to quality grade; the red circles identify the sensitivity (number of samples predicted as in the class divided by number actually in the class) and the specificity (number of samples predicted as not in the class divided by actual number not in the class) of the models. EVOO = extra virgin olive oil; VOO = virgin olive oil, LOO = lampante Olive Oil.

The results of all the models (cross and external validation), in term of probability of belonging to the correct class, are shown in Figure 2. The threshold value was fixed at 0.5, corresponding to a probability of 50%: a sample classified with a probability lower than this is considered as not correctly grouped [32].

The definition of a probability level, ranging from 50% to 100%, could be a means of identifying uncertain samples that need to be checked by sensory evaluation. In other words, the samples classified with a probability lower than the selected probability level should be submitted to panel test. These procedures would reduce the amount of the samples analyzed by the panel, but at the same time, it would insure the accuracy of the classification.

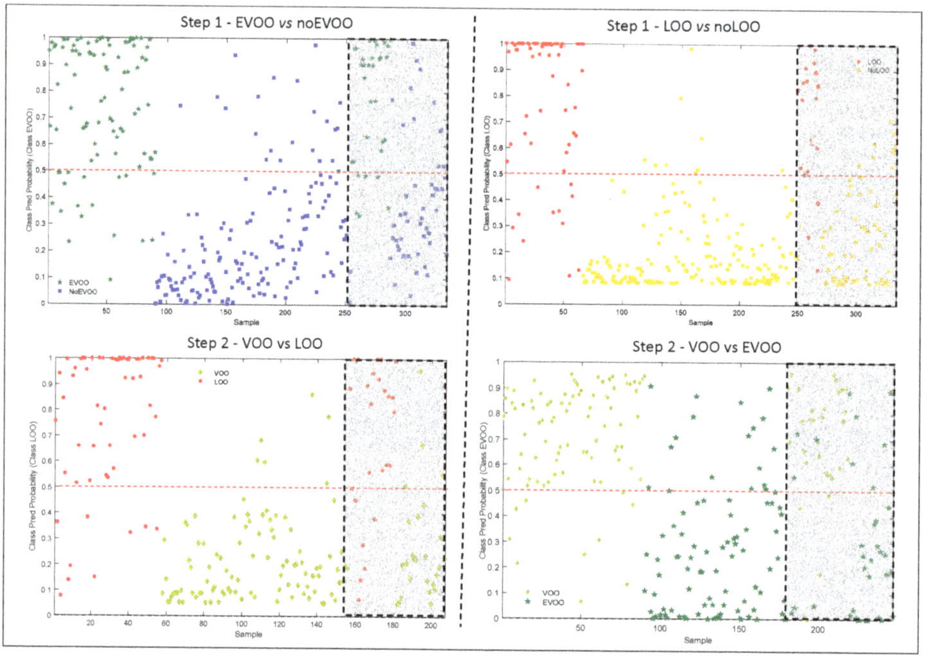

Figure 2. Class prediction probability of all samples used to develop the models, in cross and external validation (grey area). Step 1—EVOO (green star) vs. no-EVOO (blue square); step 2—VOO (yellow diamond) vs. LOO (red circle); step 1—LOO (red circle) vs. no-LOO (yellow square); step 2—VOO (yellow diamond) vs. EVOO (green star). EVOO = extra virgin olive oil; VOO = virgin olive oil, LOO = lampante olive oil.

Table 3. Results in terms of percentage and number of samples correctly classified in cross and external validation of the two classification strategies applied based on the partial least squares-discriminant analysis (PLS-DA) sequential model. EVOO = extra virgin olive oil; VOO = virgin olive oil, LOO = lampante olive oil.

1st CLASSIFICATION STRATEGY		**2nd CLASSIFICATION STRATEGY**	
1st Step: EVOO vs. no-EVOO		**1st Step: LOO vs. no-LOO**	
Cross validation EVOO: 70/90 (78%) No-EVOO: 132/164 (81%) TOTAL: 202/254 = 80%	External validation EVOO: 26/32 (81%) No-EVOO: 37/48 (77%) TOTAL: 63/80 = 79%	Cross validation LOO: 50/61 (81%) No-LOO: 172/188 (92%) TOTAL: 222/249 = 89%	External validation LOO: 17/20 (85%) No-LOO: 55/65 (85%) TOTAL: 72/85 = 85%
2nd Step: VOO vs. LOO		**2nd Step: VOO vs. EVOO**	
Cross validation VOO: 88/99 (89%) LOO: 41/57 (72%) TOTAL: 129/156 = 83%	External validation VOO: 22/26 (85%) LOO: 18/25 (72%) TOTAL: 41/51 = 78%	Cross validation VOO: 84/95 (88%) EVOO: 74/94 (78%) TOTAL: 158/189 = 84%	External validation VOO: 23/27 (85%) EVOO: 24/33 (73%) TOTAL: 47/60 = 78%

4. Conclusions

Despite the undisputed validity of the panel test, its application is time consuming and expensive. Accordingly, companies and private and public quality control labs could benefit from robust instrumental pre-classifications, which would reduce the number of samples that have to be assessed by panels, or at least prioritize their assessment.

For this reason, the development of rapid screening methods to support the official panel test, to analyze olive oils and differentiate their quality grades, is one of the challenges in the olive oil sector, as reported in the EU framework program Horizon 2020.

In this work, FGC combined with the multivariate statistical technique was applied to discriminate samples according to different quality grades (EVOO, VOO and LOO; examples of GC traces for EVOOs and LOOs are shown in Figure S1 of the Supplementary Materials). The analytical technique proposed herein for fingerprinting olive oils combined with chemometrics was effective in reducing data complexity and time to obtain a response; this rapid screening tool could be adopted for a quick pre-classification of the quality grades, e.g., by control laboratories in companies of the OO sector, before buying or blending EVOOs.

In order to propose a robust chemometric model, a large set of samples (n = 331) involving two different harvesting/production years, the most common olive cultivars, geographical origin, sensory positive attributes, and sensory defects, was analyzed. In addition, a validation protocol was adopted for evaluate the reliability of the results.

The proposed analytical fingerprinting method provided repeatable and reproducible signals with sufficient sensitivity to collect valuable information about samples.

FGC associated with the two-category sequential classification model is promising to support sensory analysis in discriminating samples of different product categories. Among the proposed classification strategy, the second (1st step: LOO vs. no-LOO; 2nd step: VOO vs. EVOO) was the best of all PLS-DA models developed with percentages of correctly classified samples ranging from 78 to 92% and from 73 to 85%, for cross and external validation, respectively.

This analytical approach is very fast, and, in fact, only around 200 s are needed to analyze a single sample. The classification model, built by using a high number of robust samples classified by sensorial analysis and representative of the commercial variability (here we used a decision tree and six panels to ensure their classification) is easily applicable in any laboratory or industry.

Future studies could be addressed to the implementation of this methodology, even in relation to an increasing interest of the food sector towards volatile compounds and more widespread use of instruments such as FGC, which are less common in quality control laboratories. An even wider sampling phase including other variables among oils, since they are natural products, could lead to a better control of classifications and would lead to implementation of this technique to a broader extent. Lastly, the use of other statistical approaches, such as nonlinear techniques, could be investigated in order to improve the results of classification.

Supplementary Materials: The following are available online at http://www.mdpi.com/2304-8158/9/7/862/s1, Table S1: Sensory results of samples from the first year. Table S2: Sensory results of samples from the second year. Table S3: available information on samples collected and evaluated during the first year of the Oleum project. Table S4: available information on samples collected and evaluated during the second year of the Oleum project. Figure S1: overlapping of the GC traces of extra virgin (EVOO) and lampante (LOO) samples.

Author Contributions: Conceptualization, S.B., A.B. and T.G.T.; Formal analysis, S.B. and C.C.; Data curation, S.B. and C.C.; Writing—original draft preparation, S.B.; Writing—review and editing, S.B., C.C., A.B., B.Q.-C.; Supervision, T.G.T. and D.L.G.-G.; Funding acquisition, T.G.T. All authors have read and agree to the published version of the manuscript.

Funding: This work is supported by the Horizon 2020 European Research project OLEUM "Advanced solutions for assuring the authenticity and quality of olive oil at a global scale", which received funding from the European Commission within the Horizon 2020 Programme (2014–2020), grant agreement No. 635690.

Acknowledgments: The information expressed in this article reflects the authors' views; the European Commission is not liable for the information contained herein. We are grateful to all producers who provided us with VOOs for this study as well as the panel members who performed sensory analysis of VOOs from each institution involved: Eurofins Analytik GmbH, Hamburg, Germany; Institute of Agriculture and Tourism, Poreč, Croatia; Institut des Corps Gras, Pessac, France; Alma Mater Studiorum-Università di Bologna; Science and Research Centre Koper, Slovenia and Ulusal Zeytin ve Zeytinyağı Konseyi, Izmir, Turkey.

Conflicts of Interest: The authors declare no conflict of interest. The funders had no role in the design of the study; in the collection, analyses, or interpretation of data; in the writing of the manuscript, or in the decision to publish the results.

References

1. Fernandes, G.D.; Ellis, A.C.; Gámbaro, A.; Barrera-Arellano, D. Sensory evaluation of high-quality virgin olive oil: Panel analysis versus consumer perception. *Curr. Opin. Food Sci.* **2018**, *21*, 66–71. [CrossRef]
2. Conte, L.; Bendini, A.; Valli, E.; Lucci, P.; Moret, S.; Maquet, A.; Lacoste, F.; Brereton, P.; García-González, D.L.; Moreda, W.; et al. Olive oil quality and authenticity: A review of current EU legislation, standards, relevant methods of analyses, their drawbacks and recommendations for the future. *Trends Food Sci. Technol.* **2019**. [CrossRef]
3. Official Journal of the European Union. *European Union, Commission Regulation 1348/2013. Amending Regulation (EEC) No 2568/91*; Official Journal of the European Union: Brussels, Belgium, 2013; Volume L338, pp. 31–67.
4. Morales, M.T.; Luna, G.; Aparicio, R. Comparative study of virgin olive oil sensory defects. *Food Chem.* **2005**, *91*, 293–301. [CrossRef]
5. Vichi, S.; Guadayol, J.M.; Caixach, J.; López-Tamames, E.; Buxaderas, S. Comparative study of different extraction techniques for the analysis of virgin olive oil aroma. *Food Chem.* **2007**, *105*, 1171–1178. [CrossRef]
6. Romero, I.; García-González, D.L.; Aparicio-Ruiz, R.; Morales, M.T. Validation of SPME–GCMS method for the analysis of virgin olive oil volatiles responsible for sensory defects. *Talanta* **2015**, *134*, 394–401. [CrossRef]
7. Inarejos Garcia, A.M.; Gómez-Alonso, S.; Fregapane, G.; Desamparados Salvador, M. Evaluation of minor components, sensory characteristics and quality of virgin olive oil by near infrared (NIR) spectroscopy. *Food Res. Int.* **2013**, *50*, 250–258. [CrossRef]
8. Cecchi, L.; Migliorini, M.; Giambanelli, E.; Rossetti, A.; Cane, A.; Melani, F.; Mulinacci, N. Headspace solid-phase microextraction–gas chromatography–mass spectrometry quantification of the volatile profile of more than 1200 virgin olive oils for supporting the panel test in their classification: Comparison of different chemometric approaches. *J. Agric. Food Chem.* **2019**, *67*, 9112–9120. [CrossRef]
9. Contreras, M.; Jurado-Campos, N.; Arce, L. A robustness study of calibration models for olive oil classification: Targeted and non-targeted fingerprint approaches based on GC-IMS. *Food Chem.* **2019**, *288*, 315–324. [CrossRef]
10. Fiorino, G.M.; Garino, C.; Arlorio, M.; Logrieco, A.F.; Losito, I.; Monaci, L. Overview on Untargeted Methods to Combat Food Frauds: A Focus on Fishery Products (2018). *J. Food Qual.* **2018**, *1*, 1–13. [CrossRef]
11. Garrido-Delgado, R.; Mar Dobao-Prieto, M.; Arce, L.; Valcárcel, M. Multi-capillary column-ion mobility spectrometry: A potential screening system to differentiate virgin olive oils. *Anal. Bioanal. Chem.* **2012**, *402*, 489–498. [CrossRef]
12. Garrido-Delgado, R.; Dobao-Prieto, M.; Arce, L.; Valcárcel, M. Determination of volatile compounds by GC-IMS to assign the quality of virgin olive oil. *Food Chem.* **2015**, *187*, 572–579. [CrossRef] [PubMed]
13. Gerhardt, N.; Birkenmeier, M.; Sanders, D.; Rohn, S.; Weller, P. Resolution-optimized headspace gas chromatography-ion mobility spectrometry (HS-GC-IMS) for non-targeted olive oil profiling. *Anal. Bioanal. Chem.* **2017**, *409*, 3933–3942. [CrossRef] [PubMed]
14. Gerhardt, N.; Schwolow, S.; Rohn, S.; Ruiz Pérez-Cacho, P.; Galán-Soldevilla, H.; Arce, L.; Weller, P. Quality assessment of olive oils based on temperature-ramped HS-GC-IMS and sensory evaluation: Comparison of different processing approaches by LDA, kNN, and SVM. *Food Chem.* **2019**, *278*, 720–728. [CrossRef] [PubMed]
15. Schwolow, S.; Gerhardt, N.; Rohn, S.; Weller, P. Data fusion of GC-IMS data and FT-MIR spectra for the authentication of olive oils and honeys—Is it worth to go the extra mile? *Anal. Bioanal. Chem.* **2019**, *411*, 6005–6019. [CrossRef]
16. Palagano, R.; Valli, E.; Cevoli, C.; Bendini, A.; Gallina Toschi, T. Compliance with EU vs. extra-EU labelled geographical provenance in virgin olive oils: A rapid untargeted chromatographic approach based on volatile compounds. *LWT Food Sci. Technol.* **2020**, *130*, 109566. [CrossRef]
17. Quintanilla-Casas, B.; Bustamante, J.; Guardiola, F.; García-GonzáLez, D.L.; Barbieri, S.; Bendini, A.; Gallina Toschi, T.; Vichi, S.; Tres, A. Virgin olive oil volatile fingerprint and chemometrics: Towards an instrumental screening tool to grade the sensory quality. *LWT-Food Sci. Technol.* **2020**, *121*, 1–8. [CrossRef]
18. Valli, E.; Panni, F.; Casadei, E.; Barbieri, S.; Cevoli, C.; Bendini, A.; García-González, D.L.; Gallina Toschi, T. An HS-GC-IMS Method for the Quality Classification of Virgin Olive Oils as Screening Support for the Panel Test. *Foods* **2020**, *9*, 657. [CrossRef]

19. Melucci, D.; Bendini, A.; Tesini, F.; Barbieri, S.; Zappi, A.; Vichi, S.; Conte, L.; Gallina Toschi, T. Rapid direct analysis to discriminate geographic origin of extra virgin olive oils by flash gas chromatography electronic nose and chemometrics. *Food Chem.* **2016**, *204*, 263–273. [CrossRef]
20. Barbieri, S.; Brkić Bubola, K.; Bendini, A.; Bučar-Miklavčič, M.; Lacoste, F.; Tibet, U.; Winkelmann, O.; García-González, D.L.; Gallina Toschi, T. Alignment and proficiency of virgin olive oil sensory panels: The OLEUM approach. *Foods.* **2020**, *9*, 355. [CrossRef]
21. Guidance for Industry, Bioanalytical Method Validation, U.S. Department of Health and Human Services, Food and Drug Administration, Center for Drug Evaluation and Research (CDER), Center for Veterinary Medicine (CMV). May 2018. Available online: https://www.fda.gov/files/drugs/published/Bioanalytical-Method-Validation-Guidance-for-Industry.pdf (accessed on 7 January 2020).
22. Commission Decision 2002/657/EC (OJ L221, p8, 17/08/2002) of 12 August 2002 Implementing Council Directive 96/23/EC Concerning the Performance of Analytical Methods and the Interpretation of Results. Available online: https://www.fsai.ie/legislation/food_legislation/veterinary_medicines/monitoring_of_residues.html (accessed on 7 January 2020).
23. Allwood, J.W.; Erban, A.; de Koning, S.; Dunn, W.B.; Luedemann, A.; Lommen, A.; Kay, L.; Löscher, R.; Kopka, J.; Goodacre, R. Inter-laboratory reproducibility of fast gas chromatography–electron impact–time of flight mass spectrometry (GC–EI–TOF/MS) based plant metabolomics. *Metabolomics* **2009**, *5*, 479–496. [CrossRef]
24. Cavanna, D.; Righetti, L.; Elliott, C.; Sumana, M. The scientific challenges in moving from targeted to non-targeted mass spectrometric methods for food fraud analysis: A proposed validation workflow to bring about a harmonized approach. *Trends Food Sci. Technol.* **2008**, *80*, 223–241. [CrossRef]
25. Tomasi, G.; Van den Berg, F.; Andersson, C. Correlation optimized warping and dynamic time warping as preprocessing methods for chromatographic data. *J. Chemom. Soc.* **2004**, *18*, 231–241. [CrossRef]
26. Horwitz, W. Evaluation of analytical methods used for regulation of foods and drugs. *Anal. Chem.* **1982**, *54*, 67–76. [CrossRef]
27. Brereton, R.G.; Lloyd, G.R. Partial least squares discriminant analysis: Taking the magic away. *J. Chemom.* **2014**, *28*, 213–225. [CrossRef]
28. Daszykowski, M.; Walczak, B.; Massart, D.L. Looking for Natural Patterns in Analytical Data. 2. Tracing Local Density with OPTICS. *J. Chem. Inf. Comput. Sci.* **2002**, *42*, 500–507. [CrossRef]
29. Barbieri, S.; Cevoli, C.; Bendini, A.; Quintanilla-Casas, B.; García-González, D.L.; Gallina Toschi, T. OLEUM project. Rapid screening tool of quality grades of virgin olive oils. FGC in tandem with chemometrics. *Università di Bologna* **2020**. [CrossRef]
30. Bishop, C.M. *Pattern Recognition and Machine Learning*; Springer: New York, NY, USA, 2006.
31. Naz, S.; Vallejo, M.; García, A.; Barbas, C. Method validation strategies involved in non-targeted metabolomics. *J. Chromatogr. A* **2004**, *1353*, 99–105. [CrossRef]
32. Pèrez, F.N.; Ferré, J.; Boquè, R. Calculation of the reliability of classification in discriminant partial least-squares binary classification. *Chemom. Intell. Lab. Syst.* **2009**, *95*, 122–128. [CrossRef]

foods

MDPI

Article

Genetic Identification of the Wild Form of Olive (*Olea europaea var. sylvestris*) Using Allele-Specific Real-Time PCR

Christina I. Kyriakopoulou and Despina P. Kalogianni *

Department of Chemistry, University of Patras, 26504 Rio Patras, Greece; xrkyriako@gmail.com
* Correspondence: kalogian@upatras.gr

Received: 28 February 2020; Accepted: 7 April 2020; Published: 9 April 2020

Abstract: The wild-type of olive tree, *Olea europaea var Sylvestris* or oleaster, is the ancestor of the cultivated olive tree. Wild-type olive oil is considered to be more nutritious with increased antioxidant activity compared to the common cultivated type (*Olea europaea* L. *var Europaea*). This has led to the wild-type of olive oil having a much higher financial value. Thus, wild olive oil is one of the most susceptible agricultural food products to adulteration with other olive oils of lower nutritional and economical value. As cultivated and wild-type olives have similar phenotypes, there is a need to establish analytical methods to distinguish the two plant species. In this work, a new method has been developed which is able to distinguish *Olea europaea var Sylvestris* (wild-type olive) from *Olea europaea* L. *var Europaea* (cultivated olive). The method is based, for the first time, on the genotyping, by allele-specific, real-time PCR, of a single nucleotide polymorphism (SNP) present in the two olives' chloroplastic genomes. With the proposed method, we were able to detect as little as 1% content of the wild-type olive in binary DNA mixtures of the two olive species.

Keywords: *Olea europaea var Sylvestris*; oleaster; olive; olive oil; real-time PCR; adulteration; SNP; DNA

1. Introduction

The wild form of the olive tree, formally named *Olea europaea var Sylvestris* or oleaster, is considered to be one of the oldest trees worldwide; it is found mainly in the Mediterranean Basin. Genetic pattering studies have shown that cultivated olive trees, i.e., *Olea europaea* L. *var Europaea*, are more similar to oleaster species, providing evidence to support the concept that oleasters are the ancestors of cultivated trees [1]. Both wild and cultivated olive oil have beneficial properties for human health, giving them high economic and nutritional value; however, this has made olive oil one of the most vulnerable agricultural products to fraud and fakery. Wild-type olive oil has higher antioxidant activity, as well as phenolic, tocopherolic and orthodiphenolic contents equal to or higher those in extra virgin cultivated olive oil [2]. Moreover, wild-type olive is a valuable natural resource due to its resistance to certain environmental and climatic conditions and diseases [3]. For the above reasons, its genetic characteristics have to be evaluated, and reliable molecular tools have to be developed for olive oil origin traceability (genetically and geographically) and wild-type olive oil identification. On the other hand, producers need accurate analytical tools for the genetic identification of their wild-type olive-related products to ensure their high added value [4].

Genetic variations between the two plant species have not been extensively explored by the research community. The analytical techniques used so far for the genetic identification of the wild form of olive tree include randomly amplified polymorphic DNA (RAPD), amplified fragment length polymorphisms (AFLPs) and intersimple and simple sequence repeats (ISSRs and SSRs), based on the chloroplastic and mitochondrial plant DNA [1]. Early research compared the genome of *Olea europaea* L. *var Europaea* to that of the wild-type olive, derived from many countries and two areas in Italy, using AFLP analysis as designed by Angiolillo et al., 1999, and Baldoni et al., 2006 [5,6]. RAPD analysis was used to distinguish oleasters from *Olea europaea* L. *var Europaea* trees on the Mediterranean islands of Corsica and Sardinia, as well as in Turkey [7,8]. Besnard et al. used RAPD markers and restriction fragment length polymorphism (RFLP) analysis based on mitochondrial and cytoplasmatic DNA to investigate the relationships among olive species and subspecies in the Mediterranean Basin and other countries in Asia and Africa. This research led to the discovery that there was a large degree of diversity among olive cultivated trees, but that they were more or less related to the local oleasters [9,10]. Moreover, ISSR and SSR markers have been utilized by many researchers to investigate the relation and differentiation of cultivated olives from wild-type olives [3,11–16]. Genome size estimation based on double-stranded DNA staining followed by flow cytometric analysis was also used for screening purposes between *Olea europaea var Sylvestris* and *Olea europaea* L. *var Europaea* species [17], while flow cytometry in combination with SSR profiles was used for the taxonomy of four olive subspecies, namely *Olea europaea ssp. cerasiformis*, *Olea europaea ssp. guanchica*, *Olea europaea var Sylvestris* and *Olea europaea* L. *var Europaea* [18].

Moreover, the wild olive has also been used for nonedible purposes in pharmacology and cosmetics to create products with specific valuable characteristics. Researches have also studied the antimicrobial activity of the wild olive against certain human bacterial pathogens [19]. Several plants, including the olive and its wild form, have also been used for the production of various food supplements [20]. Finally, phenolic extracts from wild olive leaves have been investigated for use in foodstuffs, food additives and functional food materials, due to their high antioxidant activity [21,22].

In 2017, the complete genome sequence of *Olea europaea var Sylvestris* was published by Unver et al. [23]. This will be useful, in the future, for the localization of specific genetic variations in the genome of oleasters compared to other olive subspecies.

For the first time, in this work, a single nucleotide polymorphism (SNP)-based method was developed for the detection and identification of the wild form of olive in order to distinguish it from the cultivated olive. Different olive cultivars contain different SNPs in their genome that are responsible for their unique phenotyping characteristics [24,25]. The method was based on an allele-specific, real-time PCR. The proposed method is able to detect wild-type olive DNA at levels as low as 1% in DNA derived from the cultivated olive.

2. Materials and Methods

2.1. Materials and Instrumentation

The Vent (exo-) DNA polymerase was purchased by New England Biolabs (Beverly, MA, USA). Deoxynucleoside triphosphates (dNTPs) were obtained from Kapa Biosystems (Wilmington, MA, USA). The fluorescent dye SYBR Green I 10^4 × concentrated was from Molecular Probes (Eugene, OR, USA). The primers used were from Eurofins Scientific (Brussels, Belgium) and are listed in Table 1. The size of the PCR products was 136 bp. An extra virgin olive oil sample (*Olea europaea* L. *var Europaea*) was purchased from a local market, while a certified wild-type olive oil sample (*Olea europaea var Sylvestris*) was kindly by local producer, Alexandros Karakikes, from the Olea Sylvestris estate (Agrielaio, Volos, Greece) [26].

Real-time PCR was performed using the Mini Opticon Real-Time PCR System from Biorad (Hercules, CA, USA), while the results were analyzed using the Bio–Rad CFX Manager 3.0 software.

Table 1. The primers used in the allele-specific, real-time PCR, two species-specific upstream primers and a common downstream primer, along with their melting temperatures (Tm).

Primer Name	5′–3′ Oligonucleotide Sequence	Melting Temperature *
Olea europaea var Sylvestris upstream primer	TGTCAATTTTAATCACTACTGC	62 °C
Olea europaea L. *Europaea* upstream primer	TGTCAATTTTAATCACTACTGT	61 °C
Common downstream primer	CTAGTAACTAATCCTAACATGGAA	64 °C

* according to Eurofins Scientific (Brussels, Belgium).

2.2. DNA Isolation Procedure

DNA was isolated from olive oil samples using the NucleoSpin Tissue kit from Macherey-Nagel (Düren, Germany) according to the manufacturer's instructions. The quantity and purity of the isolated DNA were determined using the Nanodrop UV/VIS Nanophotometer by Implen GmbH (Münich, Germany).

2.3. Design of the Primers

The primers used for the amplification of *Olea europaea var Sylvestris* (wild-type olive) and *var Europaea* (cultivated olive) were designed using the free online Oligo Analyzer software for primer evaluation (created by Dr. Teemu Kuulasmaa), based on the *Olea europaea var. sylvestris* NADH dehydrogenase subunit F gene, chloroplastic sequence (Accession Number: AY172114) and the *Olea europaea* L. NADH dehydrogenase subunit F (ndhF) gene chloroplastic sequence (Accession Number: DQ673278) [23].

2.4. Allele-Specific, Real-Time PCR

The allele-specific, real-time PCR reactions were conducted in a final volume of 50 μL and contained 1 × Thermopol Buffer (20 mM Tris-HCl, 10 mM $(NH_4)_2SO_4$, 10 mM KCl, 0.1% Triton® X-100 at pH 8.8), 0.5 μM of each of the upstream and downstream primers, 0.2 mM of each of the four dNTPs, 0.5 mM $MgCl_2$, 2 × SYBR Green I, one unit of Vent (exo-) DNA polymerase and 150 ng of isolated DNA. The reaction conditions involved a 95 °C incubation step for three min, followed by 45 cycles at 95 °C for 30 s, 62 °C for 30 s, 72 °C for 30 s and a final extension step at 72 °C for 10 min.

3. Results and Discussion

A new analytical method was developed for the detection and identification of *Olea europaea var Sylvestris* that refers to the wild form of the olive tree. The method was based on the detection of a specific Single Nucleotide Polymorphism (SNP) that is different in the genome of the wild olive plant. The method involves the following steps: (i) DNA isolation from olive oil samples and (ii) allele-specific, real-time PCR using an upstream primer specific to *Olea europaea var Sylvestris* or *var Europaea* species and a common downstream primer. The species-specific primers have the same 22-base sequence but differ only at the base at the 3′ end that contains the SNP of interest. The DNA sequences were amplified using a DNA polymerase that lacked the 3′ to 5′ exonuclease activity, so only the primer that was perfectly complementary to the DNA target was extended by the enzyme. The amplicons were finally detected using the DNA intercalating fluorescent dye SYBR Green I. The principle of the proposed method is illustrated in Figure 1. SYBR Green I was chosen here instead of Taqman probes in order to develop a new analytical method that could be easily transferred, with few modifications, for the detection of other SNPs that will be found in the wild olive genome in the future.

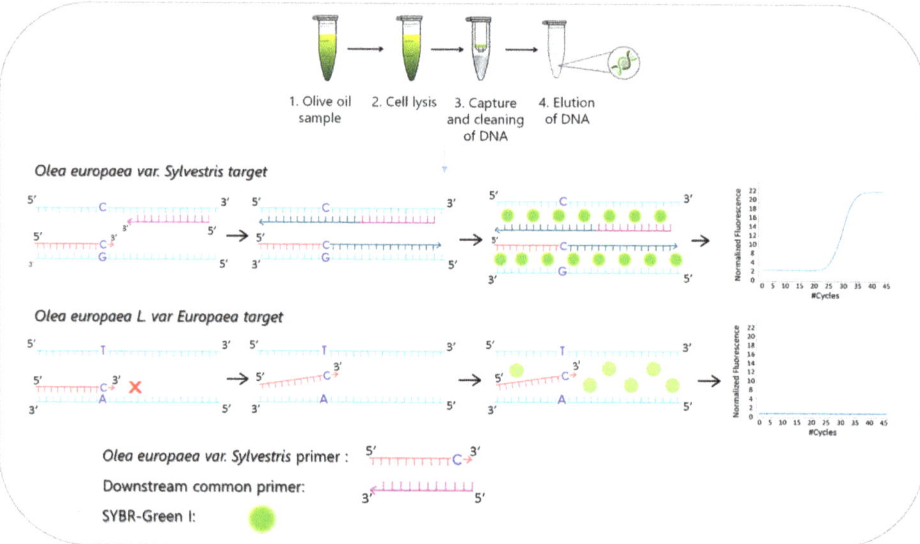

Figure 1. (**Upper panel**) Schematic illustration of the principle of the method that includes DNA isolation and purification from olive oil samples using spin cleanup columns, including the following steps: cell lysis of an olive oil sample, capture of DNA to the cleanup columns and elution of the DNA from the columns. (**Lower panel**) The allele-specific, real-time PCR. Two allele-specific upstream primers that contain the SNP of interest at their 3′ ends and one common downstream primer were used in the amplification reaction. Only the perfectly complementary upstream primer to the target was extended by the DNA polymerase, while the amplicons were detected by the DNA intercalating dye, SYBR Green I.

3.1. DNA Isolation

First, DNA was isolated from olive oil samples and its concentration was determined using a UV/VIS nanophotometer. It was found that the isolation procedure did not result in a constant DNA amount for all samples, with the DNA concentrations ranging from 8.4 to 142 ng/μL. To avoid fluctuation in the PCR yield due to different initial DNA concentrations, we decided to use the same amount (ng) of isolated DNA for all samples into the real-time PCR mixture. After amplification, the amplicons had a size of 136 bp. The quality of the isolated DNA was also determined by UV measurements; the ratios A_{260}/A_{280} were from 1174 to 1739. DNA was considered to be of high quality when the ratio A_{260}/A_{280} was above 1.8.

3.2. Optimization of the PCR Conditions

The real-time PCR conditions were initially optimized. The parameters studied were the amount of the isolated DNA, the concentration of the primers, the number of PCR cycles and the temperature of the annealing step of the reaction. At low DNA and primer concentrations, low temperature (55–60 °C) and number of cycles < 45, the PCR was not sufficiently efficient. The yield of the reaction also decreased when a high amount of initial DNA target was used. This may be attributed to the fact that the DNA isolated from olive samples has reduced quality, as it contains high amounts of PCR inhibitors that may inhibit the activity of the DNA polymerase [27]. We also observed that the highest reaction yield and specificity were obtained at an annealing temperature of 62 °C.

3.3. Specificity of the Allele-Specific Primers

The specificity of the two species-dependent upstream primers was then studied as follows: both DNA targets, *Olea europaea var Sylvestris* (wild-type olive) and *var Europaea* (cultivated olive) were subjected to two separate amplification reactions using either the upstream primer specific to the wild-type olive or the cultivated olive-specific upstream primer. As shown, in Figure 2, each primer amplified only its fully complementary DNA sequence, proving the superior specificity of the primers. To ensure that the fluorescence signals were attributed only to the specific amplicons, a melting curve analysis was also performed after each amplification reaction. The melting curve analysis revealed only one peak for each PCR product, the melting temperature (T_m) of which was 77 °C for *Olea europaea var Sylvestris* (wild-type olive) and 78 °C for *var Europaea* (cultivated olive), allowing us to distinguish between the two allele-specific DNA sequences.

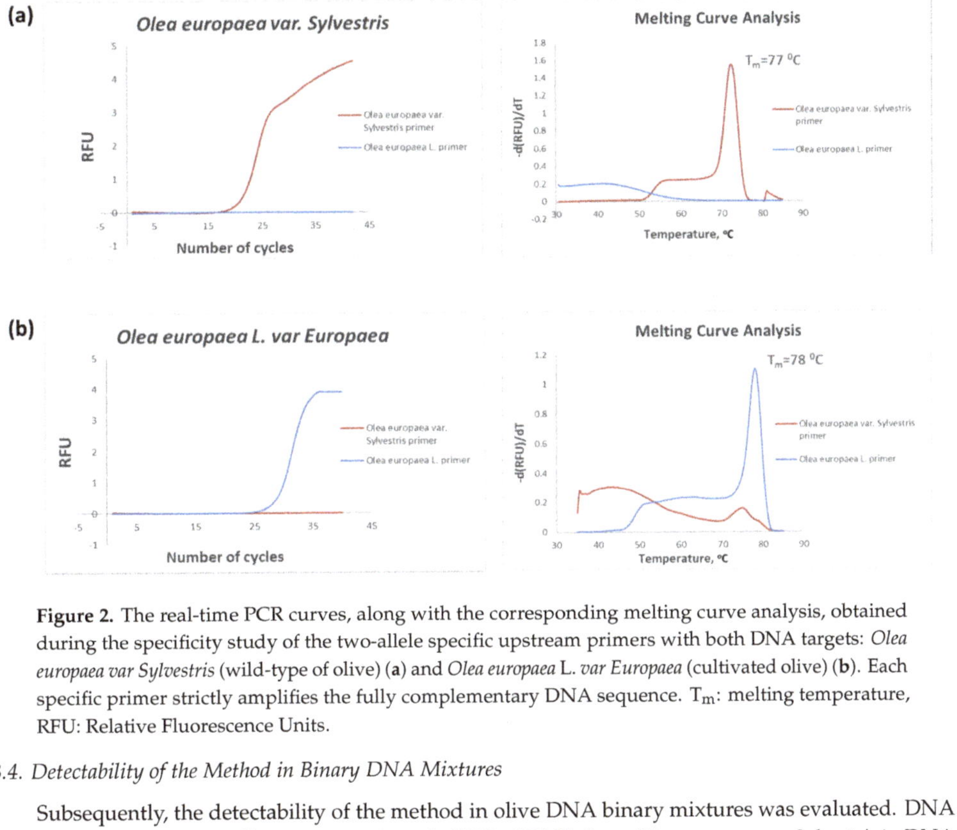

Figure 2. The real-time PCR curves, along with the corresponding melting curve analysis, obtained during the specificity study of the two-allele specific upstream primers with both DNA targets: *Olea europaea var Sylvestris* (wild-type of olive) (**a**) and *Olea europaea* L. *var Europaea* (cultivated olive) (**b**). Each specific primer strictly amplifies the fully complementary DNA sequence. T_m: melting temperature, RFU: Relative Fluorescence Units.

3.4. Detectability of the Method in Binary DNA Mixtures

Subsequently, the detectability of the method in olive DNA binary mixtures was evaluated. DNA mixtures that contained different proportions (1–50%) of DNA from *Olea europaea var Sylvestris* in DNA from *var Europaea* were prepared. An amount of 150 ng of each DNA mixture was then subjected to two separate allele-specific, real-time PCR reactions using each of the species-specific upstream primers along with the common downstream primer, respectively. A high amount of total DNA was used in the PCR in order to detect the low amount of wild olive DNA in the mixtures, e.g., for 150 ng of total DNA in the 1% mixture, only the 1.5 ng was the wild olive DNA. The results are presented in Figure 3. We were able to detect as little as 1% of DNA specific to *Olea europaea var Sylvestris* in the presence of DNA from *Olea europaea* L. *var Europaea*. The allelic ratios of the analyzed SNP for the above DNA mixtures were also calculated based on the fluorescence value at the 45th cycle of the reaction, and are presented in the same Figure. The allelic ratios for all DNA mixtures were close to the value of 0.5, as expected for a heterozygote sample.

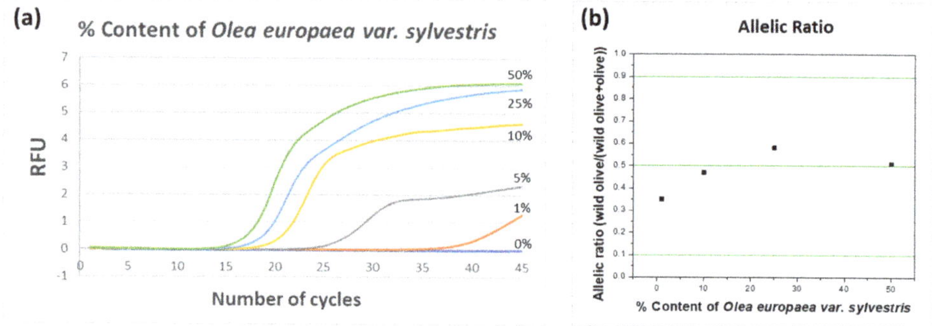

Figure 3. (**a**) The real-time PCR curves obtained for different % DNA content (0–50%) of *Olea europaea var Sylvestris* (wild-type olive) DNA in binary mixtures with *Olea europaea* L. *var Europaea* (cultivated olive) DNA. (**b**) The allelic ratios of the binary DNA mixtures calculated as the ratio of the fluorescence intensity obtained with the upstream primer specific to *Olea europaea var Sylvestris* target versus the sum of the fluorescence intensity obtained by both allele-specific primers for *Olea europaea var Sylvestris* and *var Europaea* targets. All allelic ratios were close to the value of 0.5, which corresponds to a heterozygote sample. RFU: Relative Fluorescence Units.

3.5. Reproducibility of the Method

Finally, the reproducibility of the method was determined. Two different proportions, 1% and 10%, of the above DNA mixtures, were subjected, in triplicate, to real-time PCR. The % coefficients of variation (CV) were calculated based on the obtained Cq values for all samples. The CV for the 1%-content was 10.5% and for the 10%-content was 7.5%, demonstrating the reproducibility of the method.

4. Conclusions

A new allele-specific, real-time PCR-based analytical method was developed for the detection and identification of wild-type olive oil (*Olea europaea var Sylvestris*), compared to cultivated olive oil (*Olea europaea* L. *var Europaea*). The discrimination of the two similar plant species was based on genotyping a single nucleotide polymorphism (SNP) that is differently present in the genome of the two plant species. The detection of this SNP was carried out by an allele-specific, real-time PCR that was performed using two different species-specific upstream primers that contained the analyzed SNP and a common downstream primer. Each specific primer amplified only its fully complementary DNA sequence, leading to species identification. The detection of the amplicons was accomplished using the DNA intercalating dye, SYBR Green I. With the proposed method, we were able to sucessfully distinguish between the two plant species in olive oil samples. Also, as little as 1% wild-type olive species was detected in binary DNA mixtures of the two analyzed plant species. In conclusion, the method is easy, rapid, has good detectability, is reproducible and can easily distinguish between species. The proposed method also contributes to the ability to add the higher financial value to wild-type olive-based products. In the future, the determination of different SNPs in the wild-type olive genome compared to all the known cultivated olive trees could lead to more accurate discrimination of the wild-type olive among other olive-based subspecies. The proposed method could also be applied, with some modifications, for the detection of wild olive-based ingredients in food supplements and cosmetic products. The global increase in food supplements has led to the mislabeling of these products and fraudulent practices. In both cases, the purity of the extracted DNA is more important than the

PCR yield itself, because several food additives and other ingredients may be present in the extracts, inhibiting the PCR amplification. Also, the amount of the extracted DNA may be extremely low. Thus, the DNA isolation protocols have to be properly justified to remove these inhibitors and increase the DNA recovery and the PCR yield. In some studies, however, the inability to extract DNA from some food supplements has been reported. Finally, in some products, DNA degradation may also occur due to thermal or chemical treatment, but the use of short-length amplicons can overcome this issue [28–31].

Author Contributions: Conceptualization, D.P.K.; methodology, C.I.K. and D.P.K.; software, C.I.K.; investigation, C.I.K. and D.P.K.; resources, C.I.K. and D.P.K.; data curation, C.I.K. and D.P.K.; writing—original draft preparation, C.I.K. and D.P.K.; writing—review and editing, D.P.K.; supervision, D.P.K.; project administration, D.P.K.; funding acquisition, D.P.K. All authors have read and agreed to the published version of the manuscript.

Funding: This work was funded by the project "Research Infrastructure on Food Bioprocessing Development and Innovation Exploitation—Food Innovation RI" (MIS 5027222), which is implemented under the Action "Reinforcement of the Research and Innovation Infrastructure", funded by the Operational Programme "Competitiveness, Entrepreneurship and Innovation" (NSRF 2014–2020) and co-financed by Greece and the European Union (European Regional Development Fund).

Acknowledgments: We acknowledge Alexandros Karakikes (Olea Sylvestris estate) for kindly providing us with the wild-type olive oil sample.

Conflicts of Interest: The authors declare no conflicts of interest.

References

1. Kassa, A.; Konrad, H.; Geburek, T. Molecular diversity and gene flow within and among different subspecies of the wild olive (*Olea europaea* L.): A review. *Flora Morphol. Distrib. Funct. Ecol. Plants* **2019**, *250*, 18–26. [CrossRef]
2. Bouarroudj, K.; Tamendjari, A.; Larbat, R. Quality, composition and antioxidant activity of Algerian wild olive (*Olea europaea* L. *subsp. Oleaster*) oil. *Ind. Crops Prod.* **2016**, *83*, 484–491. [CrossRef]
3. Beghé, D.; Piotti, A.; Satovic, Z.; De La Rosa, R.; Belaj, A. Pollen-mediated gene flow and fine-scale spatial genetic structure in *Olea europaea subsp. Europaea var. sylvestris*. *Ann. Bot.* **2017**, *119*, 671–679. [PubMed]
4. Bracci, T.; Busconi, M.; Fogher, C.; Sebastiani, L. Molecular studies in olive (*Olea europaea* L.): Overview on DNA markers applications and recent advances in genome analysis. *Plant Cell Rep.* **2011**, *30*, 449–462. [CrossRef] [PubMed]
5. Angiolillo, A.; Mencuccini, M.; Baldoni, L. Olive genetic diversity assessed using amplified fragment length polymorphisms. *Theor. Appl. Genet.* **1999**, *98*, 411–421. [CrossRef]
6. Baldoni, L.; Tosti, N.; Ricciolini, C.; Belaj, A.; Arcioni, S.; Pannelli, G.; Germana, M.A.; Mulas, M.; Porceddu, A. Genetic structure of wild and cultivated olives in the central Mediterranean Basin. *Ann. Bot.* **2006**, *98*, 935–942. [CrossRef]
7. Bronzini de Caraffa, V.; Maury, J.; Gambotti, C.; Breton, C.; Bervillé, A.; Giannettini, J. Mitochondrial DNA variation and RAPD mark oleasters, olive and feral olive from Western and Eastern Mediterranean. *Theor. Appl. Genet.* **2002**, *104*, 1209–1216. [CrossRef]
8. Sesli, M.; Yeğenoğlu, E.D. Determination of the genetic relationships between wild olive (*Olea europaea oleaster*) varieties grown in the Aegean Region. *Genet. Mol. Res.* **2010**, *9*, 884–890. [CrossRef]
9. Besnard, G.; Khadari, B.; Baradat, P.; Bervillé, A. Combination of chloroplast and mitochondrial DNA polymorphisms to study cytoplasm genetic differentiation in the olive complex (*Olea europaea* L.). *Theor. Appl. Genet.* **2002**, *105*, 139–144. [CrossRef]
10. Besnard, G.; Bervillé, A. Multiple origins for Mediterranean olive (*Olea europaea* L. *ssp. europaea*) based upon mitochondrial DNA polymorphisms. *Comptes Rendus l'Académie des Sci. Ser. III Sci. Vie* **2000**, *323*, 173–181. [CrossRef]
11. Vargas, P.; Kadereit, J.W. Molecular fingerprinting evidence (ISSR, inter-simple sequence repeats) for a wild status of *Olea europaea* L. (*Oleaceae*) in the Eurosiberian North of the Iberian Peninsula. *Flora* **2001**, *196*, 142–152. [CrossRef]
12. Breton, C.; Tersac, M.; Bervillé, A. Genetic diversity and gene flow between the wild olive (*oleaster, Olea europaea* L.) and the olive: Several Plio-Pleistocene refuge zones in the Mediterranean Basin suggested by simple sequence repeats analysis. *J. Biogeogr.* **2006**, *33*, 1916–1928. [CrossRef]

13. Belaj, A.; León, L.; Satovic, Z.; De la Rosa, R. Variability of wild olives (*Olea europaea subsp. europaea var. sylvestris*) analyzed by agro-morphological traits and SSR markers. *Sci. Hortic.* **2011**, *129*, 561–569. [CrossRef]
14. Belaj, A.; Muñoz-Diez, C.; Baldoni, L.; Porceddu, A.; Barranco, D.; Satovic, Z. Genetic diversity and population structure of wild olives from the north-western Mediterranean assessed by SSR markers. *Ann. Bot.* **2007**, *100*, 449–458. [CrossRef] [PubMed]
15. Abood, A.A.; Al-Ansari, A.M.; Migdadi, H.M.; Okla, M.K.; Assaeed, A.M.; Hegazy, A.K.; Alshameri, A.M.; Khan, M.A. Molecular and phytochemical analysis of wild type and olive cultivars grown under Saudi Arabian environment. *3 Biotech* **2017**, *7*, 289. [CrossRef] [PubMed]
16. Chiappetta, A.; Muto, A.; Muzzalupo, R.; Muzzalupo, I. New rapid procedure for genetic characterization of Italian wild olive (*Olea europaea*) and traceability of virgin olive oils by means of SSR markers. *Sci. Hortic.* **2017**, *226*, 42–49. [CrossRef]
17. Loureiro, J.; Rodriguez, E.; Costa, A.; Santos, C. Nuclear DNA content estimations in wild olive (*Olea europaea* L. *ssp. europaea var. sylvestris Brot.*) and Portuguese cultivars of *O. europaea* using flow cytometry. *Genet. Resour. Crop Evol.* **2007**, *54*, 21–25. [CrossRef]
18. Brito, G.; Loureiro, J.; Lopes, T.; Rodriguez, E.; Santos, C. Genetic characterisation of olive trees from Madeira Archipelago using flow cytometry and microsatellite markers. *Genet. Resour. Crop Evol.* **2008**, *55*, 657–664. [CrossRef]
19. Paudel, S.; Magrati, T.; Lamichhane, J.R.; Centre, C.H. Antimicrobial activity of wild olive crude extracts in vitro. *Int. J. Pharma Sci. Res.* **2011**, *2*, 110–113.
20. Colombo, M.L. Botanicals authentication in food, food supplements and herbal medicinal products. *Plant Biosyst.* **2016**, *150*, 22–26. [CrossRef]
21. Lafka, T.-I.; Lazou, A.; Sinanoglou, V.; Lazos, E. Phenolic extracts from wild olive leaves and their potential as edible oils antioxidants. *Foods* **2013**, *2*, 18–31. [CrossRef] [PubMed]
22. Mohamed, R.; Pineda, M.; Aguilar, M. Antioxidant capacity of extracts from wild and crop plants of the Mediterranean region. *J. Food Sci.* **2007**, *72*, S059–S063. [CrossRef]
23. Unver, T.; Wu, Z.; Sterck, L.; Turktas, M.; Lohaus, R.; Li, Z.; Yang, M.; He, L.; Deng, T.; Escalante, F.J.; et al. Genome of wild olive and the evolution of oil biosynthesis. *Proc. Natl. Acad. Sci. USA* **2017**, *114*, E9413–E9422. [CrossRef] [PubMed]
24. Kalogianni, D.P.; Bazakos, C.; Boutsika, L.M.; Targem, M.B.; Christopoulos, T.K.; Kalaitzis, P.; Ioannou, P.C. Olive oil DNA fingerprinting by multiplex SNP genotyping on fluorescent microspheres. *J. Agric. Food Chem.* **2015**, *63*, 3121–3128. [CrossRef] [PubMed]
25. Bazakos, C.; Dulger, A.O.; Uncu, A.T.; Spaniolas, S.; Spano, T.; Kalaitzis, P. A SNP-based PCR–RFLP capillary electrophoresis analysis for the identification of the varietal origin of olive oils. *Food Chem.* **2012**, *134*, 2411–2418. [CrossRef]
26. Agrielaio-olea Sylvestris. Available online: https://www.agrielaio.gr/wild-olive-oil.php (accessed on 8 April 2020).
27. Montemurro, C.; Pasqualone, A.; Rosanna Simeone, R.; Sabetta, W.; Blanco, A. AFLP molecular markers to identify virgin olive oils from single Italian cultivars. *Eur. Food Res. Technol.* **2008**, *226*, 1439–1444. [CrossRef]
28. Lo, Y.T.; Shaw, P.C. DNA-based techniques for authentication of processed food and food supplements. *Food Chem.* **2018**, *240*, 767–774. [CrossRef]
29. Costa, J.; Amaral, J.S.; Fernandes, T.J.R.; Batista, A.; Oliveira, M.B.P.P.; Mafra, I. DNA extraction from plant food supplements: Influence of different pharmaceutical excipients. *Mol. Cell. Probes* **2015**, *29*, 473–478. [CrossRef]
30. Kim, Y.S.; Yu, H.K.; Lee, B.Z.; Hong, K.W. Effect of DNA extraction methods on the detection of porcine ingredients in halal cosmetics using real-time PCR. *Appl. Biol. Chem.* **2018**, *61*, 549–555. [CrossRef]
31. Zabidi, A.R.; Fauzi, F.N.; Abd Razak, F.N.; Rosli, D.; Jamil, M.Z.M.; Wan Ibrahim, W.K.; Yahaya, N. Screening porcine DNA in collagen cream cosmetic products. *Food Res.* **2020**, *4*, 151–156. [CrossRef]

foods

MDPI

Article

Using Chemometric Analyses for Tracing the Regional Origin of Multifloral Honeys of Montenegro

Vesna Vukašinović-Pešić [1,*], Nada Blagojević [1], Snežana Brašanac-Vukanović [1], Ana Savić [2] and Vladimir Pešić [3]

1 Faculty of Metallurgy and Technology, University of Montenegro, 81000 Podgorica, Montenegro; nadab@ucg.ac.me (N.B.); snezana.v@ucg.ac.me (S.B.-V.)
2 Department of Biology and Ecology, Faculty of Sciences and Mathematics, University of Niš, 18000 Niš, Serbia; anka@pmf.ni.ac.rs
3 Department of Biology, University of Montenegro, 81000 Podgorica, Montenegro; vladopesic@gmail.com
* Correspondence: vesnav@ucg.ac.me

Received: 19 January 2020; Accepted: 14 February 2020; Published: 18 February 2020

Abstract: This is the first study of mineral content and basic physicochemical parameters of honeys of Montenegro. We examined honey samples from eight different micro-regions of Montenegro, and the results confirm that, with the exception of cadmium in samples from two regions exposed to industrial pollution, none of the 12 elements analyzed exceeded the maximum allowable level. The samples from areas exposed to industrial pollution were clearly distinguished from samples from other regions of Montenegro in the detectable contents of Pb, Cd, and Sr. This study showed that chemometric techniques might enhance the classification of Montenegrin honeys according to their micro-regional origin using the mineral content. Linear discriminant analysis revealed that the classification rate was 79.2% using the cross-validation method.

Keywords: honey; regional origin; chemometric analysis; mineral content; Montenegro

1. Introduction

Honey is a complex natural product, whose characteristics depend on the flower nectar from which it is obtained, but also on other factors such as geographical origin, bee species, season, type of processing and storage [1]. It is known that pollution and a number of different pollutants present in its foraging areas have an impact on honeybees [2] but also on nectar-providing plant species. Therefore, it is necessary to assure geographical traceability and determine the botanical origin of the foraging area of the beehive.

As stated by Karabagias and Karabournioti [3], the authentication of honey is gaining in importance and includes a number of contending parties from producers and sellers to consumers and control labs. A number of papers have shown that specific physicochemical parameters and mineral contents in combination with chemometric analyses can be a useful tool in discovering botanical and/or geographical origin of honeys that may enter the market [1,3,4].

Tracing the geographical origin of honeys can provide important information about the potential contamination of the area from which the honey production material comes. Therefore, ensuring high standards in terms of product safety leads to the need to examine the contents of essential and toxic elements in honey. Due to its bioaccumulation ability, honey can be used as an indicator of metal pollution, especially of toxic pollutants such as Pb, Cd, and As [5–7].

Due to its geographical position, climate conditions and richness of the nectar-providing plants, Montenegro provides favorable natural conditions for more intensive development of beekeeping. According to the data for 2011, the population in Montenegro was 625,266, while the honey production for that year was 394 t, and the average annual consumption 1.2 kg per person, meaning that a large part of honey consumption in Montenegro is imported [8]. Data for the last few years show an increase in honey production (627 t for 2016) but also an increase in the average annual consumption of honey per person (2.76 kg) [9].

The majority of honey on the market in Montenegro are multifloral (derived from a large number of nectar-providing plant species in the honeybees foraging area). Most of these honey types are recognizable by their local or regional origin (e.g., Katunski med (= honey), Pivski med, Piperski med.). It is worth mentioning that Montenegro and its regions are known to harbor a high number of regional floral endemics [10] that likely affect the composition and properties of honey.

There is a lack of information on the mineral content and basic physicochemical parameters of honey from the territory of Montenegro. Moreover, there is no continuity in monitoring the quality of honey, especially in areas that are exposed to the effects of potential pollution sources. Due to the high consumption of local honey in the diet, the need and obligation for its systematic characterization are highly required.

This study is aimed to investigate the mineral content and the basic physicochemical parameters of honeys from different micro-regions of Montenegro. We evaluated the usefulness of chemometric analyses for the classification of honeys according to its regional origin.

2. Materials and Methods

Twenty-four honey samples as indicated in Figure 1 were collected from eight micro-regions of Montenegro, i.e., (1) Piva, (2) Zbljevo, (3) Potrlica, (4) Mijakovići, (5) Piperi, (6) Martinići, (7) Katunska, and (8) Zeta. The Piva, Zbljevo, Potrlica and Mijakovići micro-regions are situated in the continental part of Montenegro (Alpine biogeographical region, see Figure 1) while the four other micro-regions are situated in the sub-Mediterranean part of the country belonging to the Mediterranean biogeographical region [10]. The climate in the latter region is mainly Mediterranean-Adriatic with relatively dry and warm summers (the average air temperature of the warmest month > 20 °C), but humid and mild winters (the average air temperatures varies from 6 to 9 °C), while the Alpine region has a "continental" type climate, with relatively cool and humid summers and long and harsh winters [10].

Samples were taken from individual beekeepers during the harvesting season 2015. All samples were multifloral as confirmed by the suppliers. The samples were stored in glass flasks at room temperature before analysis. Physicochemical parameters (pH, electrical conductivity (EC), free acidity (FA) and moisture) were analyzed using the Harmonized Methods of the International Honey Commission [11].

The mineral composition of honey was analyzed by inductively coupled plasma-optical emission spectrometry (ICP-OES). About 1 g of each honey sample was digested with 14 mL 65% HNO_3 and 2 mL 35% H_2O_2 on a hot plate to near dryness. The sample containing a volumetric flask was cooled at room temperature before the addition of deionized water to the mark on the flask. All samples were prepared in triplicate and their average value was assessed.

The concentration of twelve elements (Pb, Cd, Cu, Zn, Fe, Cr, Sr, Ba, Ca, Na, K, Mg) were determined by ICP-OES according to the iCAP 6000 spectrometer method.

All statistical analyses were performed using SPSS 17.0 (SPSS Statistics for Windows, Version 17.0. SPSS Inc., Chicago, IL, USA). Data were expressed as mean ± standard deviation. A Kolmogorov–Smirnov test showed that all analyzed physicochemical parameters were normally distributed, while the content of Pb, Cd, Sr and Ba in some regions exhibited significant differences from the normal distribution. The one-way analysis of variance (ANOVA) was performed on physicochemical parameters in order to determine if there any significant differences between studied micro-regions at the confidence level 0.05. The Kruskal–Wallis test was used to investigate whether the mineral contents varied significantly between the investigated micro-regions. The relationship between the mineral content and physicochemical parameters were analyzed using the Spearman's correlation analysis. For checking similarities between samples of honey of different geographical origin we used two chemometric analyses: PCA and LDA. Statistical methods based on principal component analysis (PCA) and linear discriminant analysis (LDA) have been used. The LDA was performed using R.3.5.3, while the PCA was made by using MVSP version 3.21.

Figure 1. Map of Montenegro with marked locations of honey sampling from eight micro-regions (in parentheses are given sampling location numbers): Piva (1–3), Zbljevo (4–6), Potrlica (7–9), Mijakovići (10–12), Piperi (13–15), Martinići (16–18), Katunska (19–21), and Zeta (22–24).

3. Results

The mineral content of honey samples from different geographical areas of Montenegro is presented in Table 1. The value presented for each element is the average concentration observed. A significant difference has been observed in the concentrations of Pb, Cd and Sr (p = 0.002) between studied micro-regions. In most analyzed samples the concentrations of above-listed elements were below the limit of detection except in the samples from Potrlica, Zbljevo and Mijakovići. The highest Cd concentration was observed in samples from Potrlica (0.08 ± 0.01 mg/kg). The highest concentration of Pb (0.21 ± 0.06 mg/kg) and Sr (0.12 ± 0.00 mg/kg) were recorded in samples from Zbljevo.

Table 1. Mineral content and physicochemical parameters of honey samples from studied micro-regions of Montenegro.

Regions		Cu mg/kg	Zn mg/kg	Fe mg/kg	Cr mg/kg	Ca mg/kg	Na mg/kg	K mg/kg	Mg mg/kg	Pb mg/kg	Cd mg/kg	Sr mg/kg	Ba mg/kg	EC mS/cm	pH	Moisture %	Free Acidity meq/kg
Katunska	Mean	0.78	1.61	13.00	0.90	100.49	53.62	2155.00	76.33	ND *	ND	ND	ND	0.39	4.43	16.15	25.00
	SD	0.15	0.01	1.00	0.50	3.35	6.58	115.00	14.74					0.08	0.08	0.37	7.21
Martinići	Mean	0.31	1.18	7.67	0.30	109.82	34.61	713.00	31.38	ND	ND	ND	ND	0.82	3.95	15.40	29.33
	SD	0.09	0.60	1.53	0.10	1.36	2.24	252.00	11.89					0.04	0.17	1.16	3.51
Mijakovići	Mean	0.62	1.40	3.95	0.10	80.50	46.00	2589.33	68.77	0.08	0.02	0.07	0.27	0.86	4.49	16.04	30.00
	SD	0.08	0.00	1.05	0.01	29.50	4.00	652.50	9.75	0.01	0.00	0.01	0.11	0.09	0.14	0.49	3.61
Piperi	Mean	0.53	0.84	14.67	1.13	76.70	39.93	1760.00	37.91	ND	ND	ND	ND	0.75	4.48	14.92	28.00
	SD	0.18	0.19	2.52	0.15	3.04	5.72	685.00	12.12					0.11	0.35	0.78	2.00
Piva	Mean	0.53	0.52	7.33	0.27	74.65	39.76	844.33	29.52	ND	ND	ND	ND	0.44	3.87	15.36	41.67
	SD	0.17	0.16	1.53	0.15	22.46	8.62	333.78	12.31					0.21	0.36	0.44	12.10
Potrlica	Mean	0.87	3.00	7.53	0.12	152.00	46.00	1945.33	61.83	0.12	0.08	0.08	1.44	0.68	4.32	16.05	37.00
	SD	0.22	0.61	6.05	0.07	84.26	8.54	189.72	9.22	0.02	0.01	0.05	1.67	0.20	0.19	1.43	11.14
Zbljevo	Mean	0.98	4.13	15.93	0.26	103.00	56.67	1651.33	41.00	0.21	0.07	0.12	0.62	0.63	4.11	15.40	36.33
	SD	0.19	3.30	5.71	0.06	10.15	3.51	1005.47	17.78	0.06	0.01	0.00	0.34	0.32	0.36	0.14	3.51
Zeta	Mean	0.47	0.93	11.00	0.47	48.88	63.94	1285.00	57.23	ND	ND	ND	ND	0.93	4.34	16.22	32.00
	SD	0.02	0.29	0.00	0.07	2.20	3.46	135.00	0.17					0.15	0.23	0.36	12.12
Total	Mean	0.64	1.64	10.14	0.44	93.25	47.57	1617.92	50.50	0.05	0.02	0.03	0.29	0.69	4.25	15.69	32.42
	SD	0.25	1.59	4.80	0.40	40.15	10.58	750.73	19.77	0.08	0.03	0.05	0.70	0.24	0.31	0.79	8.49
$F^{a}/\chi2^{b}$		6.59 [a]	3.17 [a]	5.08 [a]	10.65 [a]	2.64 [a]	8.68 [a]	4.61 [a]	6.63 [a]	22.87 [b]	22.57 [b]	22.06 [b]	21.62 [b]	3.79 [a]	2.69 [a]	1.17 [a]	1.42 [a]
significance		0.001 [c]	0.026 [c]	0.003 [c]	0.000 [c]	0.051 [c]	0.000 [c]	0.005 [c]	0.001 [c]	0.002 [d]	0.002 [d]	0.002 [d]	0.003 [d]	0.013 [c]	0.048 [c]	0.373 [c]	0.263 [c]

* ND = not detected. a = ANOVA F statistics, b = Kruskal-Wallis Chi-Square statistics, c = ANOVA significance, d = Kruskal-Wallis significance.

The concentrations of examined physicochemical parameters of honey samples are given in Table 1. The moisture level had similar values across studied regions and ranged from 14.92 ± 0.78% (Piperi) to 16.22 ± 0.36% (Zeta). The pH of studied honey samples varies between 3.87 and 4.49 and was lowest in samples from the Piva region (3.87 ± 0.36) and highest in honey samples from Mijakovići (4.49 ± 0.14). A significant difference has been observed in pH according to the honey regional origin ($p = 0.048$). The electrical conductivity varied from 0.39 to 0.93 mS/cm and was lowest in samples from Katunska (0.39 ± 0.08 mS/cm) and highest in honey samples from Zeta (0.93 ± 0.15 mS/cm). A significant difference has been observed in electrical conductivity according to the honey geographical origin ($p = 0.013$). Free acidity varied from 25.00 to 41.67 meq/kg and was lowest in samples from Katunska (25.00 ± 7.21 meq/kg) and highest in honey samples from the Piva region (41.67 ± 12.10 meq/kg).

Correlation analysis revealed significant correlation between contents of K (R = 0.800, significance < 0.001) and Mg (R = 0.758, significance < 0.001) from one side and pH from the other one (Table 2).

Table 2. Results of the correlation analysis between the mineral content and physicochemical parameters of analyzed honey samples from Montenegro.

		Electrical Conductivity	pH	Moisture	Free Acidity
Pb	R	−0.028	−0.052	0.013	0.224
Cd	R	−0.087	−0.030	0.080	0.259
Cu	R	−0.239	0.240	−0.043	−0.029
Zn	R	−0.329	−0.183	0.052	0.035
Fe	R	−0.379	0.044	−0.115	−0.194
Cr	R	−0.179	0.335	−0.069	−0.285
Sr	R	−0.006	0.074	0.009	0.125
Ba	R	−0.064	0.114	−0.275	−0.082
Ca	R	−0.229	−0.019	-0.031	−0.061
Na	R	0.079	0.382	0.214	−0.173
K	R	0.171	0.800 **	0.218	−0.250
Mg	R	0.159	0.758 **	0.325	−0.391

** significance < 0.01.

The first principal component explains 42.53% of the total variability and is mostly determined by Cd (R = 0.416), Pb (R = 0.398), and Cu (R = 0.352). The PC2 explains 17.97% and is mostly determined by Mg (R = −0.577), K (R = −0.54), and Na (R = −0.425). Mutual projections of factor scores and their loadings for the first two PCs are presented in Figure 2. As can be seen from the projection plot the separation of the analyzed honey samples is much clearer along the X-axis. On the one side, there are localities Piva, Piperi, Katunska, Zeta and Martinići in whose honey samples Cd, Pb, and Sr were not detected. On the other side, there are Mijakovići, Potrlica and especially Zbljevo, whose honey samples concentrations of Cd, Pb and Sr were detected.

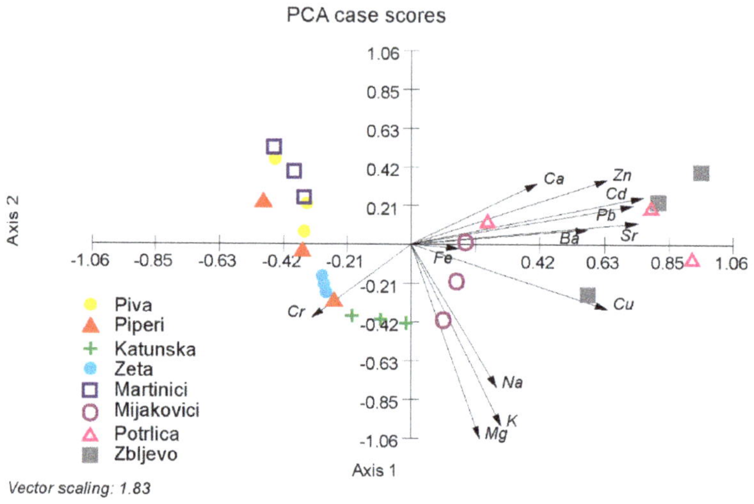

Figure 2. Principal component analysis (PCA) of the mineral content scores of analyzed Montenegrin honey samples.

LDA performed on the geographical origin revealed that the cross-validation classification was correct for 79.17% of samples (Table 3). The smallest percent of good classification was achieved in the case of honey samples from Katunska, while the highest in the case of honeys from Piperi, Zeta, Mijakovići, and Zbljevo.

Table 3. Classification of honey according to their regional origin using the linear discriminate analysis.

Predicted	Piva	Piperi	Katunska	Zeta	Martinići	Mijakovići	Potrlica	Zbljevo
Piva	66.67%	0.00%	0.00%	0.00%	33.33%	0.00%	0.00%	0.00%
Piperi	0.00%	100.00%	33.33%	0.00%	0.00%	0.00%	0.00%	0.00%
Katunska	0.00%	0.00%	33.33%	0.00%	0.00%	0.00%	0.00%	0.00%
Zeta	0.00%	0.00%	0.00%	100.00%	0.00%	0.00%	0.00%	0.00%
Martinići	33.33%	0.00%	0.00%	0.00%	66.67%	0.00%	0.00%	0.00%
Mijakovići	0.00%	0.00%	0.00%	0.00%	0.00%	100.00%	0.00%	0.00%
Potrlica	0.00%	0.00%	0.00%	0.00%	0.00%	0.00%	66.67%	0.00%
Zbljevo	0.00%	0.00%	33.33%	0.00%	0.00%	0.00%	33.33%	100.00%

The bidimensional plot (Figure 3) of the first two functions show four distinct clusters, three of them corresponding to Mijakovići, Potrlica, and Zbljevo regions, while all other regions were clustered together. The first discriminant function explains 94.2% of the total variance and it is dominated by Cd content (R = 0.95). The second discriminant function explains 4.4% of the total variance and is dominated by Pb (R = −0.55) and Sr (R = −0.59) contents.

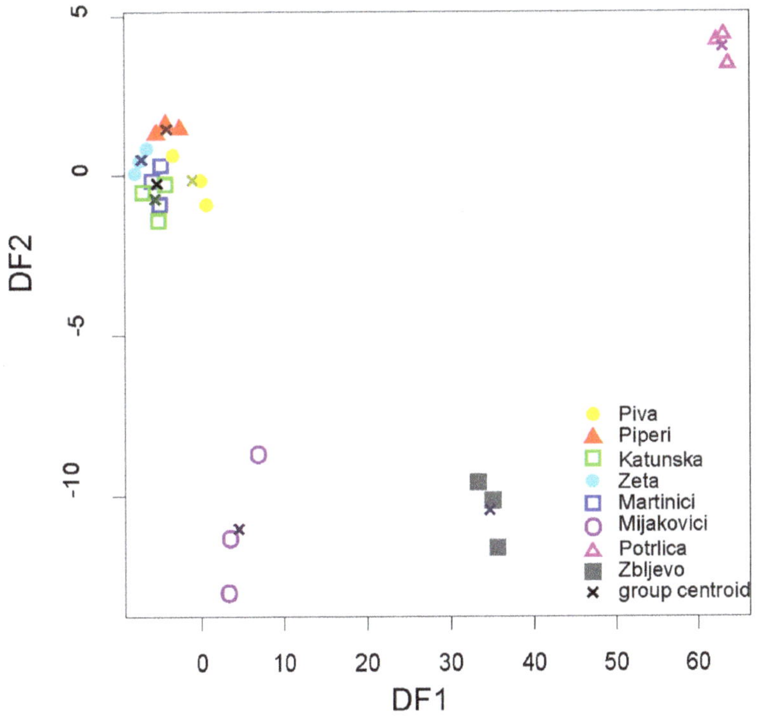

Figure 3. Linear discriminant score plot of analyzed honey samples.

4. Discussion

The values of the mineral contents have been compared with those established by the EU regulations [12]. With the exception of the concentrations of cadmium in samples from Zbljevo and Potrlica, none of the 12 elements analyzed exceeded the maximum allowable level established by the EU regulations. Our results revealed significant differences in the concentrations of Pb, Cd, and Sr between the studied geographical areas of Montenegro. The latter elements were detected only in the samples from Mijakovići, Potrlica and Zbljevo regions, which are likely under the influence of the Pljevlja Thermal Power Plant (Zbljevo, and in less extent Mijakovići) and the Pljevlja coalmine (Potrlica).

The Pb content in the analyzed honey samples varied from 80–210 μg/kg. These values were lower in comparison with the honey from Serbia (290 μg/kg [13]) and Italy (289 μg/kg [14]) but higher in comparison with those from Croatia (5.43–11.3 μg/kg [15]) and Bosnia and Hercegovina (13.4 μg/kg [16]). All these values are below the maximum allowable level established by the EU regulations (0.5 mg/kg) [12].

The values reported for Cd in this study (20–80 μg/kg) were higher in comparison with the honey from Croatia (0.69–12.8 μg/kg [15]), Bosnia and Hercegovina (0.013–22.9 μg/kg [16]), Romania (0.5–11.60 μg/kg [17]), Italy (8–18 μg/kg [14]), Spain (0.7–50 μg/kg [17]) and Serbia (0.59–30 μg/kg [1,13]). The values of Cd content in the samples from Potrlica and Zbljevo exceeds the maximum allowable level established by the EU legislation (0.05 mg/kg) [12]. As the main sources of Cd are recognized as the presence in sewage sludge and smelting from the nearby Pljevlja Thermal Power Plant (Zbljevo), or mining from the Pljevlja coalmine (Potrlica).

The Sr content in honey samples from our study varied from 0.07–0.12 μg/kg and was in the same range as those from Serbia (0.09–0.19 μg/kg [13]).

The most abundant element in honey samples was kalium, followed by Ca, Mg, Na, and Fe. In our study, we found that the content of kalium and magnesium correlated with pH. The average levels of K content ranged from 713–2589.33 mg/kg and was in the same range with those from Croatia (304.7–2824.4 mg/kg [15]), but lower in comparison with the maximum values established for honeys from Bosnia and Herzegovina (14.81–4895.73 μg/kg [16]). On the other hand, the range of concentrations of kalium in the honey from Serbia (400–1755 mg/kg [1,13]) and Slovenia (1090–1220 mg/kg [18]) were lower. The Mg content in honey samples from our study varied from 29.52 to 76.33 mg/kg. In neighboring countries the Mg content in honey varied in a similar range, as: 28.83 to 101.50 mg/kg [13] in Serbia, 2.18 to 166.04 mg/kg [16] in Bosnia and Herzegovina and from 8.02 to 59.1 mg/kg [16,19] in Croatia.

In our study, we used two chemometric analyses, PCA and LDA, respectively to test similarities between honey samples of honey of different geographical origins. Both applied methods separated the regions exposed to industrial pollution (Mijakovići, Potrlica, and Zbljevo) which are characterized by detectable content of Cd, Pb and Sr in their honey samples.

Using LDA it's possible to evaluate the capacity to correctly predict the group to which the unknown samples belong. In our study LDA analysis performed on the geographical origin revealed that the cross-validation classification was correct for 79.17% of the samples. The obtained values are in the range for those from Serbia (Zlatibor: 94.73%, Vojvodina: 70.58% [1]). On the other hand, our value was greater than those reported in the case of Romania honeys where only 21.2% were correctly classified according to their geographical origin [4].

The smallest percentage of good classification was achieved in the case of honeys from Katunska. Of the three samples from the latter region, only one was correctly classified, while the other two being misclassified as Piperi and Zbljevo, respectively. It is known that large numbers of beekeepers (especially from Katunska) in a part of the year (most often in summertime) move their bee colonies to geographically distant areas. On the other hand, the highest percentage of good classification was achieved in the case of honeys from Piperi, Zeta, Mijakovići, and Zbljevo. One cause may be that most of these sites (i.e., Zeta, Mijakovići, and Zbljevo) are more exposed to industrial pollution, resulting in increased concentration of heavy metals (Pb, Cd, and Sr showing significant difference ($p < 0.05$) between studied regions) in their honeys, which, in turn, increase the success rate of the classification of honey according to their geographical origin.

Author Contributions: Investigation, V.V.-P., S.B.-V. and N.B.; statistical analysis, A.S.; writing—original draft preparation, V.P.; writing—review and editing, V.V.-P. and V.P. All authors have read and agreed to the published version of the manuscript.

Funding: This research received no external funding.

Conflicts of Interest: The authors declare no conflict of interest.

References

1. Lazarević, K.; Trifković, J.; Andric, F.; Tešić, Ž.; Anđelković, I.; Radovic, D.; Nedić, N.; Milojković-Opsenica, D. Quality parameters and pattern recognition methods as a tool in tracing the regional origin of multifloral honey. *J. Serb. Chem. Soc.* **2013**, *78*, 1875–1892. [CrossRef]
2. Leita, L.; Muhlbachova, G.; Cesco, S.; Barbattini, R.; Mondini, C. Investigation on the Use of Honeybees and Honeybee Products to Assess Heavy Metal Contamination. *Environ. Monit. Assess.* **1996**, *43*, 1–9. [CrossRef] [PubMed]
3. Karabagias, I.; Karabournioti, S. Discrimination of clover and citrus honeys from Egypt according to floral type using easily assessable physicochemical parameters and discriminant analysis: An external validation of the chemometric approach. *Foods* **2018**, *7*, 70. [CrossRef] [PubMed]
4. Oroian, M.; Prisacaru, A.; Hretcanu, E.C.; Stroe, S.G.; Leahu, A.; Buculei, A. Heavy Metals Profile in Honey as a Potential Indicator of Botanical and Geographical Origin. *Int. J. Food Prop.* **2016**, *19*, 1825–1836. [CrossRef]

5. Paul, S.; Hossen, M.S.; Tanvir, E.M.; Afroz, R.; Hossen, D.; Das, S.; Bhoumik, N.C.; Karim, N.; Juliana, F.M.; Hua Gan, S.; et al. Minerals, Toxic Heavy Metals, and Antioxidant Properties of Honeys from Bangladesh. *J. Chem.* **2017**. [CrossRef]
6. Porrini, C.; Sabatini, A.G.; Girotti, S.; Ghini, S.; Medrzycki, P.; Grillenzoni, F.; Bortolotti, L.; Gattavecchia, E.; Celli, G. Honey bees and bee products as monitors of the environmental contamination. *Apiacta* **2003**, *38*, 63–70.
7. Staniskiene, B.; Matusevicius, P.; Budreckiene, R. Honey as an indicator of environmental pollution. *Environ. Res. Eng. Manag.* **2006**, *2*, 53–58.
8. Sector Study for the Preparation of the Strategy for the Development of Agriculture and Rural Areas 2014–2020. Union of Beekeeping Organizations of Montenegro. 2014. Available online: https://pcelarstvo.me/wpx/wp-content/uploads/2017/01/Studije-o-proizvodnji-meda.pdf?x46702 (accessed on 23 December 2019).
9. Market research of Bee Products in Montenegro. Union of Beekeeping Organizations of Montenegro. 2016. Available online: https://pcelarstvo.me/wpx/wp-content/uploads/2017/02/Istrazivanje-trzista-pcelinjih-proizvoda-u-Crnoj-Gori.pdf?x46702 (accessed on 23 December 2019).
10. Pešić, V.; Grabowski, M.; Hadžiablahović, S.; Marić, D.; Paunović, M. The Biodiversity and Biogeographical Characteristics of the River Basins of Montenegro. In *The Rivers of Montenegro*; Pešić, V., Paunović, M., Kostianoy, A., Eds.; The Handbook of Environmental Chemistry; Springer: Cham, Switzerland, 2019. [CrossRef]
11. Bogdanov, S. Harmonised Methods of the International Honey Commission, International Honey Commission, Internet Publication. 2002. Available online: http://www.apiculturacluj.com/ApiculturaCluj/italiano/Documents/IHCmethods_e.pdf (accessed on 23 December 2019).
12. Commission Regulation No 1881/2006, the Framework EU Legislation Which Sets Maximum Levels for Chemical Contaminants in Foodstuffs. Available online: https://eur-lex.europa.eu/LexUriServ/LexUriServ.do?uri=OJ:L:2006:364:0005:0024:EN:PDF (accessed on 23 December 2019).
13. Lazarević, K. Physico-Chemical Characterization and Classification of Honey from the Territory of The Republic of Serbia According to Regional and Botanical Origin Using Multivariate Chemometric Analysis. Ph.D. Thesis, Faculty of Chemistry, University of Belgrade, Belgrade, Serbia, 2016.
14. Perna, A.; Intaglietta, I.; Simonetti, A.; Gambacorta, E. Metals in Honeys from Different Areas of Southern Italy. *Bull. Environ. Contam. Toxicol.* **2014**, *92*, 253–258. [CrossRef]
15. Bilandžić, N.; Tlak Gajger, I.; Čalopek, B.; Sedak, M.; Solomun Kolanović, B.; Varenina, I.; Božić Luburić, Đ.; Varga, I.; Đokić, M. Content of Heavy Metals and Trace Elements in Different Types of Honey from Dubrovnik Country. *Vet. Stn.* **2015**, *46*, 359–368.
16. Tanković, S.; Jelušić, V.; Bilandžić, N.; Čalopek, B.; Sedak, M.; Ferizbegović, J. Concentrations of Heavy Metals and Elements in Different Types of Honey from Bosnia and Herzegovina. *Vet. Stn.* **2017**, *48*, 1–12.
17. Corredera, L.; Bayarri, S.; Pérez-Arquillué, C.; Lázaro, R.; Molino, F.; Herrera, A. Evaluation of Heavy Metals and Polycyclic Aromatic Hydrocarbons in Honeys from Different Origins. *J. Food Prot.* **2014**, *77*, 504–509. [CrossRef] [PubMed]
18. Kropf, U.; Bertoncelj, J.; Korošec, M.; Nečemer, M.; Kump, P.; Ogrinc, N.; Golob, T. Geographical origin of Slovenian multifloral and forest honey. *Apiacta* **2009**, *44*, 33–42.
19. Bilandžić, N.; Gačić, M.; Đokić, M.; Sedak, M.; Ivanec Šipušić, Đ.; Končurat, A.; Tlak Gajger, I. Major and trace elements levels in multifloral and unifloral honeys in Croatia. *J. Food Compos. Anal.* **2014**, *33*, 123–138. [CrossRef]

foods

MDPI

Article

Geographical Origin Discrimination of Monofloral Honeys by Direct Analysis in Real Time Ionization-High Resolution Mass Spectrometry (DART-HRMS)

Vincenzo Lippolis [1,*], Elisabetta De Angelis [1], Giuseppina Maria Fiorino [1], Annalisa Di Gioia [1], Marco Arlorio [2], Antonio Francesco Logrieco [1] and Linda Monaci [1]

[1] Institute of Sciences of Food Production (ISPA), National Research Council of Italy (CNR), Via G. Amendola 122/O, 70126 Bari, Italy; elisabetta.deangelis@ispa.cnr.it (E.D.A.); giuseppina.fiorino@outlook.com (G.M.F.); annalisadigioia@hotmail.it (A.D.G.); antonio.logrieco@ispa.cnr.it (A.F.L.); linda.monaci@ispa.cnr.it (L.M.)

[2] Dipartimento di Scienze del Farmaco, Università degli Studi del Piemonte Orientale "Amedeo Avogadro" (UNIUPO), Largo Donegani 2, 28100 Novara, Italy; marco.arlorio@uniupo.it

* Correspondence: vincenzo.lippolis@ispa.cnr.it; Tel.: +39-080-5929457

Received: 27 July 2020; Accepted: 28 August 2020; Published: 1 September 2020

Abstract: An untargeted method using direct analysis in real time and high resolution mass spectrometry (DART-HRMS) combined to multivariate statistical analysis was developed for the discrimination of two monofloral (chestnut and acacia) honeys for their geographical origins—i.e., Italy and Portugal for chestnut honey and Italy and China for acacia honey. Principal Component Analysis, used as an unsupervised approach, showed samples of clusterization for chestnut honey samples, while overlapping regions were observed for acacia honeys. Three supervised statistical approaches, such as Principal Components—Linear Discriminant Analysis, Partial Least Squares—Discriminant Analysis and k-nearest neighbors, were tested on the dataset gathered and relevant performances were compared. All tested statistical approaches provided comparable prediction abilities in cross-validation and external validation with mean values falling between 89.2–98.4% for chestnut and between 85.8–95.0% for acacia honey. The results obtained herein indicate the feasibility of the DART-HRMS approach in combination with chemometrics for the rapid authentication of honey's geographical origin.

Keywords: monofloral honey; direct analysis in real time (DART); high resolution mass spectrometry (HRMS); geographical origin; chemometrics

1. Introduction

Honey is a complex and high-quality natural product containing a wide range of nutritional and therapeutic properties but with a limited production and high commercial prices. Honey is defined by European Union legislation as the natural sweet substance produced by bees of *Apis mellifera* species from nectar or sugary secretions of plants, as well as from excretions of plant-sucking insects on the living parts of plants [1]. Both the European Union and Codex Alimentarius laws establish that the geographical origin, in terms of country of production, must be indicated on the label, also supplemented by specific reference to the floral or vegetable origin. Moreover, in the case of blends of honey, their origin should be declared as a "blend of EC honeys", "blend of non-EC honeys" or "blend of EC and non-EC honeys" [1,2].

Geographical and botanical origins of honey account for the peculiar chemical composition and organoleptic characteristics of the final product [3,4]. Monofloral honeys, mostly deriving from a single plant species (at least 45% of pollen grains), may considerably differ in their sensory properties with highly prominent flavor and aroma. Acacia (*Robinia pseudoacacia*) honey is one of the most consumed monofloral honeys in Europe, being appreciated for its permanently liquid state, light color, floral aroma and sweet and delicate taste [5]. Similarly, chestnut (*Castanea sativa*) honey is considered one of the most delicious and high-quality honeys, being a very good source for nectar and pollen [6,7]. For these reasons, monofloral honeys, and in particular those derived from acacia and chestnut, have recently gained consumer preferences, with an increased demand and commercial value [4]. Due to their increased commercial value, monofloral honeys are highly susceptible to fraudulent practices through mislabeling and mixing with cheaper and lower-quality honeys or with various sugar syrups.

Honey is produced in different areas of the world, with more than 2.3 million tonnes produced worldwide in 2018, with China and Turkey as main producers [8]. China is also the largest exporter of honeys in the world, while in Europe, Portugal is the country bearing the highest number of geographical protected labels on honey [9]. Honey composition is quite variable and strictly linked to its floral source and geographical origin, but external factors, including processing, packaging and storage conditions, could play an important role. Although Italy is one of the EU countries with the highest honey production [10], the market demand for honey is higher than domestic production, therefore a substantial amount of honey is imported from elsewhere in Europe and from third-world countries, in which production does not always meet the high food safety standards required. This can lead to honey mislabeled with regard to its geographical and/or botanical origin [11].

The traceability certifying the geographical origin of food products is of primary importance for traders and producers, as well as to reinforce consumer trust. The complex task of the determination of food origin is commonly applied to control products in both customs control and self-control programs of the food industry.

Melissopalynological analysis of pollen is the most used approach for the botanical and geographical origin classification of honey, as the pollen spectrum is strictly related to the environment where the nectar is collected [12]. This analysis is often complemented by other analytical methodologies, mainly based on chromatographic techniques, to assure the honey authenticity [13]. Often, the use of conventional and targeted methods is time-consuming and not sufficient to guarantee the evaluation of complex matrices, including honeys. For this reason, the development of rapid and reliable non-targeted analytical approaches, such as fingerprinting and profiling methods, is highly demanded. Indeed, these methods combined to chemometric tools allow for the detection of a high number of metabolites, leading to samples based on their pattern.

Several analytical techniques, mainly based on nuclear magnetic resonance [5,14,15], Raman and infrared spectroscopy [16,17], mass spectrometry [18–20], electronic tongue [21,22] and electronic nose [23,24], in combination with chemometrics, have been applied to discriminate the geographical origin of honey.

The use of ambient mass spectrometry (AMS) is continuously increasing in the field of metabolomic fingerprinting as a high-throughput alternative to more traditional hyphenated methods for authentication issues [25]. Among AMS techniques, direct analysis in real-time mass spectrometry (DART-MS), being simple and requiring a very limited sample preparation, has been shown to be the most promising and versatile technique, proving to be a rapid tool in the assessment of food authenticity and food quality, also thanks to the use of fast and streamlined protocols [25–27]. Such an approach offers several advantages over the conventional techniques, including direct sample analysis in open atmosphere, high sample throughput and minimal or no sample preparation requirements, the soft ionization of a wide range of both polar and apolar compounds. Several papers have been recently published demonstrating the applicability of DART-MS to assess food authenticity and detect food adulterations of olive oil [28], beer [29], wine [30], animal fat [27], milk [31] and salmon [32]. Only one paper reported the applicability of DART-MS to the discrimination of geographical origin of food—i.e., garlic produced in Czech Republic, Spain and China [33].

Regarding honey products, DART-HRMS was used as alternative approach for the determination of 5-hydroxymethylfurfural [34,35]. To the best of our knowledge, no studies based on DART-MS have been performed to date for the assessment of geographical origin of honeys.

In this context, the aim of this study was to demonstrate the feasibility of the DART-HRMS technique for the discrimination of the geographical origin of honeys. Specifically, a rapid and suitable non-targeted DART-HRMS method in combination with multivariate statistical analysis was developed and validated to discriminate two monofloral honeys varieties (i.e., chestnut and acacia) for their geographical origin (i.e., Italy and Portugal and Italy and China, respectively). Different statistical classification models were investigated and applied to the analysis of honey samples and performance results were compared.

2. Materials and Methods

2.1. Chemicals and Reagents

Methanol (HPLC grade) was purchased from Sigma-Aldrich (Milan, Italy). Ultrapure water was produced by a Milli-Q® Direct system (Merck KGaA, Darmstadt, Germany). Helium (99.9995% purity) was provided by Sapio S.r.l. (Bari, Italy). Regenerate cellulose (RC) syringe filters with 0.2 μm of porosity were purchased by from VWR International (Milan, Italy). OpenSpot (OS) Sample Cards were purchased by Ion Sense Inc. (Saugus, MA, USA).

2.2. Honey Samples

A total of 234 monofloral honey samples commonly found in marketplaces and collected in different countries with certified origins were selected for this study. Specifically, 117 chestnut honey samples were collected from Italy (39) and Portugal (78), while 117 acacia honey samples were collected from Italy (78) and China (39). The authenticity of the monofloral honeys was assessed by internal certified protocols performed by Coop Italia Soc. Cooperativa (Casalecchio di Reno, Italy) which provided samples. Only honey samples produced in seasons 2017–2018 were taken into account.

2.3. Sample Preparation

Sampling and homogenization of honey samples were performed according to AOAC 920.180 protocol [36]. For sample preparation, a rapid protocol aimed at retaining as many honey metabolites as possible—thus to obtain most comprehensive spectra applicable for discriminating between different geographical origin—was optimized for the DART-HRMS analysis. In particular, an aliquot (1 g) of homogenized honey was added to a mixture of $MeOH/H_2O$ (1:1, *v/v*), (50 mL) and the sample was vortexed for 3 min. After filtration using 0.2 μm RC syringe filter, the filtered extract was directly analyzed by DART-HRMS.

2.4. DART-HRMS Analysis

DART-HRMS analyses were carried out by using a DART ionization source SI-140-GIST (DART Thermo Ion Max Vapur Interface, Ion Sense Inc., Saugus, MA, USA) coupled to an Exactive™ monostage Orbitrap™ High Resolution mass spectrometer (Thermo Fisher Scientific, San Jose, CA, USA). An aliquot (2 μL) of the honey extract was placed onto the metallic grid of the OpenSpot® sample cards and kept at 60 °C for 5 min to facilitate solvent evaporation before its introduction into the DART source holder. The operating conditions of the DART source were: positive ion mode; helium flow of 3.2 L/min for 1 min and heated at 250 °C; discharge needle voltage kept at −6 kV; grid electrode voltage set to 250 V; distance between DART exit and MS inlet set at 5 mm. The operating conditions of DART source were set by DART-SVP controller (v. 4.0.x). The main settings of the Exactive™ mass spectrometer were the following: mass scan range of 100–600 m/z; resolution set at 25,000 (FWHM at m/z 200); microscan number of 4; Automatic Gain Control (AGC) Target of 3×10^{-6}; maximum injection time (IT) of 250 ms; capillary voltage set to 30 V; tube lens voltage set to 65 V; capillary temperature kept at 250 °C. Calibrations of the MS system were periodically performed by the direct infusion ESI-MS approach of the positive ion calibrating solution, provided by the manufacturer, in order to obtain a mass accuracy lower than 5 ppm. The MS system was controlled by using the Xcalibur™ v. 2.1 software (Thermo Fisher Scientific, San Jose, CA, USA).

To carry out the subtraction of the spectral background, a blank open spot card was acquired before analyzing each sample by DART-HRMS acquiring the relevant spectrum for 30 s.

2.5. Data Processing and Statistical Analysis

In the first step of data processing, DART-HRMS spectra acquired in the time range of 30 s were averaged and then subtracted of spectral background by using the Xcalibur™ software. Successively, for each honey sample, the full list of accurate m/z ratios and peak intensities obtained was exported and processed by MetaboAnalyst 3.0 (http://www.metaboanalyst.ca/) [37,38] for peak matching and alignment with mass tolerance of 0.25, imputation of missing values (replacing missing elements by using the half of the lowest measured peak intensity) and data filtering (by using Interquartile Range approach). Successively, after pre-processing obtained by data centering, the dataset was submitted to multivariate statistical analyses performed by V-Parvus software (release 2010, http://www.parvus.unige.it, Genova, Italy).

Principal Component Analysis (PCA) was used as an unsupervised technique to evaluate the presence of outliers. Specifically, PCA was applied to each single group of monofloral honey samples of different geographical origin, observing the relevant influence plots and excluding samples identified as extreme outliers. To establish the exact number of Principal Components (PCs) to be used to build PCA models, the Non-linear Iterative Partial Least Squares (NIPALS) algorithm was applied using V-fold of 10 (cross validation process, CV = 10). PCA was also used as exploratory technique to visualize the presence of natural sample clustering between monofloral honey samples in relation to their geographical origin [39].

Afterwards, three supervised pattern recognition techniques—i.e., Linear Discriminant Analysis (LDA), Partial Least Squares Discriminant Analysis (PLS-DA) and k-nearest neighbors (k-NN) [40], were exploited to classify monofloral honey samples on the basis of their geographical origin. For this purpose, the two data matrices were randomly split in two subsets: a modelling set (containing 60 samples) and a test set (containing 57 samples). Specifically, for each monofloral honey, a modelling set, composed by 30 samples for each geographical origin, was used to build the three different statistical models. Test sets, consisting of 9 Italian and 48 Portuguese chestnut honey samples and 48 Italian and 9 Chinese acacia honey samples, were used for the validation process.

The chemometric models of PCA-LDA was built by firstly performing PCA test to reduce the number of variables that exceeded the number of objects, thus preventing model overfitting; then the selected scores were used as classification variables for LDA [41,42]. Indeed, the number of variables should not exceed (n-g)/3, where n is the number of objects and g is the number of categories. Considering that modelling sets were composed by 60 objects (number of samples) and 2 categories (number of geographical origins) the maximum number of variables should be approximately 19.

The appropriate numbers of principal components, latent variables and k values, respectively, for PCA-LDA, PLS-DA and k-NN models were established by evaluating those determining the lowest prediction error rate in cross-validation (cross-validation segments, V = 10). This parameter guarantees to improve feature variables and, at the same time, to avoid model overfitting. Model performances for PCA-LDA, PLS-DA and k-NN, expressed as percentages, were compared with reference to their recognition ability—i.e., the ability to correctly classify samples of the modelling set—prediction ability in cross-validation (CV)—i.e., the ability to correctly classify samples of a test set generated in a V-fold cross validation—and prediction ability in external validation—i.e., the ability to correctly classify samples of the test set.

3. Results and Discussion

In the present study, the real-time mass spectrometry (DART-MS) combined with chemometric analysis, was used for the first time to the discrimination of two kind of monofloral honey samples, namely chestnut and acacia, based on their geographical origin. As for chestnut, Italian and Portuguese honey samples were compared to each other, while for acacia, Italian honeys were compared with samples from China.

Figure 1 reports four representative DART-HRMS average spectra, after blank subtraction, obtained for the chestnut honey extracts of Italian (Figure 1a) and Portuguese (Figure 1b) samples and acacia honey extracts of Italian (Figure 1c) and Chinese (Figure 1c) samples.

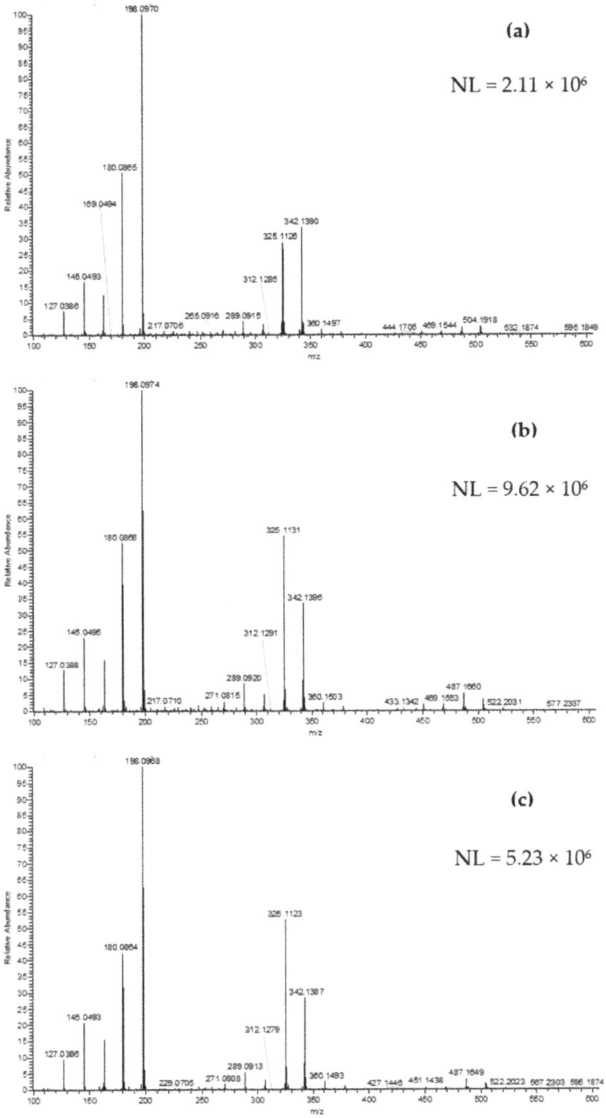

Figure 1. *Cont.*

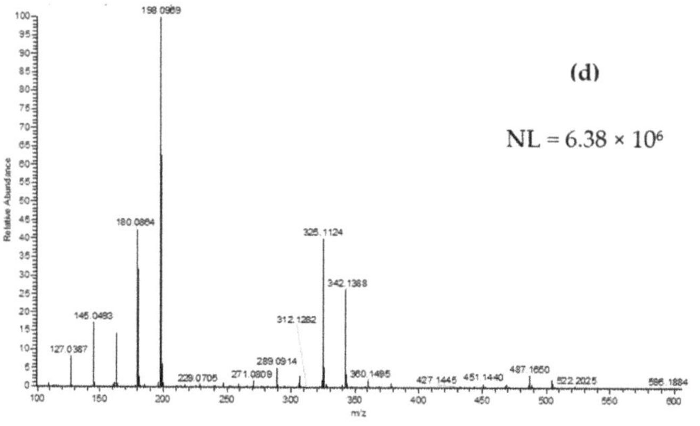

Figure 1. Representative DART-HRMS positive ion spectra acquired for the sample extracts of chestnut honeys from Italy (**a**), chestnut honeys from Portugal (**b**), acacia honeys from Italy (**c**) and acacia honeys from China (**d**). NL: Normalization level.

At first, a preliminary PCA was performed on pre-processed spectra of chestnut and acacia honey samples in order to explore the presence of outlier samples. PCA score plots highlighted that, in the case of chestnut samples, seven PCs described 96.3% of total variance for samples from Italy while nine PCs described 93.0% of total variance for samples from Portugal. In the case of acacia honey samples, PCA models showed that eight PCs described 93.7% of total variance for samples from Italy while nine PCs described 91.0% of total variance for samples from China. The absence of outliers in all classes was demonstrated using influence plots where the Mahalanobis distance was plotted versus sample residual.

Subsequently, an explorative PCA was performed using the entire data set to obtain an overview of the data distribution for each monofloral honey. Figure 2 shows the PCA score plot (PC1 vs. PC2) obtained for chestnut honey samples (Figure 2a) and for acacia honey samples (Figure 2b). A discrete visual clustering of the objects on the basis of their geographical origin was observed for chestnut honeys (PC1 and PC2 explained 88.6% and 10.2% of the total variance, respectively), while overlapping regions were observed for acacia honeys with a modest clustering for their geographical origin (with 88.4% and 9.7% of the total variance explained by PC1 and PC2, respectively). Additionally, by analyzing the score plots of the remaining PCs no visual clusterization was observed.

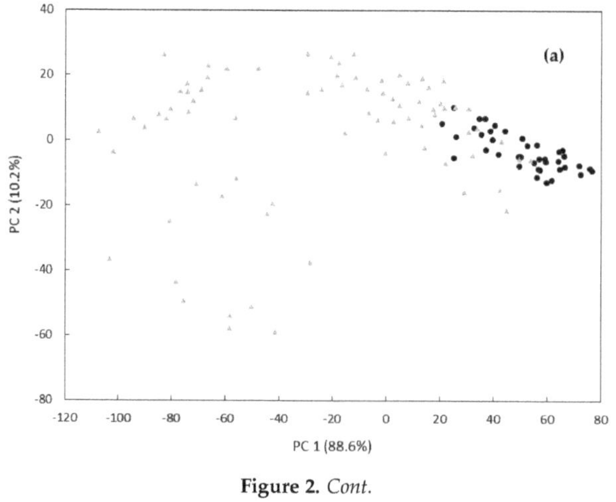

Figure 2. *Cont.*

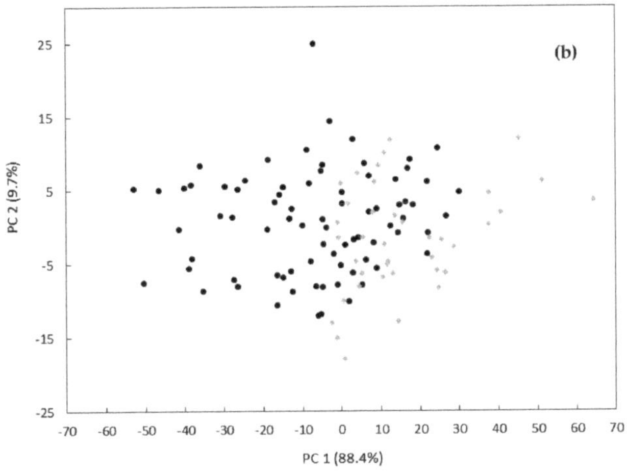

Figure 2. PC1 vs. PC2 scatter plots for monofloral chestnut (**a**) and acacia (**b**) honey samples. Geographical origins: Italy (black filled circle), Portugal (grey filled triangle), China (grey filled rhombus).

These results were confirmed by analyzing the Fisher weight (FW) values of the principal components, which measure the between-class variance/within-class variance ratio. Indeed, FW values resulted to be 2.64 for the PC1 of chestnut honeys samples and lower than 1 for all the remaining PCs of both data sets (data not shown). These results indicated that the PCA was not able to discriminate honey samples on the basis of their geographical origin; therefore, it was necessary to treat data with three different supervised discriminant techniques—i.e., PCA-LDA, PLS-DA and k-NN. These classification techniques were tested on both chestnut and acacia honey samples previously split into two subsets: a modeling set and a test set. Overall, results are indicated in Tables 1 and 2, for chestnut and acacia honeys, respectively.

Table 1. Model performances in terms of recognition, cross validation (CV) prediction abilities and external prediction to classify chestnut honeys based on their geographical origin.

	Model Performance (%)								
	Recognition Ability (Modelling)			Prediction Ability (CV [c] 10)			External Prediction		
	ITA [a]	POR [b]	Mean	ITA	POR	Mean	ITA	POR	Mean
PCA/LDA [d] (7 Principal Components)	100.0 (30/30)	96.7 (29/30)	98.4	100.0 (30/30)	96.7 (29/30)	98.4	88.9 (8/9)	91.7 (44/48)	90.3
PLS-DA [e] (10 Latent Variables)	100.0 (30/30)	96.7 (29/30)	98.4	100.0 (30/30)	93.3 (28/30)	96.7	88.9 (8/9)	89.6 (43/48)	89.2
k-NN [f] (k value of 3)	96.7 (29/30)	100.0 (30/30)	98.4	96.7 (29/30)	100.0 (30/30)	98.4	88.9 (8/9)	93.8 (45/48)	91.4

[a]: Italy; [b]: Portugal; [c]: Cross Validation; [d]: Principal Components—Linear Discriminant Analysis; [e]: Partial Least Squares—Discriminant Analysis; [f]: k-nearest neighbors.

As for LDA, PCA was used as strategy for variable reduction and to avoid model overfitting. The number of PCs (seven and nine for chestnut and acacia honeys, respectively) to be used to build the PCA-LDA models was selected on the basis of the error in prediction cross validation that has to be the lowest (CV procedure with V = 10). The PCA-LDA models provided mean value of recognition ability of 98.4% for chestnut honeys (Table 1) in both classification and CV prediction and 95.0 and 93.4% for acacia honeys (Table 2) in classification and CV prediction, respectively. The model applicability was also tested by using the test set providing mean prediction abilities of 90.3 and 89.2%, for chestnut and acacia honeys, respectively (Tables 1 and 2).

Table 2. Model performances in terms of recognition, cross validation (CV) prediction abilities and external prediction for all models built to classify acacia honeys based on their geographical origin.

	Model Performance (%)								
	Recognition Ability (Modelling)			Prediction Ability (CV [c] 10)			External Prediction		
	ITA [a]	CHI [b]	Mean	ITA	CHI	Mean	ITA	CHI	Mean
PCA/LDA [d] (9 Principal Components)	96.7 (29/30)	93.3 (28/30)	95.0	96.7 (29/30)	90.0 (27/30)	93.4	89.6 (43/48)	88.9 (8/9)	89.2
PLS-DA [e] (12 Latent Variables)	100.0 (30/30)	93.3 (28/30)	96.7	96.7 (29/30)	93.3 (28/30)	95.0	93.8 (45/48)	77.8 (7/9)	85.8
k-NN [f] (k value of 3)	100.0 (30/30)	90.0 (27/30)	95.0	93.3 (28/30)	90.0 (27/30)	91.7	91.7 (44/48)	88.9 (8/9)	90.3

[a]: Italy; [b]: Portugal; [c]: Cross Validation; [d]: Principal Components—Linear Discriminant Analysis; [e]: Partial Least Squares—Discriminant Analysis; [f]: k-nearest neighbors.

PLS-DA was applied as an alternative multivariate statistical approach of classification offering the advantage to avoid variables reduction processes. Specifically, by applying a 10-fold cross-validation, 10 and 12 latent variables (LVs) were found to produce the optimal model complexity for chestnut and acacia honey data sets, respectively. In these conditions, mean recognition rates were higher than 96.7% in both cases (Tables 1 and 2). Specifically, all Italian samples was correctly classified, while one Portuguese and two Chinese samples were not correctly assigned. The mean CV prediction rates were 96.7 and 95.0%, for chestnut and acacia honeys, respectively. In addition, mean prediction abilities of 89.2% and 85.8% for chestnut and acacia honeys samples, respectively, were obtained for the external validation procedure (Tables 1 and 2).

In the case of k-NN, the prediction error rate in cross-validation (V = 10) was calculated for each different k value. The smallest k value determining the lowest error was 3 for both data sets and therefore it was selected as the optimal value. The k-NN models provided mean recognition abilities in the range between 95.0–98.4%, while CV predictions were of 98.4 and 91.7%, for chestnut and acacia honeys samples, respectively. Finally, mean prediction abilities of 91.4% and 90.3% were obtained in the external validation for chestnut and acacia honeys, respectively.

The results herein obtained were in accordance with a similar study focused on the geographical authentication of Italian honey based on an NMR-metabolomic approach [5]. The authors developed a PLS2-DA model able to correctly discriminate 100% of Italian honeys from Eastern European ones. In another study, MIR analysis in combination with a PCA-LDA model were found able to distinguish geographical origins of monofloral honeys from Switzerland, Germany, and France, with prediction abilities ranged from 76 to 100% [17] although only a limited number of samples was used for the analysis.

In the current study, the DART-HRMS untargeted approach coupled with three supervised techniques, such as PCA-LDA, PLS-DA and k-NN, were investigated for discriminating Italian chestnut and acacia honey from Portuguese and Chinese samples. The results showed that all developed models provided acceptable and comparable prediction abilities, highlighting the robustness of the entire method, its applicability being unaffected by the statistical approach used to assess the authenticity of unknown samples. Moreover, these results demonstrated that DART-HRMS technique provides informative experimental data useful to build up appropriate models for the discrimination of monofloral honey samples on the basis of their geographical origin.

4. Conclusions

In this study, a rapid, easy-to-perform and low-cost method based on DART-HRMS analysis combined to multivariate statistical analysis was successfully developed and applied to classify monofloral honeys for their geographical origins, such as Italy and Portugal for chestnut samples and Italy and China for acacia samples. Specifically, three supervised approaches—i.e., PCA-LDA, PLS-DA and k-NN were evaluated. All tested models provided high and comparable recognition and prediction abilities in cross-validation and external validation, with mean values ranging from

89.2% and 98.4%. The performances of the proposed DART-HRMS method makes it an effective tool to assess the authenticity of honeys, for both industries of sector against unfair advantages of competitors and control bodies to fight food frauds. Future efforts will be directed to improve the current predictive models in order to discriminate honey samples from different production seasons and identify potential markers useful for developing a DART-HRMS target method aimed at honey authentication. Moreover, the use of the DART-HRMS approach, generating huge information in a single run, would be a useful tool for discriminating honey samples with similar organoleptic characteristics but different quality levels.

Author Contributions: Conceptualization: V.L., M.A., A.F.L. and L.M.; methodology, V.L. and L.M.; validation, V.L. and E.D.A.; formal analysis, G.M.F. and A.D.G.; writing—original draft preparation, V.L.; writing—review and editing, V.L., E.D.A., G.M.F., A.D.G., A.F.L., M.A. and L.M.; supervision, A.F.L. and L.M.; funding acquisition, M.A. All authors have read and agreed to the published version of the manuscript.

Funding: The present research has received funding from the European Union's Seventh Framework Programme for research, technological development and demonstration, under grant agreement No. 613688 2 "Food Integrity". The equipment used in this work was funded by project BioNet – PTP – "Biodiversità per la valorizzazione e sicurezza delle produzioni alimentari tipiche pugliesi (codice n. 73, PO Regione Puglia FESR 377 2000-2006)".

Acknowledgments: The authors thank Salvatore Cervellieri for his support in the statistical analysis and Fernando Gottardi from Coop Italia (Casalecchio di Reno, Italy) for providing the honey samples analyzed in the present study.

Conflicts of Interest: The authors declare no conflict of interest.

References

1. European Commission. Directive 2014/63/EU of The European Parliament and of the Council of 15 May 2014 amending Council Directive 2001/110/EC relating to honey. *Off. J. Eur. Comm.* **2014**, *L 164*, 1–5.
2. Food and Agriculture Organization of the United Nations. Revised Codex Standard for Honey (No. CODEX STAN 12–1981). 2001. Available online: www.fao.org/input/download/standards/310/cxs_012e.pdf (accessed on 25 July 2020).
3. Da Silva, P.M.; Gauche, C.; Gonzaga, L.V.; Costa, A.C.O.; Fett, R. Honey: Chemical composition, stability and authenticity. *Food Chem.* **2016**, *196*, 309–323. [CrossRef] [PubMed]
4. Soares, S.; Amaral, J.S.; Oliveira, M.B.P.P.; Mafra, I. A comprehensive review on the main honey authentication issues: Production and origin. *Compr. Rev. Food Sci. Food Saf.* **2017**, *16*, 1072–1100. [CrossRef]
5. Schievano, E.; Stocchero, M.; Zuccato, V.; Conti, I.; Piana, L. NMR assessment of European acacia honey origin and composition of EU-blend based on geographical floral markers. *Food Chem.* **2019**, *288*, 96–101. [CrossRef] [PubMed]
6. Castro-Vázquez, L.; Díaz-Maroto, M.C.; de Torres, C.; Pérez-Coello, M.S. Effect of geographical origin on the chemical and sensory characteristics of chestnut honeys. *Food Res. Int.* **2010**, *43*, 2335–2340. [CrossRef]
7. Persano Oddo, L.; Piro, R. Main European unifloral honeys: Descriptive sheets. *Apidologie* **2004**, *35*, S38–S81. [CrossRef]
8. Food and Agriculture Organization of the United Nations, FAOSTAT. Livestock Primary Data. 2018. Available online: http://www.fao.org/faostat/en/#data/QL/visualize (accessed on 25 July 2020).
9. European Commission. Agriculture and Rural Development DOOR. 2017. Available online: http://ec.europa.eu/agriculture/quality/door/list.html (accessed on 25 July 2020).
10. European Commission. Report from the Commission to the European Parliament and the Council on the Implementation of Apiculture Programmes. 2019. Available online: https://ec.europa.eu/info/food-farming-fisheries/animals-and-animal-products/animal-products/honey_en (accessed on 25 July 2020).
11. Aries, E.; Burton, J.; Carrasco, L.; De Rudder, O.; Maquet, A. *Scientific Support to the Implementation of a Coordinated Control Plan with a View to Establishing the Prevalence of Fraudulent Practices in the Marketing of Honey*; JRC Technical Report JRC104749; European Commission: Brussels, Belgium, 2016.
12. Von Der Ohe, W.; Oddo, L.P.; Piana, M.L.; Morlot, M.; Martin, P. Harmonized methods of melissopalynology. *Apidologie* **2004**, *35*, S18–S25. [CrossRef]
13. Pita-Calvo, C.; Guerra-Rodríguez, M.E.; Vázquez, M. Analytical methods used in the quality control of honey. *J. Agric. Food Chem.* **2017**, *65*, 690–703. [CrossRef]

14. Siddiqui, A.J.; Musharraf, S.G.; Choudhary, M.I.; Rahman, A. Application of analytical methods in authentication and adulteration of honey. *Food Chem.* **2017**, *217*, 687–698. [CrossRef]
15. Zheng, X.; Zhao, Y.; Wu, H.; Dong, J.; Feng, J. Origin Identification and Quantitative Analysis of Honeys by Nuclear Magnetic Resonance and Chemometric Techniques. *Food Anal. Methods* **2016**, *9*, 1470–1479. [CrossRef]
16. Corvucci, F.; Nobili, L.; Melucci, D.; Grillenzoni, F. The discrimination of honey origin using melissopalynology and Raman spectroscopy techniques coupled with multivariate analysis. *Food Chem.* **2015**, *169*, 297–304. [CrossRef] [PubMed]
17. Ruoff, K.; Luginbühl, W.; Künzli, R.; Iglesias, M.T.; Bogdanov, S.; Bosset, J.O.; Amadò, R. Authentication of the botanical and geographical origin of honey by midinfrared spectroscopy. *J. Agric. Food Chem.* **2006**, *54*, 6873–6880. [CrossRef] [PubMed]
18. Ozcan-Sinir, G.; Copur, O.U.; Barringer, S.A. Botanical and geographical origin of Turkish honeys by selected-ion flow-tube mass spectrometry and chemometrics. *J. Sci. Food Agric.* **2020**, *100*, 2198–2207. [CrossRef]
19. Karabagias, I.K.; Papastephanou, C.; Karabagias, V.K. Geographical differentiation of Cypriot multifloral honeys through specific volatile compounds and the use of DFA. *AIMS Agric. Food* **2019**, *4*, 149–162. [CrossRef]
20. Jandrić, Z.; Frew, R.D.; Fernandez-Cedi, L.N.; Cannavan, A. An investigative study on discrimination of honey of various floral and geographical origins using UPLC QToF MS and multivariate data analysis. *Food Control* **2017**, *72*, 189–197. [CrossRef]
21. El Hassani, N.E.A.; Tahri, K.; Llobet, E.; Bouchikhi, B.; Errachid, A.; Zine, N.; El Bari, N. Emerging approach for analytical characterization and geographical classification of Moroccan and French honeys by means of a voltammetric electronic tongue. *Food Chem.* **2018**, *243*, 36–42. [CrossRef]
22. Bougrini, M.; Tahri, K.; Saidi, T.; El Hassani, N.E.A.; Bouchikhi, B.; El Bari, N. Classification of Honey According to Geographical and Botanical Origins and Detection of Its Adulteration Using Voltammetric Electronic Tongue. *Food Anal. Methods* **2016**, *9*, 2161–2173. [CrossRef]
23. Benedetti, S.; Mannino, S.; Sabatini, A.G.; Marcazzan, G.L. Electronic nose and neural network use for the classification of honey. *Apidologie* **2004**, *35*, 397–402. [CrossRef]
24. Čačić, F.; Primorac, L.; Kenjerić, D.; Benedetti, S.; Mandić, M.L. Application of electronic nose in honey geographical origin characterisation. *J. Cent. Eur. Agric.* **2009**, *10*, 19–26.
25. Black, C.; Chevallier, O.P.; Elliott, C.T. The current and potential applications of Ambient Mass Spectrometry in detecting food fraud. *Trends Anal. Chem.* **2016**, *82*, 268–278. [CrossRef]
26. Guo, T.; Yong, W.; Jin, Y.; Liu, J.; Wang, S.; Chen, Q.; Dong, Y.; Su, H.; Tan, T. Application of DART-MS for food quality and safety assurance in food supply chain. *Mass Spectrom. Rev.* **2017**, *36*, 161–187. [CrossRef] [PubMed]
27. Vaclavik, L.; Hrbek, V.; Cajka, T.; Rohlik, B.A.; Pipek, P.; Hajslova, J. Authentication of animal fats using direct analysis in real time (DART) ionization-mass spectrometry and chemometric tools. *J. Agric. Food Chem.* **2011**, *59*, 5919–5926. [CrossRef] [PubMed]
28. Vaclavik, L.; Hrbek, V.; Cajka, T.; Hajslova, J. Ambient mass spectrometry employing direct analysis in real time ion source for olive oil quality and authenticity assessment. *Anal. Chim. Acta* **2009**, *645*, 56–63. [CrossRef] [PubMed]
29. Cajka, T.; Riddellova, K.; Tomaniova, M.; Hajslova, J. Ambient mass spectrometry employing a DART ion source for metabolomic fingerprinting/profiling: A powerful tool for beer origin recognition. *Metabolomics* **2011**, *7*, 500–508. [CrossRef]
30. Rubert, J.; Lacina, O.; Fauhl-Hassek, C.; Hajslova, J. Metabolic fingerprinting based on high-resolution tandem mass spectrometry: A reliable tool for wine authentication? *Anal. Bioanal. Chem.* **2014**, *406*, 6791–6803. [CrossRef] [PubMed]
31. Hrbek, V.; Vaclavik, L.; Elich, O.; Hajslova, J. Authentication of milk and milk-based foods by direct analysis in real time ionization-high resolution mass spectrometry (DART-HRMS) technique: A critical assessment. *Food Control* **2014**, *36*, 138–145. [CrossRef]
32. Fiorino, G.M.; Losito, I.; De Angelis, E.; Arlorio, M.; Logrieco, A.F.; Monaci, L. Assessing fish authenticity by direct analysis in real time-high resolution mass spectrometry and multivariate analysis: Discrimination between wildtype and farmed salmon. *Food Res. Int.* **2019**, *116*, 1258–1265. [CrossRef]

33. Hrbek, V.; Rektorisova, M.; Chmelarova, H.; Ovesna, J.; Hajslova, J. Authenticity assessment of garlic using a metabolomic approach based on high resolution mass spectrometry. *J. Food Compos. Anal.* **2018**, *67*, 19–28. [CrossRef]
34. Chernetsova, E.S.; Morlock, G.E. Assessing the capabilities of direct analysis in real time mass spectrometry for 5-hydroxymethylfurfural quantitation in honey. *Int. J Mass Spectrom.* **2012**, *314*, 22–32. [CrossRef]
35. Rajchl, A.; Drgová, L.; Grégrová, A.; Čížková, H.; Ševčík, R.; Voldřich, M. Rapid determination of 5-hydroxymethylfurfural by DART ionization with time-of-flight mass spectrometry. *Anal. Bioanal. Chem.* **2013**, *405*, 4737–4745. [CrossRef]
36. *AOAC Official Method. AOAC 920.180. Honey (Liquid, Strained, or Comb). Preparation of Test Sample*; 2020 Edition; AOAC: Rockville, MD, USA, 1920.
37. Xia, J.; Sinelnikov, I.V.; Han, B.; Wishart, D.S. MetaboAnalyst 3.0—Making metabolomics more meaningful. *Nucleic Acids Res.* **2015**, *43*, W251–W257. [CrossRef] [PubMed]
38. Xia, J.; Wishart, D.S. Web-based inference of biological patterns, functions and pathways from metabolomic data using MetaboAnalyst. *Nat. Protoc.* **2011**, *6*, 743–760. [CrossRef] [PubMed]
39. Jolliffe, I.T. *Principal Component Analysis*, 2nd ed.; Springer: New York, NY, USA, 2002.
40. Oliveri, P.; Downey, G. Multivariate class modelling for the verification of food-authenticity claims. *TrAC Trend Anal. Chem.* **2012**, *35*, 74–86. [CrossRef]
41. Casale, M.; Casolino, C.; Oliveri, P.; Forina, M. The potential of coupling information using three analytical techniques for identifying the geographical origin of Liguria extra virgin olive oil. *Food Chem.* **2010**, *118*, 163–170. [CrossRef]
42. Granato, D.; Putnik, P.; Kovačević, D.B.; Santos, J.S.; Calado, V.; Rocha, R.S.; Da Cruz, A.G.; Jarvis, B.; Rodionova, O.E.; Pomerantsev, A. Trends in chemometrics: Food authentication, microbiology, and effects of processing. *Compr. Rev. Food Sci. Food Saf.* **2018**, *17*, 663–677. [CrossRef]

foods

MDPI

Article

Machine Learning Approaches Applied to GC-FID Fatty Acid Profiles to Discriminate Wild from Farmed Salmon

Liliana Grazina [1], P. J. Rodrigues [2], Getúlio Igrejas [2], Maria A. Nunes [1], Isabel Mafra [1,*], Marco Arlorio [3], M. Beatriz P. P. Oliveira [1] and Joana S. Amaral [4,*]

1 REQUIMTE-LAQV, Faculdade de Farmácia, Universidade do Porto, Rua de Jorge Viterbo Ferreira, 228, 4050-313 Porto, Portugal; li_grazina@hotmail.com (L.G.); antonianun@gmail.com (M.A.N.); beatoliv@ff.up.pt (M.B.P.P.O.)

2 Research Centre in Digitalization and Intelligent Robotics (CeDRI), Instituto Politécnico de Bragança, Campus de Santa Apolónia, 5300-253 Bragança, Portugal; pjsr@ipb.pt (P.J.R.); igrejas@ipb.pt (G.I.)

3 Dipartimento di Scienze del Farmaco, Food Chemistry Unit, Università del Piemonte Orientale "A. Avogadro", Largo Donegani 2/3, 28100 Novara, Italy; marco.arlorio@uniupo.it

4 Centro de Investigação de Montanha (CIMO), Instituto Politécnico de Bragança, Campus de Sta. Apolónia, 5301-857 Bragança, Portugal

* Correspondence: isabel.mafra@ff.up.pt (I.M.); jamaral@ipb.pt (J.S.A.); Tel.: +351-220-428-500 (I.M.); +351-273-383-138 (J.S.A.)

Received: 23 September 2020; Accepted: 4 November 2020; Published: 7 November 2020

Abstract: In the last decade, there has been an increasing demand for wild-captured fish, which attains higher prices compared to farmed species, thus being prone to mislabeling practices. In this work, fatty acid composition coupled to advanced chemometrics was used to discriminate wild from farmed salmon. The lipids extracted from salmon muscles of different production methods and origins (26 wild from Canada, 25 farmed from Canada, 24 farmed from Chile and 25 farmed from Norway) were analyzed by gas chromatography with flame ionization detector (GC-FID). All the tested chemometric approaches, namely principal components analysis (PCA), t-distributed stochastic neighbor embedding (t-SNE) and seven machine learning classifiers, namely k-nearest neighbors (kNN), decision tree, support vector machine (SVM), random forest, artificial neural networks (ANN), naïve Bayes and AdaBoost, allowed for differentiation between farmed and wild salmons using the 17 features obtained from chemical analysis. PCA did not allow clear distinguishing between salmon geographical origin since farmed samples from Canada and Chile overlapped. Nevertheless, using the 17 features in the models, six out of the seven tested machine learning classifiers allowed a classification accuracy of ≥99%, with ANN, naïve Bayes, random forest, SVM and kNN presenting 100% accuracy on the test dataset. The classification models were also assayed using only the best features selected by a reduction algorithm and the best input features mapped by t-SNE. The classifier kNN provided the best discrimination results because it correctly classified all samples according to production method and origin, ultimately using only the three most important features (16:0, 18:2n6c and 20:3n3 + 20:4n6). In general, the classifiers presented good generalization with the herein proposed approach being simple and presenting the advantage of requiring only common equipment existing in most labs.

Keywords: authenticity; fish; *Salmo salar* L.; fatty acids; mislabeling; chemometrics; machine learning

1. Introduction

In recent decades, the consumption of fish has been increasingly recommended due to its health benefits, mainly related to the prevention of cardiovascular diseases [1]. In particular, fatty fishes from cold waters, such as salmon, are frequently rich in polyunsaturated fatty acids (PUFA), including the essential fatty acids linoleic (18:2n6) and α-linolenic (18:3n3), but also in several omega-3 PUFA such as eicosapentaenoic (EPA, C20:5n3) and docosahexaenoic (DHA, 22:6n3) acids. Besides being components of cell membranes, omega-3 PUFA are involved in the biosynthesis of eicosanoids and have been shown to influence health by affecting cell signaling cascades and gene expression, resulting in decreased expression of inflammatory and atherogenesis-related pathways [2,3]. Moreover, different studies showed that omega-3 PUFA play an important role in altering blood lipid profiles and associate their consumption with improved cardiovascular function and decreased risk of atherosclerosis and peripheral arterial disease [2,3].

In addition to these benefits, fish is largely consumed for its nutritional value and sensory aspects, making it one of the most traded food commodities. In this sense and considering that the world's wild fish stocks are limited, the production of farmed fish has been steadily increasing in recent last years. In fact, according to the Food and Agriculture Organization (FAO) Globefish Highlights, world fisheries capture was 92.5 million tonnes in 2017 with this figure expected to decrease to 91.3 million tonnes in 2019, by the contrary, fish capture arising from aquaculture is expected to grow from 80.1 to 86.5 million tonnes in the same period [4]. Concerning salmon, from 2000 to 2014, a much stronger increase was verified for aquaculture production (from 898,800 to 2,326,300 tonnes) compared to that of the world's capture of wild salmon (from 728,000 to 879,000 tonnes) [5]. Aquaculture allows wider consumer access to fish generally at more affordable costs, though it is known that fatty acid composition can significantly vary according to its production method (wild vs. aquaculture). Particularly for salmon, it has been reported that wild salmon generally present higher contents of valuable omega-3 PUFA [6–8]. This aspect, together with particular organoleptic characteristics, has driven several consumers to prefer wild salmon. Considering the limited availability of this type of salmon and its growing demand, prices have been increasing significantly, resulting in this product being prone to adulteration by origin mislabelling or even substitution with other lower-cost fish [9–11]. Whereas fish species authentication can be performed using well established and straightforward DNA-based methods [12], different approaches have been proposed so far to assess the origin of fish with respect to production method. These include, mainly, the use of nuclear magnetic resonance (NMR) [13,14], isotope ratio analysis [15,16], lipidic profile [17,18] or a combination of these [11,19–21]. Excellent discrimination (100%) between wild and farmed Atlantic salmon was reported by Aursand et al. [13] by applying support vector machines (SVM) to data obtained by 13C NMR. In another study of the same group, the lipid extract was analyzed by 13C NMR and by gas chromatography with flame ionization detector GC-FID for fatty acid composition to discriminate between wild and farmed Atlantic salmon and assign the origin of the aquaculture samples to the farms included in the study [19]. The application of chemometrics to the reference farmed fish showed very good results for both approaches, but, surprisingly, slightly better for GC-FID data. The use of stable isotope analysis based on isotope ratio mass spectrometry (IRMS) is also a promising approach, especially when combined with chemical composition analysis, notably fatty acids [11,15,21]. Yet, previous works have demonstrated that lipidic profile is sufficient to establish the production method of salmon samples, particularly when combined with chemometric analysis [8,19,20]. Recently, Fiorino et al. [8] analyzed the lipid extracts obtained from a total of 100 samples of farmed and wild salmon by direct analysis in real time (DART) coupled to high resolution mass spectrometry (HRMS). The proposed methodology showed to be fast and allowed a good discrimination between the two groups (wild vs. farmed), though without differentiating the geographical origin of the farmed fish. Moreover, the referred approach requires advanced and expensive equipment, which is not available in most control quality/analytical laboratories. In the present study, the fatty acid composition of the same samples of wild and farmed salmon used in the work of Fiorino et al. [8] was analyzed by GC-FID, an affordable equipment commonly available in most laboratories. Subsequently, the obtained data were submitted to advanced chemometric analysis to establish the most suitable classifier able to discriminate the origin of salmon samples (wild vs. farmed, and the geographical origin among farmed samples) with the minimum possible computational effort.

2. Materials and Methods

2.1. Samples

In this study, a total of 100 authentic salmon samples obtained in the framework of the EU-funded project FOODINTEGRITY (Working Package 18) were analyzed. The samples included 26 wild salmon captured in Canada, and 74 farmed salmon samples from aquaculture farms in Canada (25), Norway (25) and Chile (24). No information was available about the gender of each specimen, neither of the diet or farming conditions used. The samples (entire fish) were transported frozen to the laboratory (Meriex Nutriscience, Chicago, IL, USA), allowed to defrost overnight at refrigerated temperature, and filleted in a cold room (4 °C). After removing the bones and skin, the muscles were grinded and distributed in labelled glass jars containing approximately 200 g each. The jars were immediately frozen and then shipped under freezing conditions (−20 °C) to the participating laboratories in different countries. After arriving, the samples were kept at −20 °C and submitted to lipid extraction as soon as possible.

2.2. Lipid Extraction

Lipids were extracted based on the Bligh and Dyer protocol [22] with some modifications. Briefly, about 13 g of each minced fillet were added with 13 mL of NaCl (1%) and 100 µL of butylated hydroxytoluene (BHT) (0.01% in n-hexane) to avoid oxidation, and homogenized for 1 min using an Ultra-Turrax at 13,500 rpm, keeping a low temperature by immersing the tube with the sample on ice. After that, 2.5 mL of the homogenate was transferred to a new tube and added with 2.5 mL of chloroform and 5 mL of methanol, both refrigerated. The solution was mixed vigorously by vortexing for 2 min. After centrifuging (4000 rpm, 15 min at 4 °C) the upper layer was discarded, and an additional 2.5 mL of refrigerated chloroform was added. After vortexing for 30 s and centrifuging under the same conditions, the chloroformic phase was transferred into a new tube and centrifuged (4000 rpm, 5 min at 4 °C). Finally, the chloroformic phase was collected into a previously weighted vial, flushed with a nitrogen stream and stored at −20 °C until further analysis. Each sample was submitted to independent extractions (n = 3).

2.3. Fatty Acids Analysis by GC-FID

Fatty acids were methylated using acid-catalysed trans-methylation with BF_3 [23]. Firstly, the lipidic chloroformic extracts, previously stored at −20 °C, were dried under nitrogen and the tubes weighted to calculate the extraction yield. After dissolving the obtained lipids in 1 mL of n-hexane, for each sample, the volume containing 12.5 mg of lipids was transferred for a glass tube and dried under nitrogen. After adding 100 µL of BHT (0.01% in n-hexane) to prevent oxidation phenomena, fatty acid methyl esters were prepared. For that purpose, 1.25 mL of KOH (0.5 M) in methanol were added and the mixture was heated for 10 min at 100 °C after vortex-mixing vigorously. After cooling down the tubes, 1.0 mL of 14% boron-trifluoride in methanol (≥99.0% purity) (Sigma-Aldrich, Steinheim, Germany) was added to the solution, which was homogenized by vortexing, and the tubes heated again for 30 min at 100 °C. After completely cooling down the tubes in ice, 2.0 mL of n-hexane high performance liquid chromatography HPLC grade (Merck, Darmstadt, Germany) was added and the solution was vortex-mixed. Then, 1.0 mL of a saturated NaCl solution was added, followed by vigorous mixing and then by centrifuging for 5 min at 3000 rpm to obtain a clear upper phase. After that, 1.5 mL of supernatant was transferred to a new vial, added with anhydrous Na_2SO_4 and approximately 1.0 mL of FAME solution was transferred to an injection vial.

GC-FID analysis was carried out in a Shimadzu GC-2010 Plus gas chromatograph equipped with a Shimadzu AOC-20i auto-injector and a flame ionization detector (Shimadzu, Japan). FAME separation was achieved on a CP-Sil 88 silica capillary column (50 × 0.25 mm i.d., 0.20 μm, Varian, Middelburg, The Netherlands). The injector and detector temperatures were 250 °C and 270 °C, respectively. The oven parameters were set as follows: an initial temperature of 150 °C was increased at 3 °C/min to 160 °C and held for 2.0 min, then it was increased at 3 °C/min to 220 °C and held for 10 min. Helium was used as the carrier gas at a flow rate of 1 mL/min, and 1 μL of sample was injected using a split ratio of 1:50. Identification of compounds was performed by comparison of their retention times with those of authentic standards mixtures, namely 37 component FAME mix (certified reference material CRM47885) and PUFA n°.1 Marine source (standard 47,033) both from Supelco (Bellefonte, PA, USA). In addition, the fatty acid cis-11-octadecenoate (C18:1n7) was identified with and individual standard also purchased from Supelco. The results were expressed as the relative percentage of each fatty acid, calculated based on the chromatographic peak area. Each lipid extract was injected in duplicate.

2.4. *Chemometric Analysis*

2.4.1. Dataset

The data used for chemometrics resulted from the chemical analyzes, totalizing 596 instances (4 chromatograms were excluded due to injection/chromatographic system problems) that were organized into four reasonably well balanced groups, each corresponding to a class of salmon: Norway Farmed (25 salmons), Chile Farmed (24 salmons), Canada Farmed (25 salmons), Canada Wild (26 salmons). Each salmon sample was represented by a block of 6 chromatograms. The number of independent features considered was 17, corresponding to the identified fatty acids.

2.4.2. Statistical Analysis by One-Way ANOVA

The differences between groups were analysed using a one-way analysis of variance (ANOVA) followed by Tukey's honest significant difference post hoc test with $p = 0.05$. The analysis was carried out using the SPSS v. 23.0 program (SPSS v. 23.0; IBM Corp., Armonk, NY, USA).

2.4.3. Data Modelling Tools

The data modelling tools used in this work are based on the Orange 3.24 software, which, in turn, uses libraries from the Scikit-learn, Numpy and Scipy written in Python. The graphical user interface uses the cross-platform Qt framework.

Data Visualization by PCA and t-SNE

As a first approach, the possibility of separating the data by classical and linear statistical methods was evaluated. For that purpose, principal component analysis (PCA) was used to check the possibility of obtaining a separation by linear composition in a subspace of principal components based on the PCA projections. When the PCA shows data superposition among groups, it means that the possibility of separating groups in the original dimension space cannot be performed, since the mapping from the original dimension space to the principal component space is always linear [24]. A manner to overcome this issue involves using the t-distributed stochastic neighbor embedding (t-SNE) method, which is able to replicate non-linear mappings in the original data space to the lower dimension [25]. Thus, a non-linear approach by t-SNE was used to observe separations in higher dimensions when they are projected in a two-dimensional space.

Machine Learning Classifiers

Several well-known classification models were evaluated, namely k-nearest neighbors (kNN), decision tree, support vector machine (SVM), random forest, artificial neural networks, naïve Bayes and AdaBoost, whose main characteristics are described as follows:

- kNN is a method that can be used for data classification. A sample is classified by the rating vote of its neighbors present in the dataset; the result is attributed to the most common class among the k closest neighbors. If k = 1, the object is simply assigned to the class of the only nearest neighbour [26].
- Decision tree is one of the predictive modeling approaches used in machine learning. It uses a tree schema as a predictive model to move from the analysis of a sample to conclude the class that corresponds to the sample. The decision trees are built inductively from a dataset that is analyzed by measures, such as the information entropy. During the process, the dataset is divided, successively, in order to reduce the uncertainty in the classification. That division is represented by each feature (tree node) [27].
- Support-vector machines (SVM) are models of supervised learning. Given a set of training examples, each is marked as belonging to one or the other of two categories. The resulting model, after training, is a representation of the examples as points in space, mapped so that the widest possible spatial margin separates the examples of the two categories. The new examples are expected to belong to a category based on the margin side where they are located. That side is formed from the center of the separating margin. This process leads to the fact that SVM can generalize as best as possible, avoiding overfitting. In addition to performing linear mapping, SVM can efficiently perform non-linear mapping using what is called a kernel trick, implicitly mapping their inputs into high-dimensional spaces, allowing separations to happen in the high dimensional space [28].
- Random forests are a learning method that can be used for classification. They are built based on decision trees, but in this case, there is a participation of several trees that are trained with segments of the dataset and with segments of the feature set randomly selected [29]. This stochastic factor improves the generalization of the model and reduces the overfitting.
- Artificial neural networks (ANN) is a model based on a collection of connected units or nodes called "artificial neurons", which mimic neurons in a biological brain. Each connection can transmit signals from one artificial neuron to another; the magnitude of the signal is modulated by a parameter adjusted during the learning phase. Each neuron behaves like a separating hyperplane in the classification space. The association of neurons, by layers, allows obtaining conjugations of complex hyperplanes that lead to non-linear classification models. The adjustment of the parameters is made by algorithms that use the gradient descent of the error. The error is defined by the difference between the value emitted by the neural network and the desired value [30].
- Naïve Bayes, in machine learning, are probabilistic classifiers, based on the application of Bayes' theorem with evidence on the assumptions of independence between features [31]. Naïve Bayes classifiers are easy to implement using Gaussian curves and the inverse Bayes formula.
- AdaBoost (short for adaptive boosting) is based on the idea that a set of weak classifiers can result in a strong classifier. Weak classifiers are combined linearly, but modulated by coefficients that are obtained during the training. The choice of weak classifiers is made focusing on the examples that are classified with more difficulty. In this iterative process, the weak classifiers have coefficients that correspond to the classifier error on the dataset. The weak classifiers that make the least mistakes have their coefficients increased. The strong classifier aggregates all those weak classifiers according to their importance coefficients [32].

All these models are mappers with non-linear capabilities, each having different methods of statistical induction of knowledge. Thus, some may perform better for certain classification problems than others. For this reason, in this study we used a test bench formed by several models.

All these classifiers were developed/trained, in a first phase, using the 17 features present in the dataset and the obtained classification results used for the assessment of each model.

Reduction in Features

Aiming at decreasing the cost of analyses and complexity of data, accelerating the whole process, it is frequently important to reduce the number of features in chemometric analysis. On the other hand, the reduction in the number of features can enhance the generalization capability of the classifier. Therefore, after having the models parameterized using all the features, strategies were developed to explore the reduction in features. The selection of the best features was made using a ranking process that is based on the measurement of information entropy. In this case, the well-known information gain ratio criterion was applied [27]. This criterion measures the uncertainty in how the data are separated based on a specific feature; the value of the information gain ratio is calculated for each feature, representing its separation power in the dataset. The sorting of features, according to these values, establishes their ranking. This criterion is normalized regarding the number of data partitions that the usage of a given feature causes. This mechanism makes it possible to obtain a numerical criterion, independent from the classifying bias (overfitting), prompted by numerous potential partitions of information groups. Thus, in the next step, the minimum number of the best features, in that ranking, was determined, ensuring that the classification model still classifies the data accurately.

Aiming to evaluate model overfitting, assertiveness and generalization assessments of the classification models were made using both external and full-cross validation. For external validation, the test dataset was obtained by splitting the data into 20:80. For the cross-validation scheme a mechanism of leave-one-sample-out (each sample corresponding to a block of 6 chromatograms) was used. Moreover, that scheme allowed us to parameterize the models by observing the validation performance given by the average of all six-chromatogram groups. The used performance indicators were the accuracy (CA) and F1. F1 is a more revealing measure of the practical performance that a classification model presents, being more sensitive to poorly classified instances. Moreover, an assertiveness analysis was made by using confusion matrices.

During the process of feature reduction, the performance of the classifiers was tested using a successive bisection approach. Starting from a set of classification models that normally provide high assertiveness rates and using all the features sorted by the information gain ratio, the following method (Algorithm 1) was developed and applied. This algorithm allows for the optimization of the search for the minimum number of features to classify the samples, with an arbitrary minimum of 99% of accuracy.

Algorithm 1 Searching the optimal number of features

Given a set of features F of n elements with gain ratio values $F_0, F_1, F_2, \ldots, F_{n-1}$ sorted such that $F_0 > F_1 > F_2 \ldots > F_{n-1}$, and the $accuracy_m$ being the correctness classifying the dataset using the first m features.
The following algorithm is based on the binary search to find the index m in F that corresponds to the minimum index to classify the dataset properly.

1. Set L to 0 and R to $n - 1$
2. Set $m = R$, $m_{old} = m$
3. If $accuracy_m < 99\%$, stop, the classifier must use the all features
4. Set m (the middle position) to the *floor* of $\frac{L+R}{2}$
5. If $accuracy_m > 99\%$, set R to m
6. If $accuracy_m < 99\%$, set L to m
7. If abs$(m - m_{old}) > 0$, $m_{old} = m$, goto 4
8. Stop, the classifier must use the first $m + 1$ features

The algorithm is repeated independently for each of the classifiers under analysis. The minimum number of features is selected to further actions when the accuracy is ≥ 99% for at least one model. For each machine learning classifier, a trial-and-error approach was used to find the best parametrization, with classifiers being tuned at two stages. At the first stage, Algorithm 1 was applied to all the models using the maximum number of features (seventeen). This tuning aims at obtaining a good classification, concerning the dataset, for each classifier. The adjustment was done manually, in a trial-and-error fashion, changing the hyperparameters associated to each model. In this phase, to get a good functional response (selection) from Algorithm 1, it is not necessary to have a perfect tuning of the classifiers.

After this, the minimum number of features required to produce good classifications (accuracy of 99%) on a classifier are known. Thus, at the second stage, eventually, one could make new adjustments to the classifier models to improve functional performance subjected to the new subset of features selected after applying Algorithm 1. The details of the final parameters used for the best models are shown in Table 1. Figure 1 schematically describes the chemometric approaches and main process pipeline used in this work.

Table 1. Details of the parametrization used to tune each of the final classification models.

Classifier	Parameters
kNN	Number of neighbors: 3; Metric: Euclidean; Weight: Distance.
Decision tree	Limit the tree depth: 100; Do not split subsets smaller than: 2; Min. number of instances in leaves 3.
SVM	C: 15; Kernel: Radial Basis Function (RBF); g: auto.
Random forests	Number of trees: 15; Do not split subsets smaller than: 5.
ANN	Neurons in hidden layers: 300; activation: Rectified Linear Unit (Relu); solver: Adam; regularization: 0.02.
Naive Bayes	Non-applicable
AdaBoost	Number of estimators: 80; learning rate: 0,7; classification algorithm: SAMME.R; Regression loss function: Square.

kNN: k-nearest neighbors; SVM: support vector machine; ANN: artificial neural networks.

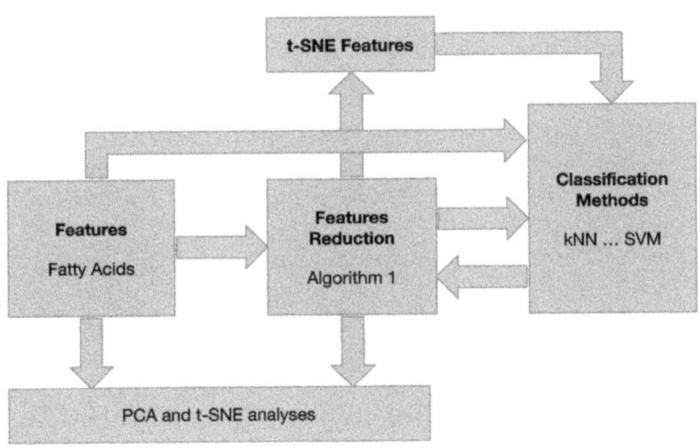

Figure 1. Main process pipeline. PCA: principal components analysis; t-SNE: t-distributed stochastic neighbor embedding.

3. Results and Discussion

3.1. Fatty Acids Composition

Figure 2 shows representative chromatograms of fatty acid analysis obtained from wild and farmed salmon samples and Table 2 presents their relative contents for the four salmon groups under evaluation, namely, wild from Canada and farmed from Canada, Chile and Norway.

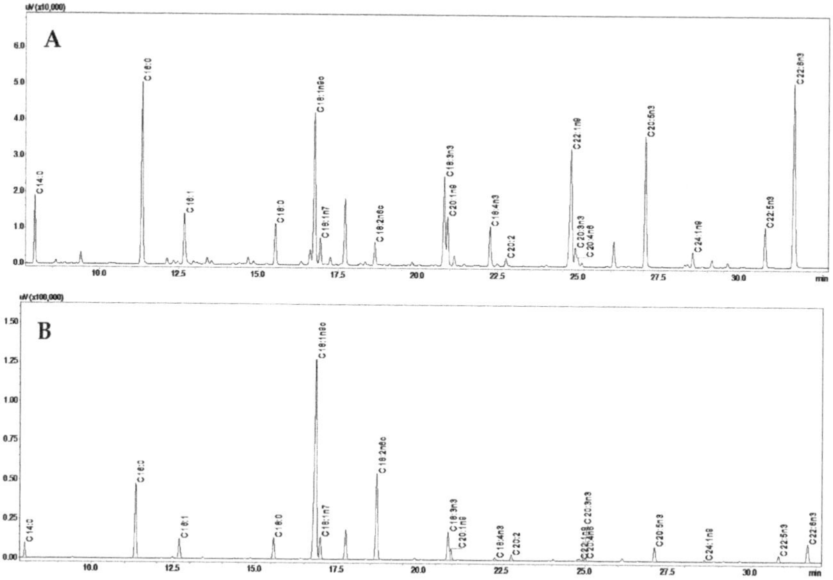

Figure 2. Chromatograms of fatty acid profiles obtained by GC-FID analysis of wild (**A**) and farmed (**B**) salmon samples from Canada.

Table 2. Fatty acid composition (relative% of the identified FAME) obtained by GC-FID analysis of lipids from the wild and farmed salmon samples of different origins. Results are given as mean ± SD of the total specimens analyzed for each group.

Fatty Acid	Wild	Farmed		
	Canada (n = 26)	Canada (n = 25)	Chile (n = 24)	Norway (n = 25)
14:0	3.86 ± 0.40 c	1.99 ± 0.47 a	1.98 ± 0.10 a	2.14 ± 0.06 b
16:0	15.75 ± 1.70 d	11.64 ± 0.83 b	12.22 ± 0.78 c	9.13 ± 0.24 a
16:1	3.73 ± 0.58 c	3.48 ± 0.80 b	2.78 ± 0.25 a	2.67 ± 0.07 a
18:0	3.61 ± 0.65 b	3.54 ± 0.22 b	3.78 ± 0.31 c	2.60 ± 0.10 a
18:1n9c	12.89 ± 2.89 a	42.15 ± 4.42 b	41.98 ± 1.58 b	43.89 ± 0.29 c
18:1n7	2.19 ± 0.28 a	3.82 ± 0.33 b	4.05 ± 0.41 c	3.81 ± 0.28 b
18:2n6c	1.76 ± 0.15 a	14.35 ± 1.52 b	16.36 ± 0.72 c	16.31 ± 0.21 c
18:3n3	6.72 ± 1.52 c	5.02 ± 0.57 b	4.47 ± 0.40 a	7.79 ± 0.27 d
20:1n9	3.57 ± 0.94 c	2.09 ± 0.65 a	2.36 ± 0.22 b	2.17 ± 0.24 a
18:4n3	2.51 ± 0.51 c	0.58 ± 0.14 a	0.61 ± 0.09 a,b	0.69 ± 0.08 b
20:2n6	0.48 ± 0.10 a	0.87 ± 0.09 b	1.04 ± 0.06 c	0.90 ± 0.06 b
22:1n11 + 22:1n9	9.30 ± 2.90 b	1.10 ± 1.11 a	0.64 ± 0.27 a	0.67 ± 0.05 a
20:3n3 + 20:4n6	1.83 ± 0.33 d	0.79 ± 0.11 c	0.60 ± 0.05 a	0.66 ± 0.02 b
20:5n3	9.12 ± 0.97 c	2.74 ± 0.42 b	2.54 ± 0.23 a	2.69 ± 0.15 ab
24:1n9	1.06 ± 0.17 c	0.24 ± 0.05 b	0.21 ± 0.02 a	0.26 ± 0.02 b
22:5n3	2.75 ± 0.30 d	1.37 ± 0.34 c	1.28 ± 0.13 b	1.17 ± 0.05 a
22:6n3	18.83 ± 2.82 d	4.23 ± 0.62 c	3.12 ± 1.11 b	2.45 ± 0.13 a
Σ SFA	23.25 ± 1.93 d	17.18 ± 1.42 b	17.98 ± 1.13 c	13.6 ± 1.98 a
Σ MUFA	32.73 ± 3.22 a	52.89 ± 1.76 c	52.02 ± 1.63 b	52.41 ± 7.55 c
Σ PUFA	43.95 ± 2.77 c	29.93 ± 0.62 a	30.00 ± 1.62 a	31.99 ± 4.62 b
n3/n6	16.75 ± 1.61 b	0.90 ± 0.24 a	0.66 ± 0.04 a	0.82 ± 0.02 a

SFA: saturated fatty acids; MUFA: monounsaturated fatty acids; PUFA: polyunsaturated fatty acids. Different letters indicate significant differences ($p < 0.05$) between groups in the statistical analysis by one-way analysis of variance (ANOVA).

Striking differences can be observed between wild and farmed salmons, namely in terms of the sum of MUFA and PUFA, ratio between omega-3 and omega-6 fatty acids, and also regarding several individual fatty acids. For the same amount of derivatized lipids, and when compared to wild, farmed salmon presented a significantly higher ($p < 0.05$) content of oleic and linoleic acids and lower contents of EPA, DHA and C22:1 isomers. In general, the obtained results are in good agreement with previous knowledge since farmed salmons are frequently described as having higher amounts of C18:1, C18:2 and C18:3 fatty acids, while wild are richer in long chain omega-3 PUFA as well as saturated fatty acids (SFA) [6,7]. Nevertheless, in the present study, similar contents of α-linolenic acid (C18:3n3) were found between the wild and farmed groups and only a slightly higher amount was verified in terms of SFA. The obtained data confirm that the consumption of wild salmon can be associated with greater health benefits due to their favorable ratio omega-3/omega-6 fatty acids. As discussed in previous papers, the differences observed are most probably related with differences in the diets of fish from the wild and in aquaculture conditions [6,17,21].

Compared to the results previously reported for the analysis of the same samples (as part of the EU-funded project FOODINTEGRITY) using a different technique, namely DART-MS, some quantitative differences can be pointed out. Namely, the content reported by Fiorino et al. [8] for 16:0 was higher in both farmed and wild groups, while the present GC-FID results show higher contents for 18:3, 18:1 (mainly for the farmed group) and 22:6 (mainly for the wild group). These dissimilarities can be due to the different techniques used, one based on mass spectrometry and normalized abundances, and the other relying on flame ionization detection and relative peak areas.

3.2. *Chemometric Analysis of the Generated Data*

3.2.1. Features Selection

The importance of each feature regarding the group separation was evaluated by applying the information gain ratio criterion, as described in the materials and methods section. Table 3 presents the ranking of features obtained based on that measurement. Subsequently, the developed algorithm (Algorithm 1) was used to determine the minimum number of features required for classifying the four groups accurately. That number was found to be six, corresponding to the following features: 16:0, 18:2n6c, 20:3n3 + 20:4n6, 14:0, 18:1n9 and 22:6n3.

Table 3. Features sorted by applying the information gain ratio criterion.

Fatty Acid	Gain Ratio
16:0	0.719
18:2n6c	0.709
20:3n3 + 20:4n6	0.675
14:0	0.615
18:1n9c	0.562
22:6n3	0.548
20:2n6	0.523
22:1n11 + 22:1n9	0.506
24:1n9	0.505
22:5n3	0.464
18:1n7	0.461
18:4n3	0.446
20:5n3	0.423
16:1	0.402
18:3n3	0.378
20:1n9	0.366
18:0	0.353

3.2.2. Data Visualization by PCA and t-SNE

As a first approach, PCA was applied to the dataset as a linear and unsupervised statistical method. This method is one of the most widespread exploratory data analysis tools, providing a fast data overview by projecting each data point onto a small number of principal components, thus reducing data dimensionality, while maintaining their variation as much as possible [24]. Moreover, this approach was used previously regarding the analysis of the same salmon samples by a distinct methodology, namely DART-MS analysis [8]. Figure 3A presents the data distribution on two principal components when all the 17 data features are used. As it can be observed, PC1 and PC2 accounted for 87.8% of the total variance and showed a clear separation between the wild samples and the farmed ones, similarly to the results reported by Fiorino et al. [8]. Although it was not possible to clearly distinguish the farmed samples according to their geographical origin, mainly due to overlapping of samples from farmed Canada and Chile groups, a better separation was achieved when compared to the results of Fiorino et al. [8] using DART-MS analyses. Interestingly, in their work, five out of the six fatty acids, exhibiting the most relevant differences between wild and farmed salmons, were in common with the ones selected by Algorithm 1. Linolenic acid (C18:3) was an exception because in the present work it ranked as the 15th position with a low information gain ratio value, thus not being relevant to distinguish the four groups using the GC-FID fatty acid profiles. Subsequently, PCA was also applied to the whole dataset, but using only the selected best six features (Figure 3B), evidencing results similar to the ones obtained with all the 17 features.

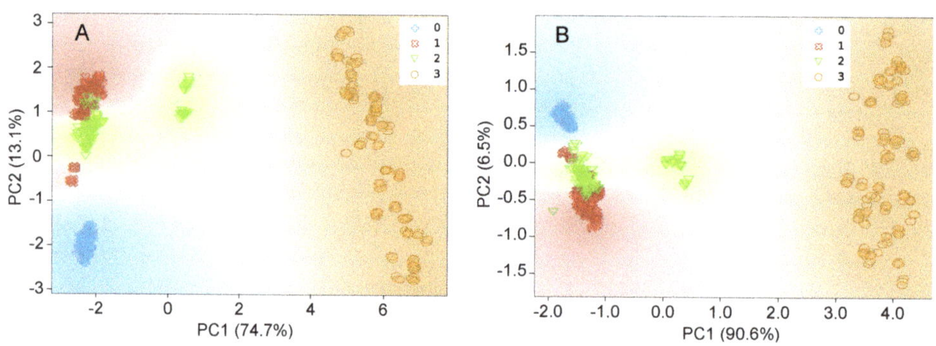

Figure 3. Scatterplot obtained for the first two principal components after applying PCA to the whole dataset using (**A**): all the 17 features, (**B**): the 6 best features (16:0, 18:2n6c, 20:3n3 + 20:4n6, 14:0, 18:1n9 and 22:6n3); 0—Norway farmed, 1—Chile farmed, 2—Canada farmed, 3—Canada wild.

The interpretation of Figure 3A,B allows drawing two conclusions: (1) most of the data are strongly explained by the first principal component regardless of the number of used features, namely all the 17 or only the best six, which confirms that most of the features are not important for the correct classification; (2) some samples of Chile farmed and Canada farmed groups are not linearly separable with data projected on a 2D subspace, thus suggesting the need for non-linear classification models. Therefore, t-SNE was applied to the dataset, first using all the 17 features, and then only the selected best six (Figure 4A,B). This method allows the projection of the original dimension on two dimensions without losing the non-linear relations presented at the high dimensional space. Thus, it is a suitable tool to perceive the separability of groups at the original dimension. As shown in Figure 4, there is no data superposition and, in general, the groups are well separated according to this method. This information suggests a good data separability when the classification models can handle non-linearities in a high dimension space.

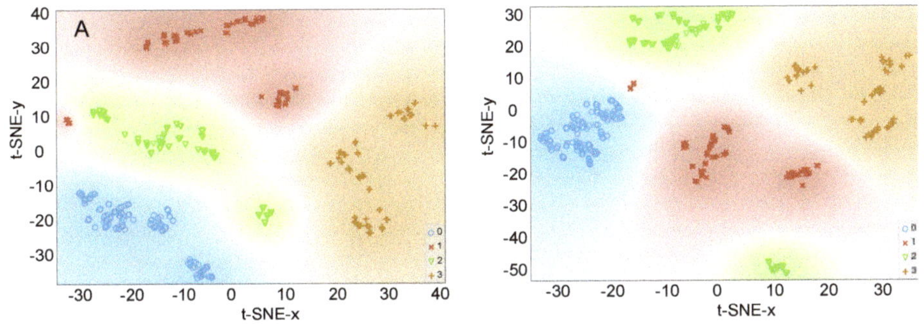

Figure 4. Scatterplot obtained after applying t-SNE to the whole dataset using (**A**): all the 17 features (**B**): only the 6 best features (16:0, 18:2n6c, 20:3n3 + 20:4n6, 14:0, 18:1n9 and 22:6n3); 0—Norway, Figure 1. Chile farmed, 2—Canada farmed, 3—Canada wild.

A good separability among groups was also observed when the number of employed input features was only six (Figure 4B). This suggests that, in the high dimension original space, the separability is achieved based on only a few features. Normally, this is an advantage for subsequently used classifiers because it promotes generalization and tends to avoid overfitting, thus strongly suggesting that new samples will be properly classified based on such classifiers.

3.2.3. Machine Learning Classifiers

In this work, a total of seven different classifiers were tested considering performance (classification accuracy) and required computational effort (evaluated as test time). Similarly, as was done for PCA and t-SNE, each classifier was first assayed using all the 17 features as inputs to the classifiers. The obtained performance is shown in Table 4, evidencing that ANN, random forest, SVM, naïve Bayes and kNN were the best models as they showed a maximum performance, allowing classifying, without error, for all of the test dataset. Nevertheless, they are closely followed by the remaining classifiers, with decision tree being the one that performed worst. In terms of performance time (test time), among the classifiers that allowed 100% accuracy (CA), naïve Bayes was the best one. This can be explained by two factors: first, one must consider that in this case the number of features exceeds the needs, thus, according to Occam's razor principle, the simpler model can achieve a good performance; second, as the model is very simple to implement, the number of required computational calculation steps is small, thus corresponding to a shorter time of performance.

Table 4. Classifiers performance, in the test dataset, using all the 17 input features.

Model	Test Time (s)	CA	F1
ANN	0.011	1.000	1.000
Naïve Bayes	0.008	1.000	1.000
kNN	0.012	1.000	1.000
SVM	0.013	1.000	1.000
Random Forest	0.011	1.000	1.000
AdaBoost	0.0064	0.991	0.991
Decision Tree	0.001	0.908	0.908

CA: accuracy; F1 score: harmonic mean of the precision and recall.

Next, the performance of classifiers was assayed with only the six best features as their inputs. As can be observed in Table 5, in this case, the ANN, SVM and kNN classifiers allowed 100% correct classification, as measured by accuracy and F1 indicators. It is possible that the elements that were not correctly classified by the remaining models do not have statistical significance to change the parameters present on the learning mechanism to the rest of classifiers. Among the best classifiers the one that presented the best computational performance was the SVM.

Table 5. Classifier performance, in the test dataset, using the selected best 6 input features.

Model	Test Time (s)	CA	F1
ANN	0.010	1.000	1.000
SVM	0.006	1.000	1.000
kNN	0.009	1.000	1.000
Random Forest	0.011	0.992	0.992
Naïve Bayes	0.003	0.992	0.992
AdaBoost	0.004	0.983	0.983
Decision Tree	0.001	0.983	0.983

Overall, the remaining classifiers were very close to the performance of ANN, SVM and kNN, despite the reduced number of features used. For this reason, it was decided to further observe the classification performance when the features are remapped by the t-SNE method as inputs for the classifiers, keeping the same parametrization for all models, as in the previous scheme. By applying Algorithm 1 and extending the processing pipeline with the t-SNE block, namely by placing that block between the features used and the classifiers, it was possible to conclude that the classification can still be performed successfully by relying on only three features, namely 16:0, 18:2n6c and the sum of 20:3n3 + 20:4n6. The obtained results are presented in Table 6, evidencing 100% accuracy of sample classification using the kNN, with only three compounds being required in this model. In this scenario, the decision tree classifier showed the worst performance, being the only one presenting an accuracy < 95%.

Table 6. Classifiers performance, in the test dataset, using the selected best 3 input features mapped by t-SNE.

Model	Test Time [s]	CA	F1
kNN	0.094	1.000	1.000
SVM	0.016	0.992	0.992
Random Forest	0.021	0.992	0.992
ANN	0.112	0.983	0.983
AdaBoost	0.020	0.967	0.967
Naïve Bayes	0.018	0.967	0.967
Decision Tree	0.001	0.925	0.925

Figure 5 shows the confusion matrices, evidencing sample classification, for the best (kNN) and worst (decision tree) models, using only the three best features, as processed by t-SNE. While the confusion matrix for the kNN model presents all samples as being correctly classified, the confusion matrix for the decision tree evidences some errors because six samples from group zero (Norway farmed) were misclassified as being from group one (Chile farmed). This shows that the inductive learning mechanism present in the decision tree was not able to classify those samples correctly, as probably happens with the remaining classifiers, except for kNN that is not based on inductive learning.

Actual \ Predicted	0	1	2	3
0	100%	0.0%	0.0%	0.0%
1	0.0%	100%	0.0%	0.0%
2	0.0%	0.0%	100%	0%
3	0.0%	0.0%	0.0%	100%

Actual \ Predicted	0	1	2	3
0	83.3%	16.7%	0.0%	0.0%
1	0.0%	94.4%	5.6%	0.0%
2	0.0%	4.2%	95.8%	0%
3	0.0%	0.0%	0.0%	100%

Figure 5. Confusion matrix (showing proportion of actual) for the decision tree model (**left**) and confusion matrix for the kNN model (**right**), both processing only three features. 0—Norway farmed, 1—Chile farmed, 2—Canada farmed, 3—Canada wild.

4. Conclusions

In general, the four evaluated groups of salmon (wild from Canada and farmed from Canada, Chile and Norway) showed different fatty acid profiles, with wild specimens presenting significantly higher contents of health beneficial omega-3 fatty acids, in particular DHA and EPA, while farmed salmon presented significantly higher ($p < 0.05$) amounts of oleic and linoleic acids. Among the three groups of farmed salmon with different geographical origins, specimens from Chile and Canada were more similar, with the ones from Norway being more distinct mainly due to their lower levels of SFA and higher levels of α-linolenic acid. The differences among farmed groups are most probably related to different types of feed used in each farm. However, information about relevant factors such as farming diet and conditions, which are known to affect the lipidic composition of fish, was not available. In this work, we demonstrated the possibility of discriminating between wild and farmed salmons, as well as differentiating the origin within farmed ones, based on the use of machine learning models applied to fatty acid composition obtained by GC-FID. Thus, compared to a previous approach reported for the same samples, namely the use of PCA applied to normalized intensities of the most abundant signals generated by DART-HRMS analysis of the lipid extracts, this method showed a higher discrimination power. Moreover, this method proved to be simple and it only requires the use of affordable equipment, commonly found in most laboratories. Nevertheless, this approach has the disadvantage of requiring a longer analysis time compared to DART-HRMS. The developed algorithm combined with the information gain ratio criterion allowed us to establish the number of optimal features, so the classification tasks can still attain a very good performance. The feature reduction offers a computational speedup during the classification process. Among the seven tested machine learning models, the best results were obtained with the k-nearest neighbors (kNN) classifier, allowing for the correct classification of all tested samples. Moreover, it was shown that using t-SNE in the processing pipeline boosts the reduction in features, while still maintaining 100% accuracy in data classification. The performance difference between the test dataset and the leave-one-sample-out cross-validation was residual, meaning a good generalization figure.

Author Contributions: Conceptualization, J.S.A., I.M., P.J.R. and G.I.; methodology, L.G., G.I. and P.J.R.; chemical analyses, L.G. and M.A.N.; writing—original draft preparation, P.J.R., G.I. and J.S.A.; writing—review and editing, I.M. and J.S.A.; supervision, I.M., M.B.P.P.O. and J.S.A.; project administration, M.A., I.M., J.S.A. All authors have read and agreed to the published version of the manuscript.

Funding: This work was supported by the European project FOODINTEGRITY (FP7-KBBE-2013-single-stage, under grant agreement No 613688) and FCT (Fundação para a Ciência e Tecnologia, Portugal) under the Partnership Agreements UIDB 50006/2020, UIDB 00690/2020 (CIMO) and UIDB/5757/2020 (CeDRI). L. Grazina and M.A. Nunes acknowledge the FCT grant SFRH/BD/132462/2017 and SFRH/BD/130131/2017 financed by POPH-QREN (subsidised by FSE and MCTES).

Acknowledgments: The authors are thankful to Emiliano De Dominicis from Mérieux-Nutrisciences for providing the salmon samples analysed in the present project.

Conflicts of Interest: The authors declare no conflict of interest.

References

1. Larsen, R.; Eilertsen, K.E.; Elvevoll, E.O. Health benefits of marine foods and ingredients. *Biotechnol. Adv.* **2011**, *29*, 508–518. [CrossRef]
2. Swanson, D.; Block, R.; Mousa, S.A. Omega-3 Fatty Acids EPA and DHA: Health Benefits Throughout Life. *Adv. Nutr.* **2012**, *3*, 1–7. [CrossRef]
3. Shahidi, F.; Ambigaipalan, P. Omega-3 polyunsaturated fatty acids and their health benefits. *Annu. Rev. Food Sci. Technol.* **2018**, *9*, 345–381. [CrossRef]
4. FAO Food and Agriculture Organization (FAO). GLOBEFISH Highlights Issue 1/2020, with Jan.–Sep. 2019 Statistics—A Quarterly Update on World Seafood Markets. 2020. Available online: http://www.fao.org/3/ca7968en/CA7968EN.pdf (accessed on 26 August 2020).
5. FAO Food and Agriculture Organization (FAO). Salmon Commodity Statistics Update. 2016. Available online: http://www.fao.org/3/a-bq971e.pdf (accessed on 26 August 2020).
6. Blanchet, C.; Lucas, M.; Julien, P.; Morin, R.; Gingras, S.; Dewailly, E. Fatty acid composition of wild and farmed Atlantic salmon (*Salmo salar*) and rainbow trout (*Oncorhynchus mykiss*). *Lipids* **2005**, *40*, 529–531. [CrossRef]
7. Lundebye, A.K.; Lock, E.J.; Rasinger, J.D.; Nøstbakken, O.J.; Hannisdal, R.; Karlsbakk, E.; Wennevik, V.; Madhun, A.S.; Madsen, L.; Graff, I.E.; et al. Lower levels of persistent organic pollutants, metals and the marine omega 3-fatty acid DHA in farmed compared to wild Atlantic salmon (*Salmo salar*). *Environ. Res.* **2017**, *155*, 49–59. [CrossRef]
8. Fiorino, G.M.; Ilario, L.; De Angelis, E.; Arlorio, M.; Logrieco, A.F.; Monaci, L. Assessing fish authenticity by direct analysis in real time-high resolution mass spectrometry and multivariate analysis: Discrimination between wildtype and farmed salmon. *Food Res. Int.* **2019**, *116*, 1258–1265. [CrossRef]
9. Kappel, K.; Schröder, U. Substitution of high-priced fish with low-priced species: Adulteration of common sole in German restaurants. *Food Control* **2016**, *59*, 478–486. [CrossRef]
10. Cline, E. Marketplace substitution of Atlantic salmon for Pacific salmon in Washington State detected by DNA barcoding. *Food Res. Int.* **2012**, *45*, 388–393. [CrossRef]
11. Thomas, F.; Jamin, E.; Wietzerbin, K.; Guérin, R.; Lees, M.; Morvan, E.; Billault, I.; Derrien, S.; Rojas, J.M.M.; Serra, F.; et al. Determination of origin of Atlantic salmon (*Salmo salar*): The use of multiprobe and multielement isotopic analyses in combination with fatty acid composition to assess wild or farmed origin. *J. Agric. Food Chem.* **2008**, *56*, 989–997. [CrossRef]
12. Fernandes, T.J.R.; Amaral, J.S.; Mafra, I. DNA barcode markers applied to seafood authentication: An updated review. *Crit. Rev. Food Sci. Nutr.* **2020**, in press. [CrossRef]
13. Aursand, M.; Standal, I.B.; Prael, A.; McEvoy, L.; Irvine, J.; Axelson, D.E. C-13 NMR pattern recognition techniques for the classification of Atlantic salmon (*Salmo salar* L.) according to their wild, farmed, and geographical origin. *J. Agric. Food Chem.* **2009**, *57*, 3444–3451. [CrossRef]
14. Masoum, S.; Malabat, C.; Jalali-Heravi, M.; Guillou, C.; Rezzi, S.; Rutledge, D.N. Application of support vector machines to 1 H NMR data of fish oils: Methodology for the confirmation of wild and farmed salmon and their origins. *Anal. Bioanal. Chem.* **2007**, *387*, 1499–1510. [CrossRef]
15. Molkentin, J.; Meisel, H.; Lehmann, I.; Rehbein, H. Identification of organically farmed Atlantic salmon by analysis of stable isotopes and fatty acids. *Eur. Food Res. Technol.* **2007**, *224*, 535–543. [CrossRef]
16. Dempson, J.B.; Power, M. Use of stable isotopes to distinguish farmed from wild Atlantic salmon, *Salmo salar*. *Ecol. Freshw. Fish* **2004**, *13*, 176–184. [CrossRef]
17. Megdal, P.A.; Craft, N.A.; Handelman, G.J. A simplified method to distinguish farmed (*Salmo salar*) from wild salmon: Fatty acid ratios versus astaxanthin chiral isomers. *Lipids* **2009**, *44*, 569–576. [CrossRef]
18. Axelson, D.E.; Standal, I.B.; Martinez, I.; Aursand, M. Classification of wild and farmed salmon using bayesian belief networks and gas chromatography-derived fatty acid distributions. *J. Agric. Food Chem.* **2009**, *57*, 7634–7639. [CrossRef]
19. Martinez, I.; Standal, I.B.; Axelson, D.E.; Finstad, B.; Aursand, M. Identification of the farm origin of salmon by fatty acid and HR C-13 NMR profiling. *Food Chem.* **2009**, *116*, 766–773. [CrossRef]
20. Molkentin, J.; Lehmann, I.; Ostermeyer, U.; Rehbein, H. Traceability of organic fish e Authenticating the production origin of salmonids by chemical and isotopic analyses. *Food Control* **2015**, *53*, 55–66. [CrossRef]

21. Aursand, M.; Mabon, F.; Martin, G.J. Characterization of farmed and wild salmon (*Salmo salar*) by a combined use of compositional and isotopic analyses. *JAOCS* **2000**, *77*, 659–666. [CrossRef]
22. Bligh, E.G.; Dyer, W.J. A rapid method of total lipid extraction and purification. *Can. J. Biochem. Physiol.* **1959**, *37*, 911–917. [CrossRef]
23. Amaral, J.S.; Casal, S.; Pereira, J.A.; Seavra, R.M.; Oliveira, M.B.P.P. Determination of sterol and fatty acid compositions, oxidative stability, and nutritional value of six walnut (*Juglans regia* L.) cultivars grown in Portugal. *J. Agric. Food Chem.* **2003**, *51*, 7698–7702. [CrossRef]
24. Jolliffe, I.T. *Principal Component Analysis*, 2nd ed.; Springer Inc.: New York, NY, USA, 2002.
25. Van der Maaten, L.J.P.; Hinton, G.E. Visualizing Data Using t-SNE. *J. Mach. Learn. Res.* **2008**, *9*, 2579–2605.
26. Altman, N.S. An introduction to kernel and nearest-neighbor nonparametric regression. *Am. Stat.* **1992**, *46*, 175–185.
27. Quinlan, J.R. Simplifying decision trees. *Int. J. Man. Mach. Stud.* **1987**, *27*, 221–234. [CrossRef]
28. Cortes, C.; Vapnik, V.N. Support-vector networks. *Mach. Learn.* **1995**, *20*, 273–297. [CrossRef]
29. Ho, T.K. Random Decision Forests. In Proceedings of the 3rd International Conference on Document Analysis and Recognition, Montreal, QC, Canada, 14–16 August 1995; pp. 278–282.
30. Rumelhart, D.E.; Hinton, G.E.; Williams, R.J. Learning representations by back-propagating errors. *Nature* **1986**, *323*, 533–536. [CrossRef]
31. Russell, S.; Norvig, P. *Artificial Intelligence: A Modern Approach*, 2nd ed.; Prentice Hall: Upper Saddle River, NJ, USA, 2003.
32. Freund, Y.; Schapire, R.E. Short Introduction to Boosting. *J. Jpn. Soc. Artif. Intell.* **1999**, *14*, 771–780.

foods

MDPI

Article

A Real-Time PCR Method for the Authentication of Common Cuttlefish (*Sepia officinalis*) in Food Products

Amaya Velasco *, Graciela Ramilo-Fernández and Carmen G. Sotelo

Instituto de Investigaciones Marinas (IIM-CSIC), Eduardo Cabello 6, 36208 Vigo (Pontevedra), Spain; graciela@iim.csic.es (G.R.-F.); carmen@iim.csic.es (C.G.S.)

* Correspondence: amayavelasco@iim.csic.es; Tel.: +34-986-23-19-30

Received: 6 February 2020; Accepted: 28 February 2020; Published: 4 March 2020

Abstract: Cephalopods are very relevant food resources. The common cuttlefish (*Sepia officinalis*) is highly appreciated by consumers and there is a lack of rapid methods for its authentication in food products. We introduce a new minor groove binding (MGB) TaqMan real-time PCR (Polymerase Chain Reaction) method for the authentication of *S. officinalis* in food products to amplify a 122 base pairs (bp) fragment of the mitochondrial COI (Cytochrome Oxidase I) region. Reference and commercial samples of *S. officinalis* showed a threshold cycle (Ct) mean of 14.40, while the rest of the species examined did not amplify, or showed a significantly different Ct ($p < 0.001$). The calculated efficiency of the system was 101%, and the minimum DNA quantity detected was 10^{-4} ng. No cross-reactivity was detected with any other species, thus, the designed method differentiates *S. officinalis* from other species of the genus *Sepia* and other cephalopod species and works for fresh, frozen, grilled, cooked and canned samples of *Sepia* spp. The method has proved to be reliable and rapid, and it may prove to be a useful tool for the control of fraud in cuttlefish products.

Keywords: Sepia; common cuttlefish; *Sepia officinalis*; real-time PCR (Polymerase Chain Reaction); species identification; food authentication; COI (Cytochrome Oxidase I)

---

1. Introduction

Cephalopods are a very diverse group of mollusks and include 28 families and more than 600 species, many of which are commercially important. As a sign of their relevance, captures of cephalopods in 2017 reached 3,772,565 t, with an estimated value of almost 8000 million dollars [1].

The common cuttlefish (*S. officinalis*) is highly appreciated by consumers around the world, and it is traded with different presentations particularly in Japan, the Republic of Korea, Italy and Spain. In the last decade, the world catches attributed to this species have registered numbers between 20,000 and 30,000 tons every year [2]. It is the species of cuttlefish with the highest commercial value.

European regulations regarding the labeling of fishery products [3,4] establish that these products must show the information about the species, with the commercial and/or scientific name depending on the type of product. Illicit substitution of one species for another may constitute economic fraud and/or misbranding violations. Furthermore, species substitution may cause potential food safety hazards to be overlooked by processors or end-users [5]. Species substitution is relatively frequent in seafood products [6], and particularly in products containing cephalopods, where several cases of species substitution have been reported [7,8].

Species belonging to the genus *Sepia* can look very similar to a non-trained consumer, especially when they are processed for the market (e.g., peeled, canned), making the visual differentiation almost impossible and increasing the possibilities of fraud. Thus, the reported cases of mislabeling in products containing *Sepia* spp. have usually been substitutions between species belonging to the same genus [9,10]. These cases can be attributed to economic fraud (e.g., substitutions between species with different commercial value) or unintentional substitutions, which can be due to similar geographic distribution of species (e.g., *S. officinalis/S. orbignyanya/S. elegans*) and/or similar morphological characteristics (e.g., juveniles of *S. officinalis/S. elegans*) which can lead to misidentification at any level of the value chain (fisheries, processors and consumers). In order to control these substitutions, a variety of genetic methods have been published for the identification of several cephalopod species. The majority of these are labor-intensive and time-consuming, such as forensically informative nucleotide sequencing (FINS), barcoding [10–14] and RFLP [8,15]. Some rapid DNA-based methods have also been published for the authentication of some cephalopod species [7,16–18], but to date, there is not any rapid technique available for the genetic identification of *S. officinalis*.

This work presents a rapid and reliable method for the authentication of *S. officinalis* in different food matrices, including processed products. Therefore, it can be a useful tool for control authorities at different levels of the value chain.

2. Materials and Methods

2.1. Sampling and DNA Extraction

In this work, 14 samples of *S. officinalis* from different locations of Spanish and Portuguese waters were used as a reference. Also, 29 individuals from 20 other cephalopod species of 11 genera from the Instituto de Investigaciones Marinas (IIM-CSIC) own tissue collection were included for the specificity assay (Table 1). All reference individuals had a known origin and were identified visually prior to the FINS identification. Additionally, 16 commercial samples were collected from supermarkets and restaurants in Galicia region (Spain) for the application to commercial products (Table 2). All tissue samples were stored at −20 °C until analysis.

Table 1. Reference samples used in this study and threshold cycle (Ct) results.

Sample Code	Species	Common Name (FAO)	Geographic Origin	Ct Mean ± SD
SOFF2	*Sepia officinalis*	Common cuttlefish	Atlantic, Northeast (FAO 27.9) Vigo	12.98 ± 0.32
SOFF3	*Sepia officinalis*	Common cuttlefish	Atlantic, Northeast (FAO 27.9) Cambados	14.05 ± 0.26
SOFF4	*Sepia officinalis*	Common cuttlefish	Atlantic, Northeast (FAO 27.9) Cambados	14.80 ± 0.04
SOFF5	*Sepia officinalis*	Common cuttlefish	Atlantic, Northeast (FAO 27.9) Cambados	16.16 ± 0.35
SOFF6	*Sepia officinalis*	Common cuttlefish	Atlantic, Northeast (FAO 27.9) Vigo	13.57 ± 0.70
SOFF7	*Sepia officinalis*	Common cuttlefish	Atlantic, Northeast (FAO 27.9) Vigo	15.37 ± 0.10
SOFF8	*Sepia officinalis*	Common cuttlefish	Atlantic, Northeast (FAO 27.9) Vigo	13.51 ± 0.07
SOFF9	*Sepia officinalis*	Common cuttlefish	Atlantic, Northeast (FAO 27)	14.11 ± 0.20
SOFF10	*Sepia officinalis*	Common cuttlefish	Atlantic, Northeast (FAO 27)	15.65 ± 0.21
SOFF11	*Sepia officinalis*	Common cuttlefish	Atlantic, Northeast (FAO 27)	13.86 ± 0.48
SOFF12	*Sepia officinalis*	Common cuttlefish	Atlantic, Northeast (FAO 27)	13.25 ± 0.39
SOFF15	*Sepia officinalis*	Common cuttlefish	Atlantic, Northeast (FAO 27.9) Algarve	12.72 ± 0.06
SOFF16	*Sepia officinalis*	Common cuttlefish	Atlantic, Northeast (FAO 27.9) Algarve	12.59 ± 0.19
SOFF17	*Sepia officinalis*	Common cuttlefish	Atlantic, Northeast (FAO 27.9) Algarve	13.98 ± 0.12
SBER 2	*Sepia betheloti*	African cuttlefish	Atlantic, Eastern Central (FAO 34)	≥40
SBER 3	*Sepia betheloti*	African cuttlefish	Atlantic, Eastern Central (FAO 34)	≥40
SORB 4	*Sepia orbygniana*	Pink cuttlefish	Atlantic, Northeast (FAO 27)	≥40
SORB 5	*Sepia orbygniana*	Pink cuttlefish	Atlantic, Northeast (FAO 27)	≥40
SPHA 1	*Sepia pharaonis*	Pharaon cuttlefish	Indian Ocean, Western (FAO 51)	≥40
LVUL 2	*Loligo vulgaris*	European squid	Western Central Atlantic (FAO 31)	27.00 ± 0.16
LVUL 1	*Loligo vulgaris*	European squid	Atlantic, Northeast (FAO 27)	26.03 ± 0.28
LVUL 5	*Loligo vulgaris*	European squid	Atlantic, Northeast (FAO 27)	29.15 ± 0.14
LVUL 3	*Loligo vulgaris*	European squid	Western Central Atlantic (FAO 31)	≥40
LVUL 4	*Loligo vulgaris*	European squid	Western Central Atlantic (FAO 31)	≥40
LVUL 6	*Loligo vulgaris*	European squid	Western Central Atlantic (FAO 31)	≥40
LVUL 7	*Loligo vulgaris*	European squid	Western Central Atlantic (FAO 31)	≥40
LVUL 8	*Loligo vulgaris*	European squid	Western Central Atlantic (FAO 31)	29.80 ± 0.40
LREY 1	*Loligo reynaudi*	Cape Hope squid	Atlantic, Southeast (FAO 47)	≥40
IILL 2	*Illex illecebrosus*	Northern Shortfin squid	Atlantic, Northwest (FAO 21)	≥40
TEBL 1	*Todaropsis eblanae*	Lesser flying squid	Atlantic, Northeast (FAO 27)	≥40
TPAC 3	*Todarodes pacificus*	Japanese flying squid	Pacific, Northwest (FAO 61)	≥40
ICOI 10	*Illex coindetii*	Southern shortfin squid	Atlantic, Northeast (FAO 27)	≥40
LGAH 9	*Loligo gahi*	Patagonian squid	Pacific, Southeast (FAO 87)	≥40
MHYA 8	*Martialia hyadesi*	Sevenstar flying squid	Atlantic, Antarctic (FAO 48)	≥40
NSLO6	*Nototodarus sloanii*	Wellington flying squid	Pacific, Southwest (FAO 81)	≥40

Table 1. *Cont.*

Sample Code	Species	Common Name (FAO)	Geographic Origin	Ct Mean ± SD
TSAG 1	*Todarodes sagittatus*	European flying squid	Atlantic, Northeast (FAO 27)	≥40
OVUL 142	*Octopus vulgaris*	Common octopus	Atlantic, Northeast (FAO 27)	≥40
OCYA 3	*Octopus cyanea*	Big blue octopus	Pacific, Western Central (FAO 71)	≥40
OCYA 4	*Octopus cyanea*	Big blue octopus	Pacific, Western Central (FAO 71)	≥40
OMIM 1	*Octopus mimus*	Changos octopus	Pacific, Southeast (FAO 87)	≥40
ECIR 143	*Eledone cirrhosa*	Horned octopus	Atlantic, Northeast (FAO 27)	≥40
DGIG 1	*Dosidicus gigas*	Jumbo squid	Pacific, Southeast (FAO 87)	≥40
AMEM 1	*Amphioctopus membranaceus*	Webfoot octopus	Indian Ocean, Western (FAO 51)	≥40

FAO: Food and Agriculture Organization. SD: Standard Deviation.

Table 2. Commercial samples used for validation. The mislabeled samples are highlighted in red.

Sample Code	Type of Processing	Type of Establishment	Species Declared	Species Identified by FINS	Ct Mean ± SD
S1	Frozen	Supermarket	*Sepia spp.*	*Sepia pharaonis*	29.77 ± 0.62
S2	Frozen	Supermarket	*Sepia spp.*	*Sepia pharaonis*	27.68 ± 0.06
S3	Frozen	Supermarket	*Sepia spp.*	*Sepia sp* (not *S. officinalis*)	≥40
S4	Frozen	Supermarket	"Sepia"	*Sepia sp* (not *S. officinalis*)	31.85 ± 0.26
S5	Canned	Supermarket	"Sepia"	*Sepia officinalis*	16.91 ± 0.47
S6	Frozen	Supermarket	*Sepia spp.*	*Sepia sp* (not *S. officinalis*)	≥40
S7	Cooked	Supermarket	*Sepia officinalis*	*Sepia officinalis*	17.88 ± 0.94
S8	Canned	Supermarket	"Sepia"	*Sepia officinalis*	15.41 ± 0.03
S10	Grilled	Restaurant	"Choco"	*Sepia officinalis*	13.70 ± 0.06
S11	Frozen	Supermarket	*Sepia aculeata*	*Sepia sp* (not *S. officinalis*)	≥40
S12	Frozen	Supermarket	*Sepiella spp.*	*Sepiella inermis*	≥40
S13	Frozen	Supermarket	*Sepia pharaonis*	*Sepia aculeata*	≥40
S14	Thawed	Supermarket	*Sepia officinalis*	*Sepia officinalis*	14.03 ± 0.28
S15	Grilled	Restaurant	"Sepia"	*Sepia bertheloti*	26.08 ± 0.11
S16	Thawed	Supermarket	*Sepia officinalis*	*Sepia officinalis*	13.43 ± 0.06
S17	Canned	Supermarket	"Sepia"	*Sepia pharaonis*	23.62 ± 0.23

FINS: Forensically Informative Nucleotide Sequencing.

A portion of 0.3 g of muscle tissue from each sample was digested at 56 °C in a thermo shaker with 860 μL of lysis buffer (1% Sodium Dodecyl Sulfate (SDS), 150 mM NaCl, 2 mM Ethylenediaminetetraacetic acid (EDTA) and 10 mM Tris-HCl at pH 8), 100 μL of guanidinium thiocyanate 5 M and 40 μL of proteinase K (20 mg/mL). After 3 h, 40 μL of extra proteinase K was added and left overnight. DNA was isolated with the Wizard DNA Clean-up System kit (Promega, Madison, WI, USA) following the manufacturer's protocol. Double-stranded DNA obtained was quantified with Qubit dsDNA BR Assay Kit (Life Technologies, Carlsbad, CA, USA) and Qubit 3.0 fluorometer (Invitrogen, Carlsbad, CA, USA). Purified DNA was stored at −20 °C until further analysis.

2.2. FINS Identification of Samples

Reference and commercial samples were authenticated by FINS (forensically informative nucleotide sequencing) in order to test the reliability of the method developed. PCR reactions were carried out in a Verity 96 wells Thermal cycler (Applied Biosystems, Foster City, CA, USA) with Illustra PuReTaq Ready-To-Go PCR Beads (GE Healthcare, Chicago, IL, USA), 1 μL of each primer (10 μM) and 100 ng of template DNA in a final volume of 25 μL. Primers designed by Folmer [19] LCO1490-5′GGTCAACAAATCATAAAGATATTGG3′ and HCO2198-5′TAAACTTCAGGGTGACCAA AAAATCA3′ were used to amplify a 750 base pairs (bp) fragment of the mitochondrial COI region, with the following thermal protocol: a preheating step of 3 min at 95 °C, followed by 35 cycles of 1 min at 95 °C, 1 min at 40 °C and 1.5 min at 72 °C, with a final extension step at 72 °C for 7 min. When amplification of COI fragment failed, the 16SVAR primers described by Chapela [11] 16SVAR-F-5′CAAATTACGCTGTTATCCCTATGG3′ and 16SVAR-R- 5′GACGAGAAGACCCTAATGAGCTTT3′ were used to amplify a 210 bp fragment of the mitochondrial 16S rDNA, with the thermal protocol as follows: a preheating step of 3 min at 95 °C, followed by 35 cycles of 40 s at 94 °C, 40 s at 50 °C and 40 s at 72 °C, with a final extension step at 72 °C for 7 min. Negative and positive controls were included in all PCR sets.

Primers designed in this study for the minor groove binding (MGB)-TaqMan assay were also used for FINS identification in 3 cases of processed commercial samples of *Sepia* spp. (cooked and canned), where both COI and 16S sets of primers failed to amplify, with the following thermal protocol: a preheating step of 3 min at 95 °C, followed by 35 cycles of 40 s at 95 °C, 40 s at 40 °C and 40 s at 72 °C with a final extension step at 72 °C for 7 min. PCR amplicons were visualized on a 2% agarose gel, using UV transillumination (BioRad, Hercules, CA, USA).

PCR products were purified with Illustra ExoProStar (GE Healthcare, Chicago, IL, USA) and sequencing reactions were performed with BigDye Terminator 1.1 (Applied Biosystems, Foster City, CA, USA), following the manufacturer's instructions. The automatic sequencing was carried out in an ABI PRISM 3130 (Applied Biosystems, Foster City, CA, USA). After automatic sequencing, F and R files were edited with Chromas and aligned with Bioedit [20] to obtain the complete sequence of the fragment. Bioedit software was also used to align the resulting sequence with reference ones from the NCBI and the IIM-CSIC sequence database, which consists of more than 2000 sequences from fish and mollusks specimens that have been collected during 30 years; most of these specimens were morphologically identified and also genetically authenticated. This alignment was imported with MEGA [21] for phylogenetic analysis. The phylogenetic model used for constructing the neighbor-joining tree was Tamura–Nei, with 1000 bootstrap replicates. The results were also authenticated with BLAST [22]. The multiple alignments and the BLAST tool were also used to check the quality and coverage of the resulting sequences.

The COI sequences obtained for reference and commercial samples of this study were uploaded to Genbank [23] (accession numbers: MN977128 to MN977135, MN977138, MN977143, MN977144, MN977146, MN977147, MN977149, MN977152, MN977154 to MN977156, MN977158, MN977159, MN977161 to MN977171, MN977173 to MN977177, MN977179 to MN977191).

2.3. RT-PCR Design

In order to find a suitable fragment to design a short and specific system, a large number of nuclear and mitochondrial cephalopod sequences from public and IIM-CSIC databases were aligned and analyzed. A fragment of the COI region was suitable for the design of an MGB-Taq-Man Primers and Probe set, complying with the requirements of showing low intraspecific variability and high interspecific variability and allowing the amplification of a short fragment (122 bp, primers included). The sequences of primers (F and R) and Probe (P) are the following (see Figure 1):

SOFI_F: 5′CTTCTCCTTACATTTAGCWGGRGTCT3′
SOFI_R: FAM-5′TACCGAYCAAGCAAATAAAGGTAGG3′-MGB
SOFI_P: 5′AGCGATTAACTTCATCA3′

2.4. Real-Time PCR Conditions and Data Treatment

Concentrations of 50, 300 and 900 nM of each primer and 25, 50, 75, 100, 125, 150, 175, 200 and 225 nM of the probe were tested in order to select the optimal reaction conditions. The combination that gave the lowest threshold cycle (Ct) value and the highest final fluorescence was selected for the subsequent assays. The selected concentrations were 300 nM of SOFI_F primer, 900 nM of SOFI_R primer and 150 nM of SOFI_P probe.

Thus, each 20 μL reaction contained 10 μL of TaqMan Fast Universal Master Mix (2X), No AmpErase UNG (Applied Biosystems, Foster City, CA, USA), 1 μL of Primer SOFI_F (6 μM), 1 μL of Primer SOFI_R (18 μM), 1 μL of Probe SOFI_P (3 μM) and 100 ng of template DNA. Reactions were amplified in a 7500 fast real-time PCR System (Applied Biosystems, Foster City, CA, USA), with the fast ramp speed protocol: 95 °C for 20 s, followed by 40 cycles of 95 °C for 3 s and 60 °C for 30 s. Samples were analyzed in triplicate, and Ct mean and standard deviation of each individual were registered.

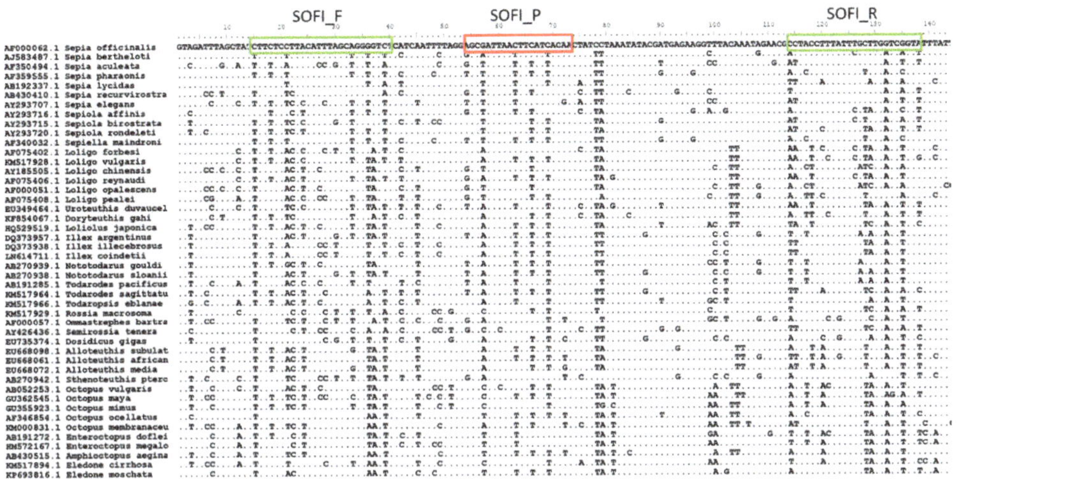

Figure 1. Multiple sequence alignment of the mitochondrial COI (Cytochrome Oxidase I) fragment, showing the position of the primers and probe designed.

3. Results

3.1. Efficiency and Detection Limit

Different quantities of template DNA of *S. officinalis*, from 10^{-5} ng to 100 ng were used for the efficiency assay. Over this range of dilutions, the response was linear with a slope of −3.13, an R^2 of 0.999 and an efficiency of 101%, following the equation: $E = 10^{-1/b} - 1$ [24]. The acceptable efficiency values range from 90% to 110%, therefore, 101% can be considered ideally optimal. The minimum quantity of DNA detected was 10^{-4} ng. The automatic threshold generated in this assay was 0.02, the value used in the subsequent analyses.

3.2. Inclusivity and Specificity

A total of 14 samples of *S. officinalis* from different locations and dates of capture were tested (Table 1), obtaining Ct data between 12.59 and 16.16, with a Ct mean of 14.04 (Figure 2A).

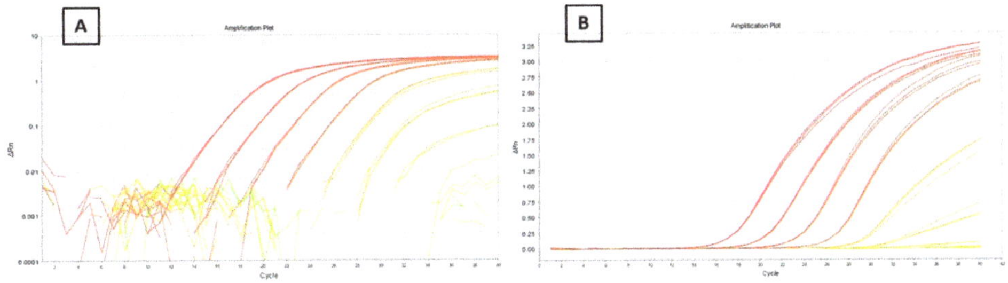

Figure 2. Amplification plots of the 10X dilution series of *Sepia officinalis* DNA (**A**): logarithmic, (**B**): linear.

In the other 19 species tested (Table 1), none of them presented any fluorescence signal with the exception of one specimen of *Loligo vulgaris*, which showed a late amplification signal (Figure 3B). In view of these results, another specificity assay was carried out with seven additional individuals of *L. vulgaris*, obtaining a Ct mean of 34.0, a result that is significantly different from the Ct of *S. officinalis* when a mean comparison test (one way ANOVA) was run ($p < 0.001$).

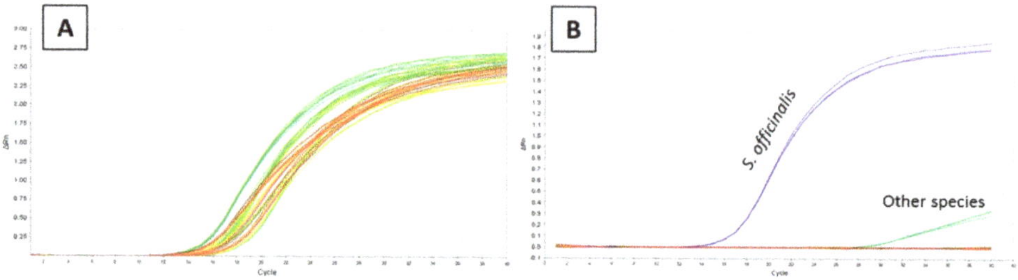

Figure 3. (**A**) Inclusivity test: amplification pattern of reference samples of *Sepia officinalis*. (**B**) Specificity test: amplification pattern of reference samples of *Sepia officinalis* and the rest of the species tested.

3.3. Application to Commercial Products

According to the Spanish regulations for the labeling of fresh, frozen and refrigerated fishery products, the commercial name "Sepia", "Choco" or "Jibia" is only accepted for products containing *S. officinalis*, while the commercial name "Sepias" can be used for all species of the genus *Sepia* [25]. In the same way, the commercial name "Jibia" or "Sepia" can be only applied to canned products containing the species *S. officinalis* [26]. Therefore, the system was also tested with 16 commercial samples labeled as "Sepia", "Choco" or "Sepias", from supermarkets and restaurants of Galicia (Spain), with different degrees of processing such as thawed, frozen, grilled, cooked and canned.

Following the above-mentioned criteria, the FINS identification results of this study revealed four cases of mislabeling regarding species (Table 2), all being substitutions between different species of the genus *Sepia*, constituting a mislabeling rate of 25%. The substitute species found were *Sepia pharaonis*, *Sepia aculeata*, *Sepia bertheloti* and a non-identified species. In four cases, it was not possible to reach the species level with the FINS identification, due to the lack of reference sequences in public databases, but authors could determine that these samples did not belong to *S. officinalis* species by analyzing the results of the neighbor-joining tree and the BLAST tool. The MGB TaqMan real-time PCR system worked in fresh and processed samples of *S. officinalis*, and the method was able to differentiate between products containing *S. officinalis* (Ct mean 15.23) and products containing other species of the Sepiidae family (Ct mean 33.82), with statistical significance ($p < 0.001$). The type of processing did not affect the Ct values, and a good differentiation was obtained both in fresh and frozen products as well as in highly processed samples, such as canned.

The Ct results obtained for both reference and commercial samples containing *S. officinalis* ranged from 12.59 to 17.88, with a Ct mean of 14.40, while the rest of species remained undetected or showed late amplification, with Ct values of 23.62 and higher and a mean Ct of 33.40 and this Ct mean resulted significantly different from the Ct of samples containing *S. officinalis* ($p < 0.001$).

4. Discussion

Results confirm TaqMan real-time PCR technique as a powerful tool for species authentication, due to its characteristics of specificity, increased with Minor Groove Binding technology (MGB probes) [27], and its sensitivity, allowing the detection of very low quantities of target DNA. Real-time PCR also allows the detection and quantification of target DNA in one step, eliminating post-PCR steps and saving labor time. The method described in this work includes these characteristics of specificity, sensitivity and fastness, since the real-time PCR analysis takes around 40 min, which means that, depending on the tissue digestion protocol, the complete analysis from the tissue sample can be carried out in 3–4 h. This feature and the reduced equipment needed, opens the possibility of the optimization of the method for analyses on-site at the different levels of the value chain, including the point of sale. The cost of the analysis (less than 5 euros per sample) is also much lower than sequencing-based methods, which makes it affordable for low-resources control units.

S. officinalis is marketed under several types of processing, including those that eliminate the characteristics for visual identification, such as peeling, cutting, cooking and canning. This makes these products vulnerable to species substitution, intentional or not. Results also show that the design of the primer set allows the amplification and authentication of the species even in samples where processing may lead to DNA degradation and/or fragmentation, such as canning. The lack of rapid methods for this task makes the control of this market laborious and costly, and this technique emerges as the only available alternative at the moment.

The sampling at supermarkets has also revealed that Spanish legislation of the commercial names in canned products needs to be updated since it has not been reviewed since 1986. Taking into account the current legislation, canned products are not obliged to show the scientific name on the labels, i.e., commercial names such as "Sepia" can be found, which correspond to several species. The authors consider that this system is no longer suitable for the current market, where the amount of cephalopod species in the market has greatly increased while different species may achieve significant differences in market price.

The Ct values obtained in this study for the target species are at the same level or lower than other recent works using the TaqMan real-time PCR technique for species identification [28,29]. The significant differences found between the data corresponding to *S. officinalis* and the other species prove that Ct values can be used to determine whether a sample contains *S. officinalis* or another cephalopod species. Results also prove the high specificity of the system, which works for the differentiation of *S. officinalis* from the other species of the genus *Sepia* with commercial importance, demonstrating the utility of the method in food control, since the reported cases of mislabeling in the family Sepiidae show substitutions between species belonging to the same genus, as shown in previous publications [9,10] and confirmed in this study. Although the system has not been tested with all the species of the genus *Sepia*, this study included those with relevance to the market. Nonetheless, further analysis could be carried out to confirm the specificity of the method with other species of the genus *Sepia* which might have some commercial relevance in certain countries. The level of mislabeling found in this work (25%) is slightly lower than those found in the aforementioned articles, but still in the range of significant mislabeling. However, the different sampling procedures do not allow an adequate comparison, therefore, authors cannot affirm that there has been a decrease in the mislabeling rates. Nevertheless, these results highlight the need for an effective tool for the control of this type of product.

5. Conclusions

As a conclusion, this work presents a rapid, non-expensive and reliable method, able to differentiate *S. officinalis* from other species of the genus *Sepia* and other cephalopod species in food samples with different levels of processing, making it useful for food control authorities in the whole food value chain.

This study also found a moderate level of mislabeling in Sepia products, which highlights the need for more efficient control of the authenticity of this type of product.

Author Contributions: Conceptualization, A.V. and C.G.S.; methodology, A.V. and G.R.-F.; software, A.V.; supervision, C.G.S.; writing—original draft, A.V.; writing—review and editing, G.R.F. and C.G.S. All authors have read and agreed to the published version of the manuscript.

Funding: This study is part of the SEATRACES project (www.seatraces.eu), funded by the EU Interreg Atlantic Area Programme (project number EAPA_87/2016).

Acknowledgments: We acknowledge the Border Control Post of Vigo (BCP Vigo), Rogério Mendes, Patricia Ramos and Marta Pérez for providing tissue samples.

Conflicts of Interest: The authors declare no conflict of interest.

References

1. FAO. *Fishery and Aquaculture Statistics*; 2017/FAO annuaire; FAO: Rome, Italy, 2019.
2. Fisheries and Aquaculture Software. FishStatJ—Software for Fishery and Aquaculture Statistical Time Series. In *FAO Fisheries and Aquaculture Department*; FAO: Rome, Italy, 2016; Updated 21 July 2016. Available online: http://www.fao.org/fishery/Rome (accessed on 6 November 2019).
3. Regulation (EU) No 1169/2011 of the European Parliament and of the Council of 25 October 2011 on the provision of food information to consumers, amending Regulations (EC) No 1924/2006 and (EC) No 1925/2006 of the European Parliament and of the Council, and repealing Commission Directive 87/250/EEC, Council Directive 90/496/EEC, Commission Directive 1999/10/EC, Directive 2000/13/EC of the European Parliament and of the Council, Commission Directives 2002/67/EC and 2008/5/EC and Commission Regulation (EC) No 608/2004. Available online: https://eur-lex.europa.eu/LexUriServ/LexUriServ.do?uri=OJ:L:2011:304:0018:0063:EN:PDF (accessed on 25 October 2011).
4. Regulation (EU) No 1379/2013 of the European Parliament and of the Council of 11 December 2013 on the common organisation of the markets in fishery and aquaculture products, amending Council Regulations (EC) No 1184/2006 and (EC) No 1224/2009 and repealing Council Regulation (EC) No 104/2000. Available online: https://eur-lex.europa.eu/LexUriServ/LexUriServ.do?uri=OJ:L:2013:354:0001:0021:EN:PDF (accessed on 11 December 2013).
5. FDA, Food and Drug Administration; U.S. Department of Health and Human Services. *Fish and Fishery Products: Hazards and Controls Guidance*, 4th ed.; Chapter 3; IFAS-University of Florida: Gainesville, FL, USA, 2019.
6. Warner, K.; Mustain, P.; Lowell, B.; Geren, S.; Talmage, S. *Deceptive Dishes: Seafood Swaps Found Worldwide*; Oceana Reports; Oceana: New York, NY, USA, 2016; Available online: www.oceana.org (accessed on 12 December 2019).
7. Espiñeira, M.; Vieites, J.M. Rapid method for controlling the correct labeling of products containing common octopus (*Octopus vulgaris*) and main substitute species (*Eledone cirrhosa* and *Dosidicus gigas*) by fast real-time PCR. *Food Chem.* **2012**, *135*, 2439–2444. [CrossRef] [PubMed]
8. Santaclara, F.J.; Espiñeira, M.; Vieites, J.M. Genetic identification of squids (families Ommastrephidae and Loliginidae) by PCR–RFLP and FINS methodologies. *J. Agric. Food Chem.* **2007**, *55*, 9913–9920. [CrossRef] [PubMed]
9. Guardone, L.; Tinacci, L.; Costanzo, F.; Azzarelli, D.; D'Amico, P.; Tasselli, G.; Magni, A.; Guidi, A.; Nucera, D.; Armani, A. DNA barcoding as a tool for detecting mislabeling of fishery products imported from third countries: An official survey conducted at the Border Inspection Post of Livorno-Pisa (Italy). *Food Control* **2017**, *80*, 204–216. [CrossRef]
10. Espiñeira, M.; Vieites, J.M.; Santaclara, F.J. Species authentication of octopus, cuttlefish, bobtail and bottle squids (families Octopodidae, Sepiidae and Sepiolidae) by FINS methodology in seafoods. *Food Chem.* **2010**, *121*, 527–532. [CrossRef]
11. Chapela, M.J.; Sotelo, C.G.; Calo-Mata, P.; Pérez-Martín, R.I.; Rehbein, H.; Hold, G.L.; Quinteiro, J.; Rey-Méndez, M.; Rosa, C.; Santos, A.T. Identification of cephalopod species (Ommastrephidae and Loliginidae) in seafood products by Forensically Informative Nucleotide Sequencing (FINS). *J. Food Sci.* **2002**, *67*, 1672–1676. [CrossRef]
12. Chapela, M.J.; Sotelo, C.G.; Pérez-Martín, R.I. Molecular identification of cephalopod species by FINS and PCR-RFLP of a cytochrome b gene fragment. *Eur. Food Res. Technol.* **2003**, *217*, 524–529. [CrossRef]
13. Sin, Y.W.; Yau, C.; Chu, K.H. Morphological and genetic differentiation of two loliginid squids, *Uroteuthis (Photololigo) chinensis* and *Uroteuthis (Photololigo) edulis* (Cephalopoda: Loliginidae), in Asia. *J. Exp. Mar. Biol. Ecol.* **2009**, *369*, 22–30. [CrossRef]
14. Undheim, E.A.; Norman, J.A.; Thoen, H.H.; Fry, B.G. Genetic identification of Southern Ocean octopod samples using mtCOI. *C. R. Biol.* **2010**, *333*, 395–404. [CrossRef] [PubMed]
15. Colombo, F.; Cerioli, M.; Colombo, M.M.; Marchisio, E.; Malandra, R.; Renon, P. A simple polymerase chain reaction-restriction fragment length polymorphism (PCR-RFLP) method for the differentiation of cephalopod mollusc families Loliginidae from Ommastrephidae, to avoid substitutions in fishery field. *Food Control* **2002**, *13*, 185–190. [CrossRef]

16. Herrero, B.; Lago, F.C.; Vieites, J.M.; Espiñeira, M. Rapid method for controlling the correct labeling of products containing European squid (*Loligo vulgaris*) by fast real-time PCR. *Eur. Food Res. Technol.* **2012**, *234*, 77–85. [CrossRef]
17. Ye, J.; Feng, J.; Liu, S.; Zhang, Y.; Jiang, X.; Dai, Z. Identification of four squid species by quantitative real-time polymerase chain reaction. *Mol. Cell. Probes* **2016**, *30*, 22–29. [CrossRef] [PubMed]
18. Ye, J.; Feng, J.; Dai, Z.; Meng, L.; Zhang, Y.; Jiang, X. Application of Loop-Mediated Isothermal Amplification (LAMP) for Rapid Detection of Jumbo Flying Squid *Dosidicus gigas* (D'Orbigny, 1835). *Food Anal. Methods* **2017**, *10*, 1452–1459. [CrossRef]
19. Folmer, O.; Black, M.; Hoeh, W.; Lutz, R.; Vrijenhoek, R. DNA primers for amplification of mitochondrial cytochrome c oxidase subunit I from diverse metazoan invertebrates. *Mol. Mar. Biol. Biotechnol.* **1994**, *3*, 294–299. [PubMed]
20. Hall, T.A. BioEdit: A user-friendly biological sequence alignment editor and analysis program for Windows 95/98/NT. *Nuceicl. Acids Symp. Ser.* **1999**, *41*, 95–98.
21. Kumar, S.; Dudley, J.; Nei, M.; Tamura, K. MEGA: A biologist-centric software for evolutionary analysis of DNA and protein sequences. *Brief. Bioinform.* **2008**, *9*, 299–306. [CrossRef] [PubMed]
22. Altschul, S.F.; Madden, T.L.; Schaffer, A.A.; Zhang, J.; Zhang, Z.; Millar, W.; Lipman, D.J. Gapped BLAST and PSIBLAST: A new generation of protein database search programs. *Nucleic Acids Res.* **1997**, *25*, 3389–3402. [CrossRef] [PubMed]
23. Clark, K.; Karsch-Mizrachi, I.; Lipman, D.J.; Ostell, J.; Sayers, E.W. GenBank. *Nucleic Acids Res.* **2016**, *44*, D67–D72. [CrossRef] [PubMed]
24. Dorak, M.T. *Real-Time PCR*; Taylor & Francis Group: New York, NY, USA, 2006.
25. *Resolución de 26 de enero de 2018, de la Secretaría General del Mar, por la que se establece y se publica el listado de denominaciones comerciales de especies pesqueras y de acuicultura admitidas en España; BOE-A-2018-2884; Boletín oficial del Estado núm. 53, de 1 de marzo de 2018*; Ministerio de Medio Ambiente, Medio Rural y Marino: Madrid, Spain, 2018; pp. 25487–25513.
26. *ORDEN de 17 de enero de 1986 sobre norma reguladora del comercio exterior de las conservas de cefalópodos;* Boletín oficial del Estado Núm. 30; Ministerio de Economía y Hacienda: Madrid, Spain, 1986; pp. 4682–4684.
27. Kutyavin, I.V.; Afonina, I.A.; Mills, A.; Gorn, V.V.; Lukhtanov, E.A.; Belousov, E.S.; Singer, M.J.; Walburger, D.K.; Lokhov, S.G.; Gall, A.A.; et al. 3′-minor groove binder-DNA probes increase sequence specificity at PCR extension temperatures. *Nucleic Acids Res.* **2000**, *28*, 655–661. [CrossRef] [PubMed]
28. Luekasemsuk, T.; Panvisavas, N.; Chaturongakul, S. TaqMan qPCR for detection and quantification of mitochondrial DNA from toxic pufferfish species. *Toxicon* **2015**, *102*, 43–47. [CrossRef] [PubMed]
29. Fernandes, T.J.R.; Costa, J.; Oliveira, M.B.P.P.; Mafra, I. Exploiting 16S rRNA gene for the detection and quantification of fish as a potential allergenic food: A comparison of two real-time PCR approaches. *Food Chem.* **2018**, *245*, 1034–1041. [CrossRef]

foods

MDPI

Article

Swordfish or Shark Slice? A Rapid Response by COIBar–RFLP

Venera Ferrito, Alessandra Raffa, Luana Rossitto, Concetta Federico, Salvatore Saccone and Anna Maria Pappalardo *

Department of Biological, Geological and Environmental Sciences, Section of Animal Biology "M. La Greca", University of Catania, Via Androne 81, 95124 Catania, Italy; vferrito@unict.it (V.F.); alessandra.raffa92@gmail.com (A.R.); lunarossa92@gmail.com (L.R.); federico@unict.it (C.F.); saccosal@unict.it (S.S.)

* Correspondence: pappalam@unict.it; Tel.: +39-0957-306051

Received: 17 September 2019; Accepted: 28 October 2019; Published: 1 November 2019

Abstract: Market transparency is in strong demand by consumers, and the authentication of species is an important step for seafood traceability. In this study, a simple molecular strategy, COIBar–RFLP (cytochrome oxidase I barcode–restriction fragment length polymorphism), is proposed to unveil commercial fraud based on the practice of species substitution in the swordfish trade. In particular, COI barcoding allowed the identification of the species *Prionace glauca*, *Mustelus mustelus*, and *Oxynotus centrina* in slices labeled as *Xiphias gladius*. Furthermore, the enzymatic digestion of COI amplicons using the *Mbo*I restriction endonuclease allowed the simultaneous discrimination of the four species. Interestingly, an intraspecific differential *Mbo*I pattern was obtained for the swordfish samples. This pattern was useful to differentiate the two different clades revealed in this species by phylogenetic analyses using several molecular markers. These results indicate the need to strengthen regulations and define molecular tools for combating the occurrence of fraud along the seafood supply chain and show that COIBar–RFLP could become a standardized molecular tool to assess seafood authenticity.

Keywords: COIBar–RFLP (cytochrome oxidase I barcode–restriction fragment length polymorphism); seafood; fraud; DNA barcoding; food authenticity

1. Introduction

Swordfish fishery is one of the most important fishing activities in the Mediterranean Sea, in particular in South Italy. Quotas have been established to combat overfishing, and fisheries have been closed over several months to protect juveniles. According to recent data from the International Commission for the Conservation of Atlantic Tuna, Italy ranks the highest in terms of swordfish catches, which amount to 45% of the total allowable in the period 2003–2016 [1]. The highest demand for fish products in general, and swordfish in particular, occurs during summer, especially in restaurants [2] but also in local markets. As a result of the high demand, the price of these large pelagic fishes is on average higher than that of small fishes [3].

With the increase in demand and price, alimentary fraud potentially increases too. This can include food mislabeling, substitution, counterfeiting, misbranding, dilution, and adulteration [4]. The mislabeling of seafood can be harmful for health, in terms of economic loss, as well as for the loss of biodiversity it may cause in the case of illegal trade of threatened species. For these reasons, European regulations have focused on traceability and, in particular, on the mandatory declaration of the species present in a product on the product label [5–10].

Despite these adequate legislative tools, the number of cases of food fraud perpetrated in the fish trade in Europe and worldwide is increasing. The results of recent investigations on this phenomenon have shown that the percentage of mislabeling was around 30% of the total samples collected [11–15].

To address this problem, researchers have increasingly asserted the importance of using molecular tools based on DNA sequencing for detecting food fraud. The most common mitochondrial (mt) genes used for this purpose have been cytochrome b, 16S rRNA, and cytochrome oxidase I (COI). Other mtDNA targets, such as the mtDNA control region (CR) [16,17], which has been the most popular molecular marker used for genetic population structure studies [18–24], have seen limited use in fish and seafood species identification. In recent years, COI has been standardized as a barcode gene for species identification in several animal taxa [25–33] including fishes [34–40]. More specifically, the high number of COI-barcode fish sequences available in the large public gene sequence databases (BOLD and GenBank) [41,42] have made this gene the most highly used gene to clearly identify fish species and cases of mislabeling of seafood products [38,43–54]. However, in the context of seafood traceability, the main goal for the implementation of these analyses is to reduce the time it takes from sampling to obtaining gene sequencing results, as well as the costs of processing.

An already well-proven technique for the identification of species is polymerase chain reaction (PCR)–restriction fragment length polymorphism (RFLP), by which the PCR product of an amplified gene is cut with different restriction endonucleases to obtain a species-specific RFLP [55–57], useful for species authentication. In this regard, the combination of DNA barcoding of COI and the consolidated method of RFLP analysis (COIBar–RFLP, cytochrome oxidase I barcode–restriction fragment length polymorphism) has been successfully used to discriminate several fish species belonging to the Engraulidae, Merluccidae, Soleidae, and Acipenseridae families in processed seafood products [49,52–54,58]. It should be noted that the time and cost of execution of the COIBar–RFLP are lower than those of DNA sequencing (about 7 h and 10 euros per sample *vs* 24 h and 17 euros per sample, respectively).

Focusing on swordfish adulteration problems, the most commonly used species for fraudulent substitution are elasmobranches, including some species of shark. It should be noted that the market of shark meat is very wide for both fresh and frozen foods also in Italy, and cases of mislabeling have been frequently recorded for these products imported from all over the world [59–61].

Therefore, food fraud occurs due to an economic return when using shark meat. However, the substitution of a more valuable fish, such as swordfish, with shark meat leads to an even more serious fraud in economic terms. In the last decades, several studies have been carried out to detect the rate of mislabeling of different seafood products, and in some cases, shortfin mako (*Isurus oxyrinchus*) and blue shark (*Prionace glauca*) have been found to be sold as swordfish [36,37,62,63].

On the basis of the considerations above, the aim of this work is to extend the use of COIBar–RFLP to investigate the identity of swordfish products in the south of Italy and to discriminate swordfish (*Xiphias gladius*) from other fish species to detect fraudulent actions, such as species substitution, which represent the most common fraud in seafood. First, we sequenced the conventional COI barcode in a large number of samples collected in local fish markets and supermarkets, labeled as swordfish slices. Subsequently, the COIBar–RFLP procedure was applied on reference samples of the COI-barcoded species to obtain a species-specific restriction enzyme pattern. Finally, this pattern was used for swordfish slice authentication.

2. Materials and Methods

2.1. Sampling

Fresh and frozen slices of swordfish were acquired in 2010 and 2018 from local fish markets and supermarkets of south Italy for a total of 35 samples. Another 10 samples from the local harbor were collected and identified on the basis of morphological traits [64,65] and used to construct a reference COI-barcode library. The samples collected in 2010 had already been processed [37] and were used in this study only for the application of COIBar–RFLP. The remaining samples, preserved at room temperature in 1.5 mL labeled tubes filled with 95% ethanol, were processed for DNA barcoding and COIBar–RFLP (Table 1). DNA samples were deposited as vouchers at the Department of Biological, Geological, and Environmental Science, Section of Animal Biology, in Catania, Italy.

Table 1. Samples examined in this study.

Code	DNA Sample Voucher	Declared Species	GenBank Access N°	Species Matched by BLAST	Matched Accession From BLAST	% Identity With 100% Coverage
XIP1	DBGES18-125	*Xiphias gladius* *	JN083390	*Xiphias gladius*	JN083390	100
XIP2	DBGES18-129	*Xiphias gladius* *	MN447670	*Xiphias gladius*	JN049558	99.71
MUS1	DBGES18-133	*Mustelus mustelus* *	MN447688	*Mustelus mustelus*	JN641215	99.39
MUS2	DBGES18-134	*Mustelus mustelus* *	MN447689	*Mustelus mustelus*	JN641214	99.24
OXY1	DBGES18-111	*Oxynotus centrina* *	MN447691	*Oxynotus centrina*	JF834320	98.96
OXY2	DBGES18-130	*Oxynotus centrina* *	MN447692	*Oxynotus centrina*	JF834320	98.66
PRI1	DBGES18-121	*Prionace glauca* *	MN447694	*Prionace glauca*	KJ146044	99.85
PRI2	DBGES18-138	*Prionace glauca* *	MN447695	*Prionace glauca*	MH719984	99.83
SCY1	DBGES18-144	*Scyliorhinus canicula* *	MN457949	*Scyliorhinus canicula*	KJ205311	99.70
SCY2	DBGES18-146	*Scyliorhinus canicula* *	MN457950	*Scyliorhinus canicula*	KJ205311	99.85
X1	DBGES10-045	*Xiphias gladius*	JN083389	*Xiphias gladius*	JN083389	100
X2	DBGES10-048	*Xiphias gladius*	JN083387	*Xiphias gladius*	JN083387	100
X3	DBGES10-049	*Xiphias gladius*	JN049559	*Xiphias gladius*	JN049559	100
X4	DBGES10-054	*Xiphias gladius*	JN083397	*Xiphias gladius*	JN083397	100
X5	DBGES10-055	*Xiphias gladius*	JN083387	*Xiphias gladius*	JN083387	100
X6	DBGES10-058	*Xiphias gladius*	JN049559	*Xiphias gladius*	JN049559	100
X7	DBGES10-059	*Xiphias gladius*	JN049559	*Xiphias gladius*	JN049559	100
X8	DBGES10-060	*Xiphias gladius*	JN083387	*Xiphias gladius*	JN083387	100
X9	DBGES10-068	*Xiphias gladius*	JN083393	*Xiphias gladius*	JN083393	100
X10	DBGES10-074	*Xiphias gladius*	JN049559	*Xiphias gladius*	JN049559	100
X11	DBGES10-077	*Xiphias gladius*	JN083393	*Xiphias gladius*	JN083393	100
X12	DBGES10-080	*Xiphias gladius*	JN083387	*Xiphias gladius*	JN083387	100
X13	DBGES10-083	*Xiphias gladius*	MN447696	*Prionace glauca*	MH719984	99.83
X14	DBGES10-089	*Xiphias gladius*	JN083386	*Xiphias gladius*	JN083386	100
X15	DBGES10-095	*Xiphias gladius*	JN083387	*Xiphias gladius*	JN083387	100
Y1	DBGES18-132	*Xiphias gladius*	MN447671	*Xiphias gladius*	JN049558	99.71
Y2	DBGES18-135	*Xiphias gladius*	MN447672	*Xiphias gladius*	JN083390	99.71
Y3	DBGES18-136	*Xiphias gladius*	MN447673	*Xiphias gladius*	JN083387	99.71
Y4	DBGES18-137	*Xiphias gladius*	MN447674	*Xiphias gladius*	JN049558	99.71
Y5	DBGES18-139	*Xiphias gladius*	MN447675	*Xiphias gladius*	JN083387	99.85

Table 1. *Cont.*

Code	DNA Sample Voucher	Declared Species	GenBank Access N°	Species Matched by BLAST	Matched Accession From BLAST	% Identity With 100% Coverage
Y6	DBGES18-141	*Xiphias gladius*	MN447676	*Xiphias gladius*	JN083387	99.71
Y7	DBGES18-143	*Xiphias gladius*	MN447690	*Mustelus mustelus*	JN641214	99.39
Y8	DBGES18-144	*Xiphias gladius*	MN447677	*Xiphias gladius*	JN049558	99.85
Y9	DBGES18-151	*Xiphias gladius*	MN447678	*Xiphias gladius*	JN083387	99.71
Y10	DBGES18-152	*Xiphias gladius*	MN447693	*Oxynotus centrina*	JF834320	98.07
Y11	DBGES18-153	*Xiphias gladius*	MN447679	*Xiphias gladius*	JN083387	99.12
Y12	DBGES18-154	*Xiphias gladius*	MN447680	*Xiphias gladius*	JN049558	99.85
Y13	DBGES18-161	*Xiphias gladius*	MN447681	*Xiphias gladius*	JN083390	99.41
Y14	DBGES18-162	*Xiphias gladius*	MN447682	*Xiphias gladius*	JN083387	98.97
Y15	DBGES18-172	*Xiphias gladius*	MN447697	*Prionace glauca*	MH194484	99.39
Y16	DBGES18-173	*Xiphias gladius*	MN447683	*Xiphias gladius*	JN049558	99.27
Y17	DBGES18-174	*Xiphias gladius*	MN447684	*Xiphias gladius*	JN083387	99.85
Y18	DBGES18-175	*Xiphias gladius*	MN447685	*Xiphias gladius*	JN083390	99.27
Y19	DBGES18-182	*Xiphias gladius*	MN447686	*Xiphias gladius*	JN049558	99.71
Y20	DBGES18-187	*Xiphias gladius*	MN447687	*Xiphias gladius*	JN083387	99.56

DBGES (Department of Biological, Geological and Environmental Sciences). BLAST (Basic Local Alignment Search Tool). * = species of the samples identified by using morphological features.

2.2. DNA Barcoding

Genomic DNA was extracted from 25 mg of tissue using a commercial kit based on silica purification (DNeasy tissue kit, Qiagen, Hilden, Germany) following the manufacturer's guidelines. All samples were analyzed by amplifying a portion of about 650 bases of the COI gene in a 20 μL reaction mixture also containing the M13 tailed primers (VF2_t1 and FishR2_t1) described in Ivanova et al. [66] to improve the sequencing quality of the PCR products and following the PCR conditions reported by Pappalardo et al. [54]. All PCR products were checked by 0.8% agarose gel electrophoresis, visualized with SYBR® Safe (Thermo Fisher, Waltham, MA USA), displayed through a Safe Imager TM 2.0 Blue Light Transilluminator (Thermo Fisher, Waltham, MA USA), and then purified with the QIAquick PCR purification kit (Qiagen, Hilden, Germany). Sanger sequencing, using M13 primers, was subsequently conducted by Genechron in both forward and reverse directions to generate the DNA barcodes [67].

The sequence chromatograms were checked visually and assembled. Multiple-sequence alignment was carried out by the online version of MAFFT v.7 [68]. Ambiguous sequences were trimmed, and primer sequences were cut. The sequences were carefully checked for the presence of nuclear mitochondrial pseudogenes or NUMTs (nuclear mitochondrial DNA sequences), which could be easily coamplified with orthologous mtDNA sequences [69]. The EMBOSS Transeq tool [70] was used to translate the nucleotide sequences to amino acids to check for premature stop codons and to verify that the open reading frames were maintained in the protein-coding locus. To confirm the identity of the amplified sequences, we conducted BLAST (Basic Local Alignment Search) searches in GenBank with default parameters [71]. All sequences obtained from the present study were published in the National Center for Biotechnology Information database (NCBI), and their GenBank accession numbers are reported in Table 1. After the BLAST search, six shark species sequences (HM909857, JF493927, KF899461, KI709900, JF493694, JN641217) downloaded from GenBank were added to our dataset to construct a phylogenetic tree. We used jModelTest v 2.1.10 [72] to select the best-fitting substitution model for our sequences according to the corrected Akaike information criterion. A maximum likelihood (ML) tree by using a GTR + I + G model was implemented in MEGA v 6.0 (Biodesign Institute, Arizona, MA, USA) [73]. The evaluation of the statistical confidence of nodes was based on 1000 non-parametric bootstrap replicates [74].

2.3. COIBar–RFLP

The selection of the most suitable restriction enzymes to discriminate swordfish from other shark species (*Mustelus mustelus*, L., 1758, *Oxinotus centrina* (L., 1758), *P. glauca* (L., 1758), *Scyliorhinus canicula* L., 1758) was performed through "Remap" [75]. The in silico analysis was preliminarily carried out using a total of 10 COI barcode sequences (of about 650 bases) of the examined species, downloaded from public databases (GenBank and BOLD) [41,42]. Five different restriction enzymes were tested to scan all validated sequences and to detect the expected size of the digested products: *Hpa*II (C*CGG), *Hinf*I (G*ANTC), *Mbo*I (*GATC), *Rsa*I (GT*AC), and *Hind*III (A*AGCTT). Finally, a total of 49 COI sequences were analyzed by Remap to test for evidence of intraspecific variation at the recognition site of the restriction endonuclease suitable for simultaneous discrimination of the examined species (Figure 1, Table 2).

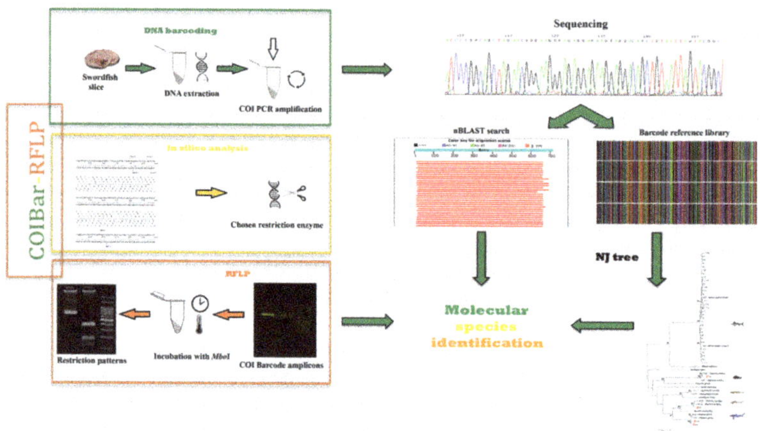

Figure 1. Flow chart of COIBar–RFLP (cytochrome oxidase I barcode–restriction fragment length polymorphism) or species discrimination. DNA barcoding steps: DNA isolation from swordfish slices and barcode region PCR amplification. In silico analysis steps: search for an appropriate restriction enzyme. RFLP steps: incubation of barcode amplicons with *Mbo*I to obtain the COIBar–RFLP pattern. ML, maximum likelihood; nBLAST, nucleotide Basic Local Alignment Search Tool.

Table 2. In silico analysis of swordfish and shark COI sequences scanned by Remap and using MboI as restriction endonuclease.

Species	Sequence Number	Genbank Accession Number	Sequence Size (bases)	Restriction Fragment Size (base pair)
Xiphias gladius	10	JN083387	682	≈ 145 - 270 - 220
		JN083389	682	≈ 145 - 270 - 220
		JN049558	682	≈ 145 - 270 - 220
		JF952886	652	≈ 145 - 265 - 220
		HQ024928	652	≈ 145 - 265 - 220
		HQ024927	652	≈ 145 - 265 - 220
		KR086931	652	≈ 145 - 265 - 220
		GU324195	652	≈ 145 - 265 - 220
		DQ107625	655	≈ 145 - 265 - 200
		DQ107623	655	≈ 145 - 265 - 220
Mustelus mustelus	10	JN641215	676	≈ 80 - 390 - 171
		JN641214	679	≈ 80 - 390 - 169
		JN641213	672	≈ 80 - 390 - 167
		JN641212	666	≈ 80 - 390 - 156
		JN641211	681	≈ 80 - 390 - 170
		KJ768265	652	≈ 80 - 390 - 142
		KJ768266	652	≈ 80 - 390 - 142
		JN641208	656	≈ 75 - 390 - 156
		JN641209	669	≈ 80 - 390 - 160
		JN641210	664	≈ 80 - 390 - 160

Table 2. *Cont.*

Species	Sequence Number	Genbank Accession Number	Sequence Size (bases)	Restriction Fragment Size (base pair)
Oxynotus centrina	9	KT307360	648	≈ 510 - 95
		KT307361	620	≈ 480 - 95
		KT307362	648	≈ 510 - 95
		KT307363	648	≈ 510 - 95
		KT307364	648	≈ 510 - 95
		JF834320	672	≈ 505 - 100
		KY176547	642	≈ 495 - 105
		GU805137	637	≈ 500 - 95
		GU805138	648	≈ 510 - 95
Prionace glauca	10	JN312505	652	≈ 85 - 405 - 70
		JN312504	652	≈ 85 - 405 - 70
		JN312503	652	≈ 85 - 405 - 70
		KP193446	652	≈ 85 - 405 - 70
		KP193455	652	≈ 85 - 405 - 70
		KP193350	652	≈ 85 - 405 - 70
		KP193339	652	≈ 85 - 405 - 70
		KP193159	652	≈ 85 - 405 - 70
		KC015834	652	≈ 85 - 405 - 70
		KC015833	652	≈ 85 - 405 - 70
Scyliorhinus canicula	10	JN641243	675	≈ 85 - 400 - 100
		JN641242	676	≈ 85 - 400 - 100
		JN641241	652	≈ 70 - 400 - 100
		JN641240	680	≈ 85 - 400 - 100
		JN641239	675	≈ 85 - 400 - 100
		JN641238	671	≈ 85 - 400 - 100
		JN641237	680	≈ 85 - 400 - 100
		JN641236	674	≈ 85 - 400 - 100
		JN641234	672	≈ 85 - 400 - 100
		JN641233	680	≈ 85 - 400 - 100

Afterwards, the COI-barcode PCR products obtained from *X. gladius* and shark samples were digested with the selected restriction enzymes. For each endonuclease, a 15 μL reaction volume containing 13 μL of unpurified PCR product, 1 μL of digestion buffer (1X), and 1 μL of each endonuclease (10 U each) was prepared. The reaction mixtures were incubated at an optimum temperature of 37 °C for 1 h. The digested amplicons were then separated on a 3% agarose gel using Trackit TM 100 bp DNA ladder (Invitrogen) as a size standard. The restriction pattern obtained from the validated samples was exploited to unequivocally identify the unknown commercial slices.

3. Results

3.1. DNA Barcoding

The length range of the obtained COI sequences was between 669 bases and 681 bases. Each of them was a functional mitochondrial sequence without stop codons. NUMTs generally smaller than 600 bases were not sequenced [71]. Five species were identified in all examined samples: *X. gladius* (Xiphiidae), *P. glauca* (Charcarinidae), *M. mustelus* (Triakidae), *S. canicula* (Scyliorinidae), and *O. centrina* (Oxynotidae). The sequences obtained from morphologically validated species were compared with the sequences retrieved from GenBank through a BLAST search. The identity percentage between the COI query sequences and their top-match sequences ranged from 98.07% to 100% (Table 1). The ML tree (Figure 2) showed the relationship between the sequences of several unidentified samples and the reference barcode sequences. High bootstrap values (>60%) supported the nodes connecting the sequences of the same species in the tree. The samples of *X. gladius* clustered into two main clades (named clade I and II), as already found by Pappalardo et al. [36,37]. Only one case of mislabeling

(1 out of 15) was found in the samples examined in 2010 (6.7%), while 15% (3 out of 20) of mislabeling was found in the samples collected during 2018 (Table 1). Swordfish was substituted with *P. glauca* (2 products), *M. mustelus* (1 product), and *O. centrina* (1 product).

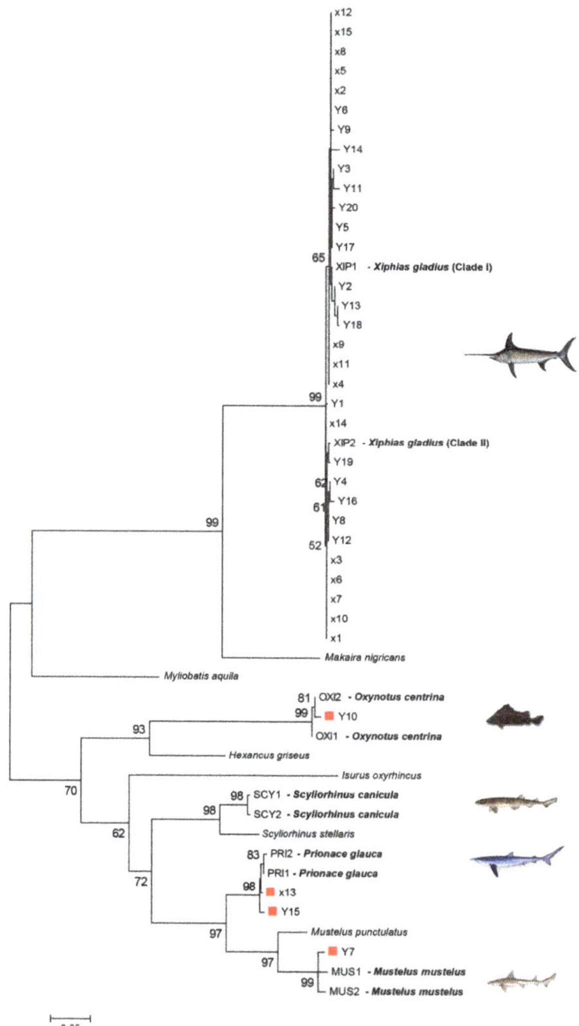

Figure 2. Maximum likelihood (ML) tree showing the relationships of unknown samples sequences (X and Y) to validated reference barcode sequences. The numbers above the nodes represent bootstrap analyses after 1000 replicates. Bootstrap values greater than 60% are shown. The red square indicates swordfish mislabeled samples. Scale bar refers to a distance of 0.05 nucleotide substitutions per site.

3.2. COIBar–RFLP

The preliminary in silico analysis using "Remap" showed that the *Mbo*I enzyme produced a species-specific pattern useful to discriminate simultaneously all examined species. No intraspecific variation of the *Mbo*I recognition sites was detected for any species tested by "Remap", with the exception of the *X. gladius* digestion pattern (Table 2). Figure 3 highlights both the size of the undigested COI amplicon, of about 750 bp, and the *Mbo*I differential restriction pattern obtained for each species:

one fragment of 510 bp was obtained for *O. centrina*; two fragments of 110 and 400 bp and of 150 and 400 bp were obtained, respectively, for *P. glauca* and *S. canicula*; finally, three fragments of 120, 180, and 390 bp were obtained for *M. mustelus*. The negative control is not shown in the figure. The enzymatic digestion of *X. gladius* amplicons produced two different patterns (Figure 4) corresponding to clades I and II, already described in this species. In particular, three fragments of 170, 220, and 240 bp were detected for clade I and three fragments of 170, 220, and 280 bp were found for clade II. On the basis of this intraspecific pattern, the swordfish sample shown in Figure 3 belongs to clade I.

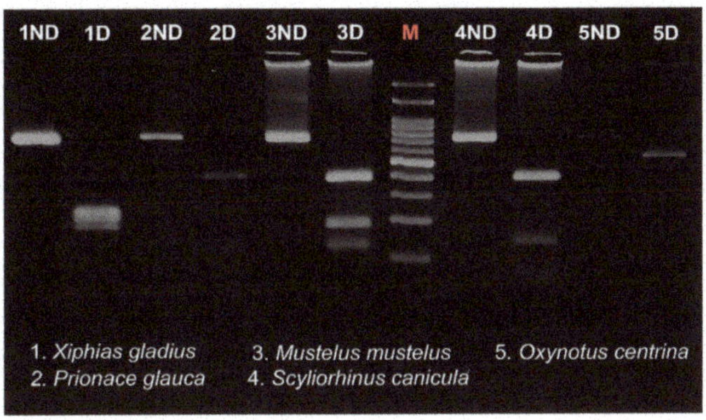

Figure 3. Example of COIBar–RFLP identification of swordfish and shark species on a 3% agarose gel by restriction by *Mbo*I of the cytochrome oxidase I amplicons. Bands smaller than 100 bp were not considered. The 5ND and 5D bands differ in intensity because they were obtained from two different PCR amplifications. ND = not digested, D = digested. M = molecular weight marker (100 bp DNA ladder, biotechrabbit GmbH, Berlin, Germany).

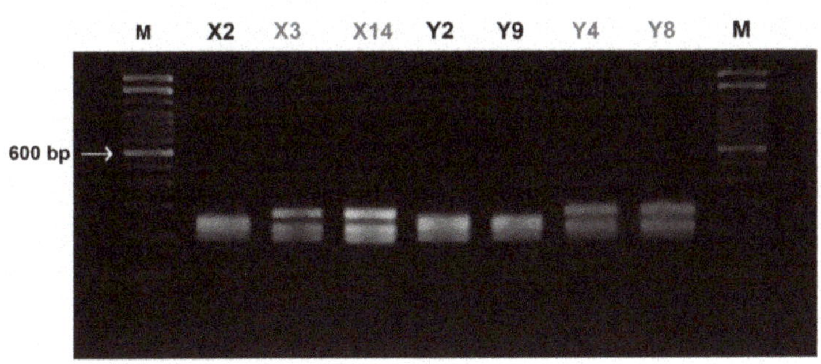

Figure 4. COIBar–RFLP patterns obtained from *Mbo*I digestion of *X. gladius* samples. Clade I (black code), Clade II (grey code). M = molecular weight marker (Trackit ™ 100 bp DNA ladder, Invitrogen). For X and Y codes see Table 1.

4. Discussion

The results obtained in this study once again confirm the efficacy of COIBar–RFLP in discriminating fish species in commercial products and also highlight the fraudulent practice of species substitutions in seafood products, consisting in the use of less valuable shark species in place of swordfish. The *Mbo*I endonuclease restriction enzyme produced species-specific restriction patterns of the COI amplicons useful to differentiate *X. gladius* from shark species. Another interesting result proving the sensitivity of this methodology is the intraspecific differential *Mbo*I pattern obtained for the swordfish samples. This pattern was useful to discriminate the two different clades revealed in this species by phylogenetic analyses using several molecular markers [36,37,76–78]. COI DNA-barcoding showed that 15% of the swordfish samples purchased in local fish markets during 2018 was mislabeled, with an evident economic loss for the consumers. This percentage was at least two times higher than that recorded in 2010, demonstrating that despite the current European legislation focused on consumer protection against fraud, fraud remains frequent and widespread. In this context, there is no doubt that molecular tools are very useful and effective to fight commercial fraud and that DNA-based methods have become increasingly important for seafood authentication. However, while the practice of commercial fraud in the seafood market is a global concern, to date there is no standardized global methodology to expose this practice. Firstly, all states have not yet incorporated into their legislation the use of molecular methods to combat commercial fraud; this is true for Italy, for example. Secondly, significant differences among countries have been found in methods used by accredited laboratories for food authenticity [79]. Thirdly, together with the classic methods (protein- and DNA sequence-based methods), new and sophisticated methods are being developed to identify seafood species [80]. It is evident that the first two issues can be solved only by adopting a common global policy to fight food fraud. The European legislation, for example, could require, rather than only suggest, the application of DNA analysis in the context of seafood traceability [81], also indicating the most useful methodology to be used across European laboratories. In this regard, the features that molecular methods should have for a rapid authentication of species in seafood products can be debated. To be effective for routine activities carried out by local food safety and quality authorities, from the traceability of the catch to the labeling of the products, effectiveness in terms of cost and time-saving and correctness of species identification should be a priority. Among the classic methods, the protein-based methods, such as isoelectric focusing of sarcoplasmic proteins, are still used as official methods for fish species identification [82], but the DNA-sequencing methods, and the DNA-barcoding methodology in particular, have become more common in laboratories specialized in food authentication ([3] and literature therein). Increasingly, new methodologies are emerging for species identification, such as qPCR, DNA microarrays, high-resolution melting analysis, mass spectrometry, high-throughput sequencing, and the recently developed handheld testing devices [80], all of them suitable and effective in terms of cost and time consumption.

However, these new methodologies require, in some cases, extensive technical equipment and specific skills by the operators and need to be standardized for use as official methods. Furthermore, the application of these methods is limited to a few cases of species authentication, while wide databases of reference samples are needed for their validation as official methods. The methodological approach we propose, COIBar–RFLP, although it cannot substitute DNA sequencing in general, takes advantage of large databases of reference DNA sequences of fish species and of the positive results from several study cases for species of relevant commercial interest under various food matrices [49,52–54,58]. COIBar–RFLP successfully and simultaneously discriminated the fish species analyzed in these studies, through the banding pattern obtained after digestion with only one endonuclease restriction enzyme. This simple, robust, easy-to-perform, and cost-effective strategy can potentially cover a wide range of species and provide a versatile tool to monitor the mislabeling of fish products. However, it should be noted that poor enzyme storage, as well as the processing conditions, could compromise the advantages of the methodology in terms of expected time of processing and misleading results. In a recent investigation on the methodological approach performed in 45 European laboratories,

Griffiths et al. [79] revealed that PCR–RFLP was used in 40% of the laboratories involved in seafood authentication; this result suggests that this method could become a standardized molecular tool to assess seafood authenticity.

5. Conclusions

The efficacy of COIBar–RFLP was tested for species authentication on slices labeled as swordfish. The illegal practice of species substitution was observed, with the species *P. glauca*, *M. mustelus*, and *O. centrina* being sold in place of swordfish. These results indicate the need to strengthen regulations and to define molecular tools to fight the occurrence of fraud along the seafood supply chain, from the traceability of the catch to the labeling of the products, and to achieve market transparency, which is highly demanded by the consumers. Finally, the future perspectives of COIBar–RFLP rest on the need to build a database of COI restriction patterns to be used for unequivocal species identifications.

Author Contributions: A.M.P. and V.F. conceived and designed the experiments; A.M.P., A.R., and L.R. performed the experiments; A.M.P., C.F., and S.S. analyzed the data; A.M.P. and V.F. wrote the paper.

Funding: This work was supported by the Annual Research Plan 2016-2018 of the Department of Biological, Geological and Environmental Sciences, University of Catania (Grants #22722132134).

Acknowledgments: We thanks the University of Catania for the economic support.

Conflicts of Interest: The authors declare no conflict of interest.

References

1. SCRS Report. *Report of the Standing Committee on Research and Statistics (SCRS) International Commission for the Conservation of Atlantic Tunas-Biennial Period 2016–2017*; SCRS: Madrid, Spain, 2017.
2. ISMEA Osservatorio consumi extra-domestici. Analisi & Ricerche n. 3 2005. Available online: http://www.ismea.it (accessed on 25 June 2019).
3. Cosmina, M.; Demartini, E.; Gaviglio, A.; Mauracher, C.; Prestamburgo, S.; Trevisan, G. Italian Consumers' Attitudes Towards Small Palegic Fish. *New Medit.* **2012**, *1*, 52–57.
4. Reilly, A. *Overview of Food Fraud in the Fisheries Sector*; FAO In Fisheries and Aquaculture Circular; FIAM: Rome, Italy, 2018.
5. European Union. Regulation (EC) No. 178/2002 of the European parliament and of the Council of 28 January 2002 laying down the general principles and requirements of food law, establishing the European food safety authority and laying down procedures in matters of food safety. *Off. J. Eur. Union* **2002**, *31*, 1–24.
6. European Union. Regulation (EC) No. 1169/2011 of the European parliament and of the council of 25 October 2011 on the provision of food information to consumers. *Off. J. Eur. Union* **2011**, *304*, 18–63.
7. European Union. Regulation (EC) No. 1379/2013 of the European parliament and of the council of 11 December 2013 on the common organisation of the markets in fishery and aquaculture products. *Off. J. Eur. Union* **2013**, *354*, 1–21.
8. Altalex IT. Art 440. Codice Penale Italiano, Libro II Titolo VI Capo II. 2017. Available online: http://www.altalex.com/documents/news/2014/06/03/dei-delitti-contro-l-incolumita-pubblica (accessed on 10 April 2019).
9. Decreto Legislativo 27 Gennaio 1992, n. 109 Attuazione delle Direttive n. 89/395/CEE e n. 89/396/CEE Concernenti L'etichettatura, la Presentazione e la Pubblicità Dei prodotti Alimentari. Available online: http://www.comune.jesi.an.it/MV/leggi/dlvo109-92.htm (accessed on 27 January 1992).
10. Decreto Legislativo 6 settembre 2005, n. 206-codice del consumo, a norma dell'art. 7 della legge 29 luglio 2003, n. 229. Available online: https://www.camera.it/parlam/leggi/deleghe/testi/05206dl.htm (accessed on 6 September 2005).
11. Benard-Capelle, J.; Guillonneau, V.; Nouvian, C.; Fournier, N.; Le Loët, K.; Dettai, A. Fish mislabeling in France: Substitution rates and retail types. *PeerJ* **2015**, *2*, e714. [CrossRef]
12. Christiansen, H.; Dettai, A.; Heindler, F.M.; Collins, M.A.; Duhamel, G.; Hautecoeur, M.; Steinke, D.; Volckaert, A.M.; Van de Putte, A.P. Diversity of mesopelagic fishes in the southern ocean—A phylogeographic perspective using DNA barcoding. *Front. Ecol. Evol.* **2018**, *6*, 120. [CrossRef]

13. Do, T.D.; Choi, T.J.; Kim, J.; An, H.E.; Park, Y.J.; Karagozlu, M.Z.; Kim, C.B. Assessment of marine fish mislabelling in South Korea's markets by DNA barcoding. *Food Control* **2019**, *100*, 53–57. [CrossRef]
14. Garcia-Vazquez, E.; Perez, J.; Martinez, J.L.; Pardiñas, A.F.; Lopez, B.; Karaiskou, N.; Triantafyllidis, A. High level of mislabelling in Spanish and Greek hake markets suggests the fraudulent introduction of African species. *J. Agric. Food Chem.* **2011**, *59*, 475–480. [CrossRef]
15. Von Der Heyden, S.; Barendse, J.; Seebregts, A.J.; Matthee, C.A. Misleading the masses: Detection of mislabelled and substituted frozen fish products in south Africa. *ICES J. Mar. Sci.* **2010**, *67*, 176–185. [CrossRef]
16. Quinteiro, J.; Vidal, R.; Izquierdo, M.; Sotelo, C.G.; Chapela, M.J.; Pérez-Martín, R.I.; Rehbein, H.; Hold, G.L.; Russell, V.J.; Pryde, S.E.; et al. Identification of hake species (*Merluccius* genus) using sequencing and PCR-RFLP analysis of mitochondrial DNA control region sequences. *J. Agric. Food Chem.* **2001**, *49*, 5108–5114. [CrossRef]
17. Kumar, G.; Kocour, M.; Kunal, S.P. Mitochondrial DNA variation and phylogenetic relationships among five tune species based on sequencing of D-loop region. *Mitochondr. DNA Part A* **2016**, *27*, 1976–1980.
18. Rocco, L.; Ferrito, V.; Costagliola, D.; Marsilio, A.; Pappalardo, A.M.; Stingo, V.; Tigano, C. Genetic divergence among and within four Italian populations of *Aphanius fasciatus* (Teleostei, Cyprinodontiformes). *Ital. J. Zool.* **2007**, *74*, 371–379. [CrossRef]
19. Pappalardo, A.M.; Ferrito, V.; Messina, A.; Patarnello, T.; De Pinto, V.; Guarino, F.; Tigano, C. Genetic structure of the killifish *Aphanius fasciatus* Nardo 1827 (Teleostei, Cyprinodontidae), results of mitochondrial DNA analysis. *J. Fish Biol.* **2008**, *72*, 1154–1173. [CrossRef]
20. Ferrito, V.; Pappalardo, A.M.; Canapa, A.; Barucca, M.; Doadrio, I.; Olmo, E.; Tigano, C. Mitochondrial phylogeography of the killifish *Aphanius fasciatus* (Teleostei, Cyprinodontidae) reveals highly divergent Mediterranean populations. *Mar. Biol.* **2013**, *160*, 3193–3208. [CrossRef]
21. Cuttitta, A.; Patti, B.; Maggio, T.; Quinci, E.M.; Pappalardo, A.M.; Ferrito, V.; De Pinto, V.; Torri, M.; Falco, F.; Nicosia, A.; et al. Larval population structure of *Engraulis encrasicolus* in the strait of Sicily as revealed by morphometric and genetic analyses. *Fish Ocean* **2015**, *24*, 135–149. [CrossRef]
22. Pedrosa-Gerasmio, I.R.; Agmata, A.B.; Santos, M.D. Genetic diversity, population genetic structure, and demographic history of *Auxis thazard* (Perciformes), *Selar crumenophthalmus* (Perciformes), *Rastrelliger kanagurta* (Perciformes) and *Sardinella lemuru* (Clupeiformes) in Sulu-Celebes sea inferred by mitochondrial DNA sequences. *Fish Res.* **2015**, *162*, 64–74.
23. Machado, V.N.; Willis, S.C.; Teixeira, A.S.; Hrbek, T.; Farias, I.P. Population genetic structure of the Amazonian black flannelmouth characin (Characiformes, Prochilodontidae: *Prochilodus nigricans* Spix & Agassiz, 1829): Contemporary and historical gene flow of a migratory and abundant fishery species. *Environ. Biol. Fishes* **2017**, *100*, 1–16.
24. Duong, T.; Uy, S.; Chheng, P.; So, N.; Thi Tran, T.; Nguyen, N.T.; Pomeroy, R.; Egna, H. Genetic diversity and structure of striped snakehead (*Channa striata*) in the lower Mekong basin: Implications for aquaculture and fisheries management. *Fish Res.* **2019**, *218*, 166–173. [CrossRef]
25. Hebert, P.D.N.; Cywinska, A.; Ball, S.; deWaard, J.R. Biological identifications through DNA barcodes. *Proc. R. Soc. Lond. B* **2003**, *270*, 313–321. [CrossRef]
26. Hebert, P.D.N.; Ratnasingham, S.; deWaard, J.R. Barcoding animal life: Cytochrome c oxidase subunit 1 divergence, among closely related species. *Proc. R. Soc. Lond. B* **2003**, *270*, S96–S99. [CrossRef]
27. Hebert, P.D.N.; Stoeckle, M.Y.; Zemlak, T.S.; Francis, C.M. Identification of birds through DNA barcodes. *PLoS Biol.* **2004**, *2*, e312. [CrossRef] [PubMed]
28. Hajibabaei, M.; Smith, M.A.; Janzen, D.H.; Rodriguez, J.J.; Whitfield, J.B.; Hebert, P.D.N. A minimalist barcode can identify a specimen whose DNA is degraded. *Mol. Ecol. Notes* **2006**, *6*, 959–964. [CrossRef]
29. Hajibabaei, M.; Singer, G.A.C.; Hebert, P.D.N.; Hickey, D.A. DNA barcoding: How it complements taxonomy, molecular phylogenetics and population genetics. *Trends Genet.* **2007**, *23*, 167–172. [CrossRef] [PubMed]
30. Ward, R.D.; Holmes, B.H.; White, W.T.; Last, P.R. DNA barcoding Australasian chondrichthyans: Results and potential uses in conservation. *Mar. Fresh Res.* **2008**, *59*, 57–71. [CrossRef]
31. Vitale, D.G.M.; Viscuso, R.; D'Urso, V.; Gibilras, S.; Sardella, A.; Marletta, A.; Pappalardo, A.M. Morphostructural analysis of the male reproductive system and DNA barcoding in *Balclutha brevis* Lindberg 1954 (Homoptera, Cicadellidae). *Micron* **2015**, *79*, 36–45. [CrossRef] [PubMed]

32. Federico, C.; Lombardo, D.; La Porta, N.; Pappalardo, A.M.; Ferrito, V.; Lombardo, F.; Saccone, S. Rapid molecular identification of necrophagous Diptera by means of variable-length of intron sequences in the wingless gene. *J. Forensic Legal Med.* **2018**, *56*, 66–72. [CrossRef] [PubMed]
33. Conti, E.; Mulder, C.; Pappalardo, A.M.; Ferrito, V.; Costa, G. How soil granulometry, temperature and water predict genetic differentiation in namibian *Ariadna* spiders and explain their behavior. *Ecol. Evol.* **2019**, *9*, 4382–4391. [CrossRef]
34. Ward, R.D.; Zemlak, T.S.; Innes, B.H.; Last, P.R.; Hebert, P.D.N. DNA barcoding Australia's fish species. *Philos. Trans. R. Soc. Lond. B* **2005**, *360*, 1847–1857. [CrossRef]
35. Steinke, D.; Zemlak, T.S.; Hebert, P.D.N. Barcoding Nemo: DNA-based identifications for the ornamental fish trade. *PLoS ONE* **2009**, *4*, e6300. [CrossRef]
36. Pappalardo, A.M.; Guarino, F.; Reina, S.; Messina, A.; De Pinto, V. Swordfish COI-DNA barcoding as a suitable tool for species and stock identification. *FEBS J.* **2011**, *278*, 440.
37. Pappalardo, A.M.; Guarino, F.; Reina, S.; Messina, A.; De Pinto, V. Geographically widespread swordfish barcode stock identification: A case study of its application. *PLoS ONE* **2011**, *6*, e25516. [CrossRef] [PubMed]
38. Pappalardo, A.M.; Ferrito, V. DNA barcoding species identification unveils mislabeling of processed flatfish products in southern Italy markets. *Fish Res.* **2015**, *164*, 153–158. [CrossRef]
39. Pappalardo, A.M.; Cuttitta, A.; Sardella, A.; Musco, M.; Maggio, T.; Patti, B.; Mazzola, S.; Ferrito, V. DNA barcoding and COI sequence variation in Mediterranean lanternfishes larvae. *Hydrobiologia* **2015**, *745*, 155–167. [CrossRef]
40. Pappalardo, A.M.; Federico, C.; Sabella, G.; Saccone, S.; Ferrito, V. A COI nonsynonymous mutation as diagnostic tool for intraspecific discrimination in the European anchovy *Engraulis encrasicolus* (Linnaeus). *PLoS ONE* **2015**, *10*, e0143297. [CrossRef] [PubMed]
41. Benson, D.A.; Cavanaugh, M.; Clark, K.; Karsch-Mizrachi, I.; Ostell, J.; Pruitt, K.D.; Sayers, E.W. GenBank. *Nucleic Acids Res.* **2018**, *46*, D41–D47. [CrossRef] [PubMed]
42. Ratnasingham, S.; Hebert, P.D.N. BOLD: The barcode of life data system. *Mol. Ecol. Notes* **2007**, *7*, 355–364. [CrossRef] [PubMed]
43. Dawnay, N.; Ogden, R.; McEwing, R.; Carvalho, G.R.; Thorpe, R.S. Validation of the barcoding gene COI for use in forensic genetic species identification. *Forensic Sci. Int.* **2007**, *173*, 1–6. [CrossRef]
44. Espiñeira, M.; González-Lavín, N.; Vieites, J.M.; Santaclara, F.J. Development of a method for the genetic identification of flatfish species on the basis of mitochondrial DNA sequences. *J. Agric. Food Chem.* **2008**, *56*, 8954–8961. [CrossRef]
45. Cawthorn, D.M.; Steinman, H.A.; Corli Witthuhn, R. Establishment of a mitochondrial DNA sequence database for the identification of fish species commercially available in south Africa. *Mol. Ecol. Res.* **2011**, *11*, 979–991. [CrossRef]
46. Armani, A.; Castigliego, L.; Guidi, A. Fish frauds: The DNA challenge. CAB reviews 071. *Anim. Sci. Rev.* **2012**, *7*, 227–238.
47. Armani, A.; Guardone, L.; La Castellana, R.; Gianfaldoni, D.; Guidi, A.; Castigliego, L. DNA barcoding reveals commercial and health issues in ethnic seafood sold on the Italian market. *Food Control* **2015**, *55*, 206–214. [CrossRef]
48. Khaksar, R.; Carlson, T.; Schaffner, D.W.; Ghorashi, M.; Best, D.; Jandhyala, S.; Traverso, J.; Amini, S. Unmasking seafood mislabeling in U.S. markets: DNA barcoding as a unique technology for food authentication and quality control. *Food Control* **2015**, *56*, 71–76. [CrossRef]
49. Ferrito, V.; Bertolino, V.; Pappalardo, A.M. White fish authentication by COIBar-RFLP: Toward a common strategy for the rapid identification of species in convenience seafood. *Food Control* **2016**, *70*, 130–137. [CrossRef]
50. Nagalakshmi, K.; Annam, P.K.; Venkateshwarlu, G.; Pathakota, G.B.; Lakra, W.S. Mislabelling in Indian seafood: An investigation using DNA barcoding. *Food Control* **2016**, *59*, 196–200. [CrossRef]
51. Ferrito, V.; Pappalardo, A.M. Seafood species identification by DNA barcoding, a molecular tool for food traceability. *Biodivers. J.* **2017**, *8*, 65–72.
52. Pappalardo, A.M.; Copat, C.; Ferrito, V.; Grasso, A.; Ferrante, M. Heavy metal content and molecular species identification in canned tuna: Insights into human food safety. *Mol. Med. Rep.* **2017**, *15*, 3430–3437. [CrossRef]

53. Pappalardo, A.M.; Federico, C.; Saccone, S.; Ferrito, V. Differential flatfish species detection by COIBar-RFLP in processed seafood products. *Eur. Food Res. Technol.* **2018**, *244*, 2191–2201. [CrossRef]
54. Pappalardo, A.M.; Petraccioli, A.; Capriglione, T.; Ferrito, V. From fish eggs to fish name: Caviar species discrimination by COIBar-RFLP, an efficient molecular approach to detect fraud in the caviar trade. *Molecules* **2019**, *24*, 2468. [CrossRef]
55. Cocolin, L.; D'Agaro, E.; Manzano, M.; Lanari, D.; Comi, G. Rapid PCR-RFLP method for the identification of marine fish fillets (seabass, seabream, umbrine, and dentex). *J. Food Sci.* **2000**, *65*, 1315–1317. [CrossRef]
56. Hold, G.L.; Russell, V.J.; Pryde, S.E.; Rehbein, H.; Quinteiro, J.; Rey-Mendez, M.; Sotelo, C.G.; Pérez-Martin, R.I.; Santos, A.T.; Rosa, C. Validation of a PCR-RFLP based method for the identification of salmon species in food products. *Eur. Food Res. Technol.* **2001**, *212*, 385–389. [CrossRef]
57. Hsieh, H.S.; Chai, T.; Hwang, D.F. Using the PCR-RFLP method to identify the species of different processed products of billfish meats. *Food Control* **2007**, *18*, 369–374. [CrossRef]
58. Pappalardo, A.M.; Ferrito, V. A COIBar-RFLP strategy for the rapid detection of *Engraulis encrasicolus* in processed anchovy products. *Food Control* **2015**, *57*, 385–392. [CrossRef]
59. Dent, F.; Clarke, S. *State of the Global Market for Shark Products*; Food and Agriculture Organization FAO: Rome, Italy, 2015.
60. Blanco, M.; Pérez-Martín, R.I.; Sotelo, C.G. Identification of shark species in seafood products by forensically informative nucleotide sequencing (FINS). *J. Agric. Food Chem.* **2008**, *56*, 9868–9874. [CrossRef] [PubMed]
61. Barbuto, M.; Galimberti, A.; Ferri, E.; Labra, M.; Malandra, R.; Galli, P.; Casiraghi, M. DNA barcoding reveals fraudulent substitutions in shark seafood products: The Italian case of palombo (*Mustelus* spp.). *Food Res. Int.* **2010**, *43*, 376–381. [CrossRef]
62. Filonzi, L.; Chiesa, S.; Vaghi, M.; Nonnis Marzano, F. Molecular barcoding reveals mislabelling of commercial fish products in Italy. *Food Res. Int.* **2010**, *43*, 1383–1388. [CrossRef]
63. Herrero, B.; Lago, F.C.; Vieites, J.M.; Espiñeira, M. Authentication of swordfish (*Xiphias gladius*) by RT-PCR and FINS methodologies. *Eur. Food Res. Technol.* **2011**, *233*, 195–202. [CrossRef]
64. Tortonese, E. Fauna d'Italia, 'Osteichtyes': Pesci ossei. *Calderini Bologna* **1975**, *11*, 133–139.
65. Whitehead, P.J.P.; Bauchot, M.L.; Hureau, J.C.; Nielsen, J.; Tortonese, E. *Fishes of the North-Eastern Atlantic and the Mediterranean*; UNESCO: Paris, France, 1986.
66. Ivanova, N.V.; Zemlak, T.S.; Hanner, R.H.; Hebert, P.D.N. Universal primer cocktails for fish DNA barcoding. *Mol. Ecol. Notes* **2007**, *7*, 544–548. [CrossRef]
67. GENECHRON. Available online: http://www.genechron.it/index.php/sanger-sequencing (accessed on 17 November 2018).
68. Katoh, K.; Rozewicki, J.; Yamada, K.D. MAFFT online service: Multiple sequence alignment, interactive sequence choice and visualization. *Brief. Bioinform.* **2019**, *20*, 1160–1166. [CrossRef]
69. TRANSEQ. Available online: http://www.ebi.ac.uk/Tools/st/emboss_transeq (accessed on 26 July 2019).
70. GENBANK. Available online: http://www.ncbi.nlm.nih.gov/genbank/ (accessed on 12 September 2019).
71. Zhang, D.X.; Hewitt, G.M. Nuclear integrations: Challenges for mitochondrial DNA markers. *Trends Ecol. Evol.* **1996**, *11*, 247–251. [CrossRef]
72. Darriba, D.; Taboada, G.L.; Doallo, R.; Posada, D. jModelTest 2: More models, new heuristics and parallel computing. *Nat. Methods* **2012**, *9*, 772. [CrossRef] [PubMed]
73. Tamura, K.; Stecher, G.; Peterson, D.; Filipski, A.; Kumar, S. MEGA6: Molecular evolutionary genetics analysis version 6.0. *Mol. Biol. Evol.* **2013**, *30*, 2725–2729. [CrossRef] [PubMed]
74. Felsenstein, J. Confidence limits on phylogenies: An approach using the bootstrap. *Evolution* **1985**, *39*, 783–791. [CrossRef] [PubMed]
75. REMAP EMBOSS. Available online: http://emboss.bioinformatics.nl (accessed on 26 July 2019).
76. Alvarado-Bremer, J.R.; Baker, A.J.; Mejuto, J. Mitochondrial DNA control region sequences indicate extensive mixing of swordfish (*Xiphias gladius*) populations in the Atlantic ocean. *Can. J. Fish Aquat. Sci.* **1995**, *52*, 1720–1732. [CrossRef]
77. Alvarado-Bremer, J.R.; Mejuto, J.; Greig, T.W.; Ely, B. Global population structure of the swordfish (*Xiphias gladius* L.) as revealed by analysis of the mitochondrial DNA control region. *J. Exp. Mar. Biol. Ecol.* **1996**, *197*, 295–310. [CrossRef]
78. Bradman, H.; Grewe, P.; Appleton, B. Direct comparison of mitochondrial markers for the analysis of swordfish population structure. *Fish Res.* **2011**, *109*, 95–99. [CrossRef]

79. Griffiths, A.M.; Sotelo, C.G.; Mendes, R.; Perez Martin, R.I.; Schröder, U.; Shorten, M.; Silva, H.A.; Verrez-Bagnis, V.; Mariani, S. Current methods for seafood authenticity testing in Europe: Is there a need for harmonisation? *Food Control* **2014**, *45*, 95–100. [CrossRef]
80. Verrez-Bagnis, V.; Sotelo, C.G.; Mendes, R.; Silva, H.; Kappel, K.; Schröder, U. Methods for seafood authenticity testing in Europe. In *Bioactive Molecules in Food, Reference Series in Phytochemistry*; Mérillon, J.-M., Ramawat, K.G., Eds.; Springer: Berlin, Germany, 2018; pp. 1–55.
81. EUR-Lex Council Regulation (EC) No. 1224/2009 of 20 November. *Off. J. Eur. Union* **2009**, *343*, 50.
82. Kappel, K.; Schröder, U. Substitution of high-priced fish with low-priced species: Adulteration of common sole in German restaurants. *Food Control* **2016**, *59*, 478–486. [CrossRef]

foods

MDPI

Article

Comparison of Targeted (HPLC) and Nontargeted (GC-MS and NMR) Approaches for the Detection of Undeclared Addition of Protein Hydrolysates in Turkey Breast Muscle

Liane Wagner [1], Manuela Peukert [1], Bertolt Kranz [1], Natalie Gerhardt [2], Sabine Andrée [1], Ulrich Busch [2] and Dagmar Adeline Brüggemann [1,*]

[1] Department of Safety and Quality of Meat, Max Rubner-Institut, Federal Research Institute of Nutrition and Food, E.-C.-Baumann-Straße 20, 95326 Kulmbach, Germany; liane.wagner@mri.bund.de (L.W.); manuela.peukert@mri.bund.de (M.P.); bertolt.kranz@mri.bund.de (B.K.); sabine.andree@mri.bund.de (S.A.)

[2] Department of Food Analysis, Bavarian Health and Food Safety Authority (LGL), NMR, Veterinärstr 2, 85764 Oberschleißheim, Germany; natalie.gerhardt@lgl.bayern.de (N.G.); ulrich.busch@lgl.bayern.de (U.B.)

* Correspondence: dagmar.brueggemann@mri.bund.de; Tel.: +49-9221-803-249

Received: 29 June 2020; Accepted: 6 August 2020; Published: 8 August 2020

Abstract: The adulteration of fresh turkey meat by the undeclared addition of protein hydrolysates is of interest for fraudsters due to the increase of the economic gain by substituting meat with low cost ingredients. The aim of this study was to compare the suitability of three different analytical techniques such as GC-MS and ^{1}H-NMR with HPLC-UV/VIS as a targeted method, for the detection of with protein hydrolysates adulterated turkey meat. For this, turkey breast muscles were treated with different plant- (e.g., wheat) and animal-based (e.g., gelatin, casein) protein hydrolysates with different hydrolyzation degrees (15–53%: partial; 100%: total), which were produced by enzymatic and acidic hydrolysis. A water- and a nontreated sample (REF) served as controls. The data analyses revealed that the hydrolysate-treated samples had significantly higher levels of amino acids (e.g., leucine, phenylalanine, lysine) compared with REF observed with all three techniques concordantly. Furthermore, the nontargeted metabolic profiling (GC-MS and NMR) showed that sugars (glucose, maltose) and/or by-products (build and released during acidic hydrolyses, e.g., levulinic acid) could be used for the differentiation between control and hydrolysates (type, degrees). The combination of amino acid profiling and additional compounds gives stronger evidence for the detection and classification of adulteration in turkey breast meat.

Keywords: ^{1}H-NMR; GC-MS; HPLC-UV/VIS; metabolomics; food fraud; protein hydrolysate; free amino acid contents; ProHydrAdd

1. Introduction

Meat is an important supplier of high-quality nutrients such as proteins, minerals and vitamins. Since it is sold on the market for a low price, which does not cover the increasing production cost, it is in focus for food fraud. Adulterators look for opportunities to increase the economic gain. Possibilities would be to misrepresent, use illegal supply chains and/or manipulate the food product, e.g., replace/substitute some, or all, premium quality materials with lower-grade, cheaper cuts of meat, meat from other species or nonmeat components (e.g., water, additives) [1–3]. The fraud can influence the consumers' satisfaction and confidence (religious, moral, cultural), but worse it can be extremely dangerous for human health, e.g., causing illness, provoke allergies or even causing death (e.g., melamine scandal) [1,4]. Therefore, it is necessary to have reliable analytical methods to detect adulteration.

The water binding capacity of meat is strongly related to the rate of early postmortem metabolism and the ultimate pH value. It is lowest at a pH of 4.9–5.4, but will increase with increasing or decreasing of the pH [5]. The ultimate pH of poultry breast muscle is 5.67–5.69 and therefore, the water uptake capacity is high [6]. A simple exposure to just water can lead to weight gain of the meat and consequently increase the economic gain. For a fair market competition, the detection of such fraudulent practices is required. The traditional method to determine extraneous water in meat is to analyze the water/protein ratio [7]. In an untreated sample is the water/protein ratio of chicken as well as turkey breast muscle <3.40, of chicken legs between 4.05–4.30 and of turkey legs between 3.80–4.05 (Commission regulation No 543/2008) [8]. If water is added, the water/protein ratio will be higher.

For more than a decade an undeclared addition of protein hydrolysates to poultry meat or meat products could be observed. This way, the analytical protein content rises, masking the water addition [3]. Protein hydrolysates consist mainly of amino acids and possibly peptides. They are cheaper, better soluble and harder to detect than protein additions. Besides amino acids, hydrolysates contain additional compounds like carbohydrates, fatty acids and/or side products, which are formed during the hydrolyzation process. A suitable method is the detection of the free amino acid contents using high performance liquid chromatography with ultraviolet-visible detection (HPLC-UV/VIS). This technique is well established in laboratories and is used as an alternative to the official method in Germany (§64 LFGB: Determination of free amino acids in meat using gas chromatography with flame ionization detection (GC-FID)) [9].

Alternative analytical approaches such as nontargeted metabolomics provide an entire profile (chromatogram, spectrum, fingerprints, etc.) of a suspicious sample [10,11]. These methods have gained more and more in importance in recent years due to strong technical improvements. With these promising and valuable high-throughput tools such as mass spectrometry (MS) based techniques or nuclear magnetic resonance (NMR) spectroscopy, it is possible to identify and quantify small organic molecules with molecular weights of less than 1.5 kDa, including carbohydrates, peptides, nucleotides, lipids and amino acids [12]. In order to analyze such a broad spectrum of metabolites with diverse properties and concentrations (over several magnitudes) in just one single sample, it is important to have techniques, which are robust and sensitive. The major techniques are MS coupled to a chromatographic separation and NMR. The high sensitivity and selectivity make MS a powerful tool to detect molecular masses and fragmentation patterns for chemical structure identification. It is possible to profile myriads of metabolites due to the different combinations of separation, ionization and detection technique. On the other hand, NMR spectroscopy provides characteristic information on the metabolic profile by analyzing small amounts of one sample in a nondestructive, quantitative and short-time period way with convenient sample preparation [13]. However, independent of the used approach (targeted or nontargeted), large reference datasets are required to account for the natural variation in different products due to multiple influencing factors such as feeding or storage conditions (duration and temperature) [3,14].

Chemometrics is a powerful multivariate data analysis tool that reduces a huge amount of generated data by (1) grouping or ordering unknown samples with similar characteristics (qualitative) and (2) ascertaining adulterant analytes in sample (quantitative) or (3) for assessing their quality or authenticity [14]. Chemometrics, and for this purpose used clustering, principal component analyses (PCA) as well as regression analyses, is a routine complement for MS- and NMR-based metabolomics [15].

The objective of this study was to compare the suitability of three different analytical methods such as GC-MS and ^{1}H-NMR (nontargeted approaches) and HPLC-UV/VIS as a targeted method, for the detection of adulterated turkey breast muscle with protein hydrolysates. HPLC-UV/VIS was selected due to the fact that it requires inexpensive equipment frequently available in every control laboratory. The first part focuses on the identification of specific markers such as amino acids, which are detectable with all three methods. The second part (nontargeted approaches) focuses on the identification of additional markers as well as possible classification of the added hydrolysates.

2. Materials and Methods

2.1. Chemicals

Sodium hydroxide, 5-sulfo salicylic acid, sodium chloride (NaCl, 99%), hydrochloric acid (HCl), sodium acetate, ninhydrin, hydrindantin-dihydrat, methanol and chloroform were obtained from Merck (Darmstadt, Germany). Ethylenediaminetetraacetic acid (EDTA, 99%) was purchased from AppliChem (Darmstadt, Germany) and tris(hydroxymethyl)aminomethane (TRIS Pufferan, Tris) from Carl Roth (Karlsruhe, Germany). The reagents (eluent buffers A, B, C, D, E, F; cleaning solution W, derivatization reagent R, sampling dilution buffer and autosampler solution) used for the amino acid (AA) determination were bought from membraPure (Henningsdorf, Germany). The AA standards (AA standards physiological: acidics–neutrals–basics), the amino acid mix solution (certified) and L-asparagine, L-glutamine, S-(2-aminoethyl)-L-cysteine hydrochloride (thialysine), L-norleucine, isopropanol, gelatin hydrolysate enzymatic, HyPep® 4601 protein hydrolysate from wheat gluten, protein hydrolysate N-Z-amine® AS, casein from bovine milk, gelatin from porcine skin and gluten from wheat were bought from Sigma-Aldrich (Steinheim, Germany). 2-Methoxyethanol was obtained from Riedel-deHaen (Seelze, Germany). For GC-MS derivatization methoxyamine hydrochloride (MAH) and pyridine were purchased from Sigma-Aldrich (Steinheim, Germany), and *N*-methyl-*N*-(trimethylsilyl)trifluoroacetamide plus 1% chlorotrimethylsilane (MSTFA + 1% TMCS) were purchased from Thermo Fisher (Dreieich, Germany). For NMR analyses reagents such as deuterium oxide (D_2O) containing 0.05 wt% TSP (sodium-3-(trimethylsilyl)-2,2,3,3-tetradeuteriopropionate) and maleic acid were obtained from Sigma-Aldrich (Steinheim, Germany) and D_2O (99.96%) was obtained from VWR (Ismaning, Germany).

2.2. Sampling and Adulteration of Turkey Breast

Meat. Three female turkeys (BUT Big 6, *Meleagris gallopavo*) with an average weight of 10.3 kg and in average 112 days old were collected at 4 °C from a slaughterhouse in Germany directly after slaughter. The *Musculus pectoralis superficialis* was taken after dissection [16]. After sampling, all meat pieces were immediately frozen in liquid nitrogen and stored at −20 °C until further use.

Protein hydrolyzation. The protein powders (1.0 g of casein from bovine milk, gelatin from porcine skin or gluten from wheat) were hydrolyzed at 150 °C in 8 mL of 6 M aqueous HCl solution for 1 h (total hydrolyzation: TH, degree of hydrolyzation: 100%). After that, neutralization with solid 1.0 M NaOH was performed. The final concentration of amino acids was 55.6 g/L (~0.5 M).

The protein hydrolysate powders (gelatin hydrolysate enzymatic, HyPep® 4601 protein hydrolysate from wheat gluten, protein hydrolysate N-Z-amine® AS (casein)) were used without further hydrolyzation (partial hydrolyzation: PH). All three bought protein hydrolysates were peptones (enzymatic hydrolysis with pepsin (E.C.3.4.4.1) or acidic hydrolysis). The degree of hydrolyzation was analyzed photometrically as the following describes. From each sample, 250 µL (aqueous AA solution) were added to 250 µL of 4 M acetate buffer (pH 5.5) and mixed. After that, 75 µL of 1 M NaOH and 250 µL ninhydrin solution (174 mg Ninhydrin + 28 mg hydrindantin-dihydrat in 10 mL 2-methoxyethanol) were added. The reaction took place at 95 °C for 20 min and was stopped by cooling the mixture in an ice bath (0 °C) for 20 min. 300 µL of each mixture was diluted with 1000 µL isopropanol solution (50% *v*/*v* in water) and measured at room temperature at 570 nm (photometer DU 640, Beckman Coulter GmbH, Krefeld, Germany). The degree of hydrolyzation (DH) was calculated by comparing the free amino groups of the samples with a solution of the same not hydrolyzed protein (0% DH) and total hydrolyzed protein (100% DH). That implies a hydrolyzation degree for gelatin, wheat and casein of 15% ± 3%, 16% ± 2% and 53% ± 2%, respectively.

Hydrolysate injection. About 1 mL solution (0.5 M) hydrolyzed protein or commercially available protein hydrolysate or water was injected per g turkey breast meat across and along the muscle fiber. The samples were frozen at −80 °C and lyophilized for GC-MS and ^{1}H-NMR analyses. All samples were stored at −80 °C until further use.

Sample code. The sample codes were chosen as followed: reference sample without injection is called REF, an additional control sample injected with water is called water. The different protein hydrolysates (gelatin (G), wheat (W) and casein (C)) with different hydrolyzation degree (partial (PH) or total (TH)) are indicated with the following codes: GPH, WPH and CPH for partial, respectively, and GTH, WTH and CTH for total hydrolyzation, respectively. This means, for example, that the sample GPH contained protein hydrolysate gelatin and was partially hydrolyzed.

2.3. *Amino Acid Analysis Using HPLC-UV/VIS*

2.3.1. Sample Preparation for Amino Acid Analysis

The frozen turkey samples (2 g each) were homogenized with an Ultra-Turrax T10 (12,000 rpm, IKA Werke GmbH und CO KG, Staufen, Germany) in 5 mL 0.025 M EDTA/0.100 M Tris buffer (pH 8.0) at −20 °C. As internal standards, L-norleucine and L-thialysine were used with a final concentration of 133 μM and 88 μM, respectively. The proteins and longer peptides were precipitated with 30% *v*/*v* of 15% 5-sulfosalicylic acid over 30 min at 4 °C at a resulting pH of about 2.2 (pH less than 2.5 is recommended for cation exchange chromatography). The samples were centrifuged (15 min, 4 °C, 6827× *g*), filtered (45 μm) and then frozen at −20 °C until further use. The centrifugation and filtration were repeated directly before the analysis.

2.3.2. Amino Acid Content Determination

The samples were analyzed in duplicates with internal (L-norleucine- and L-thialysine-solution, see Section 2.3.1) and external AA standards. The external standards were worked up similar to the meat samples. A regularly calibration of the amino acid analyzer was done with a certified AA standard (amino acid mix solution [17]). For the cation exchange chromatography, a column with 3 μm beads with a separation over a pH-range from 2.9 to 10.4 and ninhydrin post column derivatization was performed. An injection volume of 20 μL (pH 2.2) and a flow rate of 180 μL/min were applied. For the spectrophotometric analysis two photometers with wavelengths of 440 nm and 570 nm were used for detection of the free amino acids. The limit of detection (LOD) and quantification (LOQ) were defined as three and ten times the signal to noise ratio of the external standard solution, respectively. All measured contents of the free amino acids (FAA) were above the LOQ (0.13 mg/100 g–0.33 mg/100 g, depending on the different AA).

2.4. *GC-MS-Based Metabolomics Analyses*

2.4.1. Sample Preparation for Metabolomics Study of White Breast Meat

The sample set consisted of study samples, mix samples and blanks. The mix samples were prepared by combining an aliquot from each muscle powder and served for normalization. All sample types were extracted in the same manner. Study samples were extracted in triplicates. In a first step 20 mg of the dried, homogenous meat powder was extracted with 600 μL ice-cold 80% methanol containing 10 internal standards using a bead mill homogenizer (Minilys, Bertin Technologies SAS, Montigny-le-Bretonneux, France) for 2 times 30 s and an ultrasonic bath for 2 min. The raw extract was centrifuged at 15,000× *g* for 20 min at 4 °C. In a second step the pellet was re-extracted with 600 μL ice-cold methanol:chloroform (2:1 *v*/*v*) according to the first extraction step. Both supernatants were combined, mixed and centrifuged at 15,000× *g* for 20 min at 4 °C. 100 μL of the supernatant were transferred into 2 mL glass vials containing a 200 μL glass insert and evaporated in a vacuum centrifuge (Christ Speedvac RVC 2¨C18 CD plus, Germany). The dried samples were stored under protective argon atmosphere at −80 °C until analysis.

2.4.2. GC-MS Measurements and Data Processing

Prior to measurement samples were derivatized by methoximation and trimethylsilylation. For methoximation samples were shaken in 30 µL of a 20 mg/mL solution of MAH in pyridine at 50 °C for 1 h. Subsequently, 70 µL MSTFA+1%TMCS were added and samples shook at 70 °C for 1 h. GC-MS analysis was performed on a Shimdazu GCMS QP2010 instrument (Shimadzu, Duisburg, Germany) equipped with an OPTIC-4 injector (GL Sciences, Eindhoven, The Netherlands). A 1 µL aliquot of each sample was injected in a 1:7 split ratio. A series of n-alkanes (C7-C30) was used as a retention time standard. Analytes were separated on a 30 m Rxi-5SIL MS column containing a 10 m Integra-Guard column (Restek, 0.25.mm i.d., 0.25 µm film thickness), and with a linear temperature gradient starting from 80 °C to 300 °C with 5 °C/min and a final 5 min hold at 320 °C. Masses between 60 and 600 *m*/*z* were scanned.

For data analysis only those features were selected that were not present in blank samples. The annotation of compounds was performed using the NIST 14 library database implemented in GCMSsolution (Shimadzu, Duisburg, Germany). The retention times and selected masses used for relative quantification are listed in Supplementary Table S1. Samples were normalized according to the calculated means of the mix samples for each feature to reduce the impact of device maintenance (instrument tuning, liner exchange and septum exchange during the measurement batch). Briefly, signal intensities of the mix-samples within one measurement period (between device maintenance) were averaged as well as for the whole measurement batch. Using these means a correction factor was determined between the total means and the means for each measurement period between device maintenances. These correction factors were then applied to the corresponding features of study samples and calibration samples.

2.4.3. Amino Acid Quantification

For AA quantification a calibration curve consisting of a reference standard mixture and 8 final concentrations in the range 1.96–250 pmol/µL was applied. These standard mixes were spiked into aliquots from pooled REF samples (50 µL of standard mixture into 50 µL of REF mixture) to reduce for the impact of the sample matrix. The calibration samples were prepared in duplicates. Calibration samples were evaporated in a vacuum centrifuge and prepared for GC-MS measurement according to Section 2.4.2. For quantification, the calibration samples were at first normalized and then the mean values (ion counts) of selected quantitative ions from calibration-reference samples were subtracted from calibration samples containing the standard mixtures. Calibration coefficients were R^2 > 0.99 with acceptance of alanine which had an $R^2 = 0.97$). The retention times and selected masses used for quantification are listed in Supplementary Table S2.

2.5. *^1H NMR-Based Metabolomics Analyses*

2.5.1. Sample Preparation for Metabolomics Study of White Breast Meat

The same dried, homogenous breast muscle samples as used for GC-MS were prepared for NMR metabolomics as described previously by Wagner et al. [18] with slight modifications. In brief, 20 mg of lyophilized, grinded, homogeneous muscle powder was extracted using first ice-cold methanol, then ice-cold chloroform and finally ice-cold water (400 µL of each solvent). The samples were vortexed for 1 min and stored on ice for 10 min between each step, and then stored at 4 °C overnight for separation and finally after centrifugation (2000× *g*, 4 °C, 30 min) the aqueous phase was collected in a new tube. The collected samples (750 µL) were dried using a vacuum centrifuge. The dried samples were redissolved with 550 µL D_2O, 25 µL MilliQ water and 25 µL D_2O containing 0.05 wt% TSP as internal standard for quantification and chemical shift reference.

2.5.2. ^{1}H NMR Spectroscopy, Data Processing and Identification of the Signals

All samples were analyzed with a Bruker 400 MHz spectrometer (Bruker BioSpin GmbH, Rheinstetten, Germany). For the aqueous white breast muscle, a noesygppr1d pulse program at 25 °C with 64 scans, a spectral width of 8224 Hz collected into 65,536 data point and acquisition time of 3.98 s and an interscan relaxation delay of 4 s was used. ^{1}H-^{1}H COSY, ^{1}H-^{1}H TOCSY and ^{1}H-^{13}C HSQC were obtained on one representative muscle sample for metabolite identification purposes.

All data were processed using Bruker Topspin 3.6.0 software (Bruker), Fourier-transformed after multiplication by line broadening of 0.30 Hz and subsequently referenced to standard peak TSP at 0.00 ppm. After spectral phase and baseline were corrected, each NMR spectrum was integrated using Matlab R2017b (Mathworks, Natick, MA, USA) into 0.01 ppm integral regions (buckets) between 8.60 ppm and 0.80 ppm (area between 4.75 ppm and 4.80 ppm corresponding to the water signal was excluded). Each muscle spectrum region was scaled to the intensity of internal standard (TSP) for quantitative measurements. Afterwards, the signals were identified using ChenomX NMR Suite 8.4 library (ChenomX Inc., Edmonton, AB, Canada), the Human Metabolome Database (www.hmdb.ca) and previous literature [18–20] and confirmed with 2D-NMR in case of multiplicity. For quantification (profiling approach), 86 metabolites were identified by overlapping with standard spectra (Supplementary Table S3) and their concentrations (µmol/mg) were calculated using ChenomX NMR Suite 8.4 library after accounting for overlapping signals. The absolute concentrations were presented as mg/100 g wet weight (Supplementary Table S4).

2.6. Statistical Analysis

Multivariate data analyses were performed for the GC-MS data, the NMR spectral data (buckets) and the absolute concentrations of the metabolites (profiling approach) using the Simca-P+ software (version 13.0; Umetrics, Umeå, Sweden). All variables were centered and "pareto-scaled" (Par) (GC-MS data and NMR spectral data) or "unit variance" (UV)-scaled (NMR data, absolute concentrations). Principal component analysis (PCA) was used to screen the data and search for outliers. Outliers were determined using PCA-Hotelling T^2 Ellipse (95% confidential interval (CI)).

All statistical calculations such as one-way analysis of variance (ANOVA) and Dunnett's test were done in JMP (13.1.0, SAS Institute Inc., Cary, NC, USA).

The amino acid contents (alanine, leucine, methionine, phenylalanine, proline, serine, tyrosine, histidine, lysine and glutamate) determined by HPLC, GC-MS and NMR are presented as mg/100 g wet weight, henceforth referred to as mg/100 g. All data presented are mean ± standard deviation and differences were considered significant when $p < 0.01$.

3. Results and Discussion

The aim of this study was to compare different analytical techniques in respect to their performance for the detection of undeclared protein hydrolysates in fresh turkey breast. For this purpose, a traditional HPLC-UV/VIS approach focusing on the detection of free proteinogenic amino acids was compared with two nontargeted metabolic profiling techniques, GC-MS and ^{1}H-NMR. Additionally, both nontargeted approaches were compared for their suitability in the detection of the adulterated turkey breast muscle with protein hydrolysates.

3.1. Capability of Amino Acid Profiling for the Detection of Added Protein Hydrolysates

The contents of the ten free amino acids (FAA) alanine, leucine, methionine, phenylalanine, proline, serine, tyrosine, histidine, lysine and glutamate out of the 20 proteogenic AA were analyzed by all methods. Those ten AA were selected as exemplary free amino acids with different properties, e.g., aliphatic (alanine, leucine, proline), aromatic (phenylalanine, tyrosine), acidic (glutamate), basic (lysine, histidine), hydroxylic (serine) and sulfur-containing (methionine). The results of five groups (REF and addition of water, gelatin-, wheat- and casein-hydrolysates) were compared depending on the hydrolyzation degree (PH or TH).

The contents of FAA for the addition of partial and total protein hydrolysates are shown in Figures 1 and 2, respectively. All these data are also summarized in the Supplementary Table S5. The reference samples (REF) were not modified and variations were therefore only occurring through the analytical error and natural variations of the FAA contents. The natural FAA contents depend on several conditions, e.g., the gender of the birds [21] or special feed additives [22].

It is obvious that the amount of FAA contents determined by HPLC-UV/VIS (in mg/100 g) and GC-MS as well as ^{1}H-NMR (in mg/100 g) are quite different. The FAA contents determined by GC-MS and ^{1}H-NMR are on average 3.8 times (range: 0.6-fold to 7.0-fold) and 3.3 times (range: 0.1-fold to 10.2-fold) higher, respectively, compared with the contents determined by HPLC-UV/VIS. The differences can be arising from the different sample preparations. One possibility could be that the homogenization method (HPLC-UV/VIS) was not able to dissolve all FAA. The homogenization method for HPLC did not contain any specific extraction step with solvents, whereas for GC-MS and ^{1}H-NMR the samples were extracted using mixtures of water, methanol and chloroform. Moreover, it is possible that the extraction method (GC-MS and ^{1}H-NMR) led to protein hydrolysis. However, the different quantification procedures could also have caused these differences. Comparative experiments (e.g., with protease inhibitors) clearly demonstrated that no protein hydrolysis occurred by using the homogenization method (manuscript in preparation). This method was also used to determinate 18 of the 20 proteinogenic FAA contents of chicken breast meat (manuscript in preparation) and the contents were in agreement with Rikimaru and Takahashi [23].

The addition of water to the turkey breast muscle resulted in tendentially lower contents of FAA, but statistical significance was not reached by using the Dunnett's test (Figures 1 and 2 and Supplementary Table S5). The reduced mean values in water treated samples could be explained as dilution or even wash-out effect. When the amount of water injected to the sample exceeds its water binding capacity, some endogenous compounds (e.g., FAA) might be washed out.

3.1.1. Comparison of Partial Hydrolyzed Wheat-, Gelatin- and Casein-Hydrolysates

Partial enzymatic hydrolysates from gelatin (GPH, hydrolyzation degree 15% ± 3%), wheat (WPH, hydrolyzation degree: 16% ± 2%) and casein (CPH, hydrolyzation degree 53% ± 2%) were added to the meat samples, respectively (Figure 1). WPH and GPH were only slightly hydrolyzed and therefore most of the protein was converted to peptides. Therefore, the amount of FAA in these two hydrolysates was lower compared to CPH, which was hydrolyzed to a higher hydrolyzation degree.

Hence, nearly no significant differences were found for the FAA contents of GPH and WPH related to the REF. Only the content of free lysine (^{1}H-NMR, from 5.26 mg/100 g ± 0.45 mg/100 g (REF) to 26.9 mg/100 g ± 3.37 mg/100 g) for GPH, as well as the free methionine (HPLC-UV/VIS) and free leucine contents (^{1}H-NMR) for WPH showed significant differences (Supplementary Table S5).

Contrary to this, the CPH showed clearly significant different FAA contents compared to the REF. As determined with HPLC-UV/VIS-method, the FAA contents of leucine, methionine, phenylalanine and histidine were highly significant different ($p < 0.001$) and for serine significant different ($p < 0.01$). For example, the FFA content increased for leucine from 1.97 mg/100 g ± 0.33 mg/100 g (REF) to 19.9 mg/100 g ± 2.39 mg/100 g (CPH). The FAA contents analyzed with GC-MS showed significant differences for five of the ten listed FAA. Leucine was also highly increased (6.97 mg/100 g ± 2.78 mg/100 g for REF to 111 mg/100 g ± 26.4 mg/100 g (CPH), $p < 0.001$), as well as methionine, phenylalanine and lysine. The histidine content was increased significantly ($p < 0.01$). The analysis with ^{1}H-NMR showed seven increased FAA contents: leucine, methionine, phenylalanine, proline and lysine increased highly ($p < 0.001$), whereas alanine and serine increased significantly ($p < 0.01$). For example, leucine increased from 4.26 mg/100 g ± 1.25 mg/100 g (REF) to 78.6 mg/100 g ± 19.6 mg/100 g (CPH).

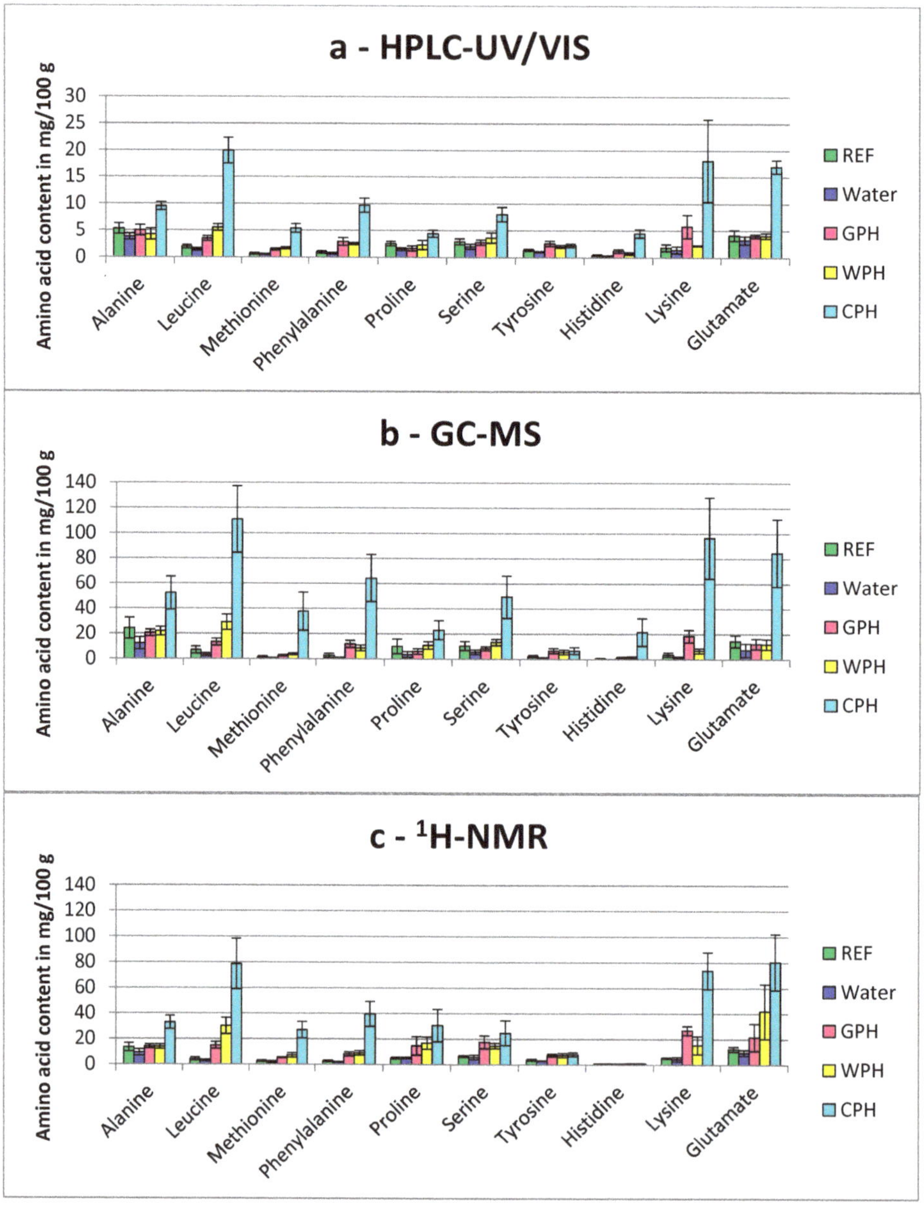

Figure 1. Free amino acids contents (mean ± standard deviation) of turkey breast meat samples treated with and without the addition of partial protein hydrolysates or water and analyzed via: (**a**) HPLC-UV/VIS (**b**) GC-MS (**c**) ^{1}H-NMR. Sample codes: REF: Reference, Water: injected with water, GPH: partial hydrolysate gelatin, WPH: partial hydrolysate wheat; CPH: partial hydrolysate casein.

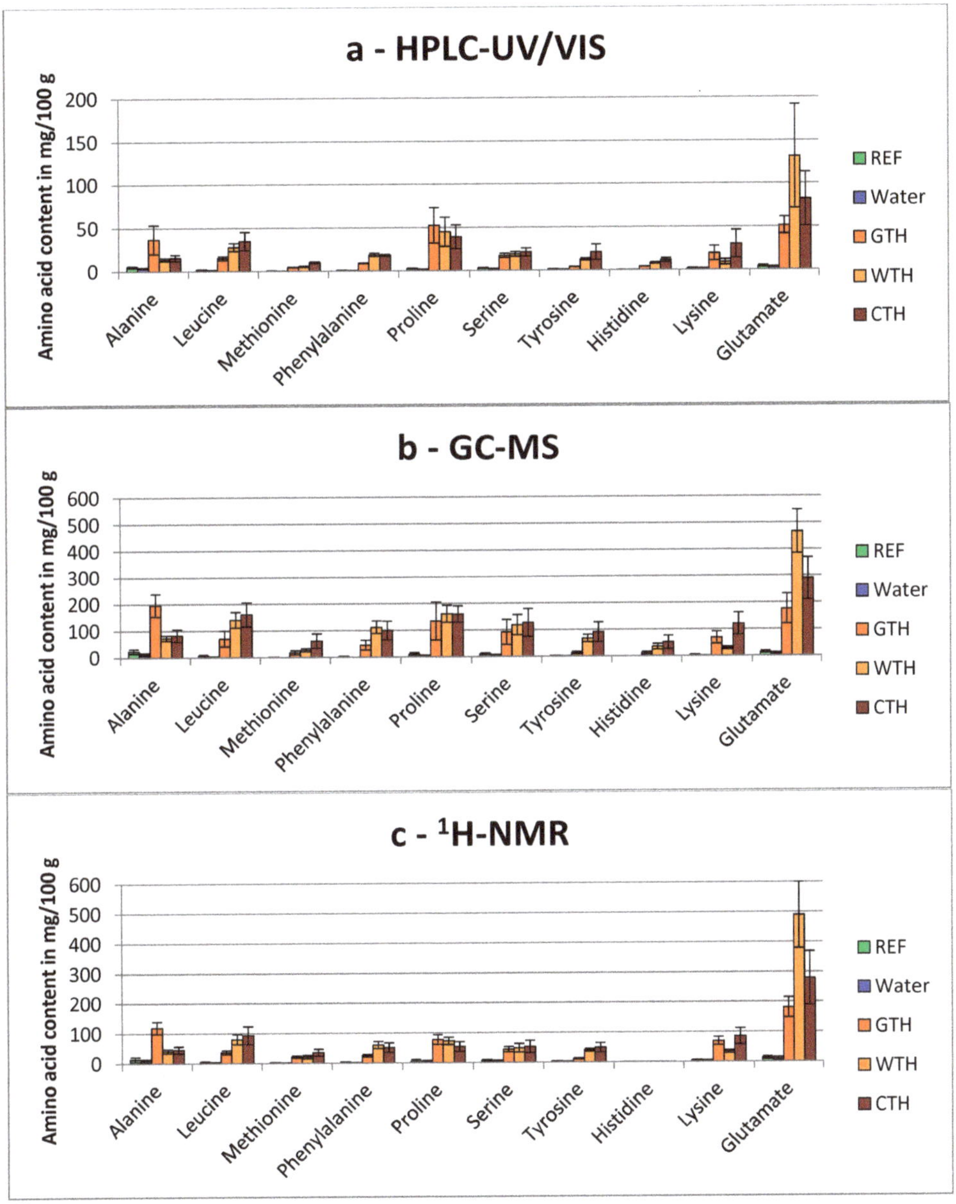

Figure 2. Free amino acids contents (mean ± standard deviation) of turkey breast meat samples treated with and without the addition of total protein hydrolysates or water and analyzed via: (**a**) HPLC-UV/VIS (**b**) GC-MS (**c**) ^{1}H-NMR. Sample codes: REF: Reference, Water: injected with water, GTH: total hydrolysate gelatin, WTH: total hydrolysate wheat; CTH: total hydrolysate casein.

For a clear proof, several FAA contents should differ significantly from the reference sample. In this way other reasons for different FAA contents (e.g., feed supplementation with AA) can be excluded with higher probability. It can be concluded in this study, that in case of partial hydrolysate treatment the detection of fraud could not be ensured by using only these ten FAA contents.

3.1.2. Comparison of Total Hydrolyzed Wheat-, Gelatin- and Casein-Hydrolysates

All the total hydrolyzed proteins (GTH, WTH, CTH) showed a hydrolyzation degree of 100% and were therefore only composed of AA. As expected, the FAA contents of all hydrolysate-treated samples were increased dramatically compared to the REF (Figure 2). The analysis with HPLC-UV/VIS revealed two significant and seven highly significant increases of FAA contents for GTH, eight highly significant increased FAA contents for WTH and nine rises of the amounts of FAA for CTH ($p < 0.001$). For example, the content of proline changed from 2.56 mg/100 g ± 0.42 mg/100 g (REF) to 52.5 mg/100 g ± 20.4 mg/100 g (GTH), 44.9 mg/100 g ± 16.8 mg/100 g (WTH) and 39.2 mg/100 g ± 13.8 mg/100 g (CTH), respectively. With GC-MS, only highly significant changes were found: for GTH seven, for WTH nine and for CTH all ten FAA contents were increased. As for HPLC-UV/VIS, the proline content was raised obviously: from 10.0 mg/100 g ± 5.81 mg/100 g to 134 mg/100 g ± 71.4 mg/100 g (GTH), 161 mg/100 g ± 31.8 mg/100 g (WTH) and 160 mg/100 g ± 30.5 mg/100 g (CTH), respectively. The same results were also found for the ^{1}H-NMR analysis: one significant and eight highly significant increases for GTH (proline: from 5.06 mg/100 g ± 0.78 mg/100 g (REF) to 77.5 mg/100 g ± 16.3 mg/100 g), and nine highly significant differences for WTH (proline: 72.8 mg/100 g ± 12.1 mg/100 g) and CTH (proline: 53.8 mg/100 g ± 16.1 mg/100 g). Therefore, only tyrosine (GTH), alanine and lysine (WTH) as well as alanine (CTH) showed no significant differences determined by HPLC-UV/VIS. For the GC-MS-method, only methionine, tyrosine, histidine (GTH) and lysine (WTH) were not significantly different. The analysis by ^{1}H-NMR revealed also nearly exclusive significant differences with only few exceptions (tyrosine, histidine for GTH, histidine for WTH and CTH).

Depending on the hydrolysate type, different AA were more affected. The addition of GTH resulted in higher levels of alanine, whereas WTH showed higher levels of glutamate and CTH higher levels of leucine, methionine, tyrosine and lysine (Figure 2 and Supplementary Table S5). The latter AA might be used as an indicator for animal-based protein origins whereas glutamate could indicate plant-based protein origins.

It was shown in this study that in case of total hydrolysate treatment a general detection of fraud is possible.

3.1.3. General Aspects Regarding the Detection of Free Amino Acids in Treated Breast Muscles

All three methods used (HPLC-UV/VIS, GC-MS, ^{1}H-NMR) showed comparable results. Although the FAA contents of the first method were about three- to fourfold lower compared to the other two methods, the validity was given. This is due to the fact that all samples were compared to the corresponding REF, determined with the same method. Hence, the method of sample preparation plays an important role for absolute quantities, whereas regarding the differentiation between REF and hydrolyzed-treated samples (rations are kept independently of the sample preparation) the method has no impact.

It can be concluded that the differentiation of hydrolysate addition depends on the degree of hydrolyzation. If breast muscles were treated with low degree hydrolysates, the additional injected FAA might not induce a significant increase over the range of natural variation. It was shown that a high hydrolyzation degree significantly increased the free AA content of several AA independently of which analytical technique (HPLC-UV/VIS, GC-MS or ^{1}H-NMR) was used.

Specific FAA profiles might be used for a tentative classification of the origin of the hydrolysate type (e.g., plant-based vs. animal-based protein origins). Nevertheless, a clear classification and identification of the protein used for hydrolyzation was not possible. Thus, it is of interest whether further information about additional compounds might be helpful for the detection and classification of the hydrolysates. For this, the following hypotheses for section two were postulated: (1) Original protein sources are not clean and contain additional compounds, which can be introduced into the breast meat. (2) Acidic hydrolysis leads to formation of byproducts and these compounds are also possible to be found in the breast meat. (3) Additional metabolites can be washed out from the breast meat and (4) therefore, information from metabolite profiling might be of interest and was included in the analysis.

3.2. Metabolomics Approaches to Obtain Additional Information Regarding Hydrolysate-Treated Samples Independently of the Hydrolyzation Degree

The detection of hydrolysate treatment in turkey breast muscle by amino acid profiling largely depends on the hydrolyzation degree. Our results clearly indicated that an addition of total hydrolyzation increases the free amino acid content tremendously (Figure 2) so that a detection with all three presented methods was possible. However, the lower the hydrolyzation degree the more uncertain is the validity of amino acid profiles between the natural variation and the differentiation due to hydrolysate treatment. Therefore, we applied two nontargeted metabolite profiling approaches (GC-MS and ^{1}H-NMR) to test for their suitability in the detection of hydrolysate treatment in turkey breast muscle. Both approaches allow to detect additionally several metabolites besides the proteogenic amino acids like carbohydrates, organic acids, lipids, et cetera. For both techniques PCA was used to check for the variation of metabolite profiles between controls and treatments and between the different types of hydrolysates (Figure 3).

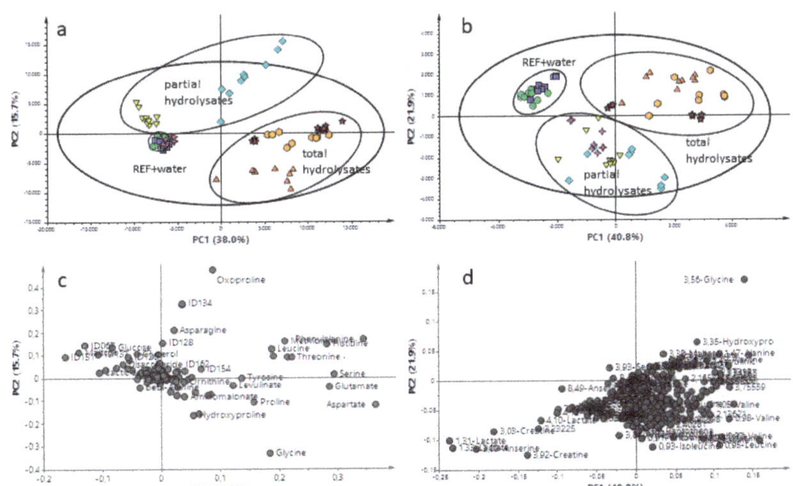

Figure 3. The score (**a**,**b**) and the loading (**c**,**d**) plots of all data of the principal component analysis (PCA) based on the metabolic fingerprint from adulated turkey breast meat with different protein hydrolysates (type and degrees) analyzed by GC-MS (**a**,**c**) and ^{1}H-NMR (**b**,**d**). REF (●), Water (■), GPH (✦), WPH (▽), CPH (◇), GTH (▲), WTH (⬢) and CTH (★). GC-MS: The first component is explained by 38.0% and the second component by 15.7% of the variation (model parameters: $R^2X = 98.4\%$, $Q^2 = 72.8\%$, 17 components). ^{1}H-NMR: The first and second components explained 40.8% and 21.9% of variation, respectively (model parameters: $R^2X = 98.5\%$, $Q^2 = 93.4\%$, 16 components).

With GC-MS a total of 129 features were considered for PCA. The first (horizontal) and second (vertical) components explained 38.0% and 15.7% of the variation, respectively, with $R^2X = 98.4\%$, $Q^2 = 72.8\%$ (Figure 3a). A clear separation between the controls (REF and water-treated control) and five of the hydrolysate-treated sample groups was observed. GPH did not segregate from the controls. WPH varied only in PC2 direction whereas CPH, with 53% hydrolyzation degree, stronger differentiated in PC1 direction. The total hydrolysate treated sample groups showed most variation in PC1 direction.

In addition, the ^{1}H-NMR spectra obtained were compared by PCA (PC1 vs. PC2) (Figure 3b). The first component (horizontal) which is explained by 40.8% of spectral variation clearly separates controls (REF and water-treated control, left) with total hydrolyzed-treated samples (right) and partial hydrolyzed-treated samples (middle). The second component explained 21.9% of variation and separates the controls and total hydrolyzed-treated samples (top) from partial hydrolyzed (bottom) samples. The model parameters were the following: $R^2X = 98.5\%$, $Q^2 = 93.4\%$, 16 components. In order to identify metabolic changes, the absolute concentrations of 86 metabolites were quantified through a profiling approach from ^{1}H-NMR spectra.

The total hydrolysate treated samples clearly separated from the controls in PC1 observed with both techniques. Interestingly, with GC-MS analysis the wheat and casein origins showed higher similarities to each other compared to gelatin (GTH). Whereas for NMR analyses, higher similarities were observed between GTH and WTH. Obviously, independently of the analytical technique, there is a clear separation between gelatin (GTH) and casein (CTH). From the loading plot (Figure 3c,d), it can be deduced that proteinogenic amino acids particularly contribute to the differentiation of the total hydrolysate treated samples and the controls (in PC1), which is in accordance to the results presented in Section 3.1. Nevertheless, besides the amino acids other compounds could be identified which play an additional role for the variation in PC1 such as hydroxyproline, levulinic acid, ornithine or glycerol (the complete feature tables are presented in Supplementary Tables S3 and S4). Compounds, which were additionally detected by ^{1}H-NMR were pyruvate and acetate. Further compounds detected by GC-MS were 5-hydroxylysine, 3-MCPD or aminomalonic acid among several nonidentified molecular features. These additional compounds represent characteristics of the protein origin or are byproducts formed during the acidic hydrolyzation process. In addition to the amino acid profiles these byproducts might contribute to a better classification of the protein sources.

The wheat protein source contained higher amounts of sugars, which was also observed for the WPH treated breast samples (see below). During the acidic hydrolysis of the protein source, the sugars contained therein such as maltose, saccharose, glucose or fructose are converted to levulinic acid in presence of hydrochloride and under high temperature [24,25]. Thus, the high levels of levulinic acid detected in our analyses could be used to differentiate plant-based hydrolysates and as a marker for acidic hydrolyzation treatment. Nevertheless, other plant-based hydrolysates need to be tested for their carbohydrate content in contrast to animal-based protein sources.

The total protein hydrolysate from gelatin contained higher amounts of AA derivates such as hydroxyproline and hydroxylysine. Gelatin is the denatured form of collagen, which is one of the most abundant proteins in meat, ranging between 2 and 4 mg/g in chicken breast meat [26]. Most abundant amino acids of collagen are glycine, proline, glutamate and hydroxyproline [27]. Hydroxyproline is specific to collagen and its concentration in collagen is rather constant with ~12% [28]. Therefore, hydroxyproline is used to estimate the connective tissue content [29,30]. Regarding the treatment of turkey breast meat with protein hydrolysates, the hydroxyproline content can serve as a marker for animal-based protein sources such as gelatin. Hydroxylysine is another modified amino acid, which is unique to collagens. Similar to hydroxyproline, this amino acid becomes posttranslational hydroxylated and subsequently glycosylated forming the α-helical structure of collagens [31]. Therefore, 5-hydroxylysine might serve as an additional indicator of gelatin hydrolysate treatment. Aminomalonic acid that was most abundant in GTH followed by CTH treated samples represents an amino acid derivative, whose origin is suspected to be related to protein oxidation processes [32] and to play a role in the serine-glycine interconversion [33]. According to our results, the acidic hydrolysis process might increase the formation of aminomalonic acid in dependence of the protein source, namely the glycine-rich gelatin.

For casein, the second tested animal derived hydrolysate, it was not that a particular molecule was strongly increased, but the combination of several molecular features could hint towards this treatment. In addition to the amino acid profile, the casein treated samples had higher levels of 3-MCPD and a number of not-identified molecular features (Supplementary Table S1).

Interestingly, the acidic treatment of the different protein sources led to the formation of 3-MCPD. This compound can be found in numerous foodstuffs and is described to be present in acidic hydrolysates of proteins [34]. Depending on the remaining lipids in the original protein sources, different amounts of 3-MCPD and 3-MCPD fatty acid esters might be formed and injected into the breast meat. Thus, 3-MCPD represents an additional marker for acidic hydrolysis, similar to levulinic acid.

From the score plots of the partial hydrolysate treated breast muscles, we observed a clear separation of all three sample groups using ^{1}H-NMR technique, whereas by GC-MS analysis the GPH group largely overlapped with the controls. The WPH treated samples were in a medium distance to the controls, and the highest variation to the controls was observed for CPH treated samples. Those observations are in accordance with the different degrees of hydrolyzation in the partial hydrolysates with GPH having a hydrolyzation degree of 15% and CPH having a hydrolyzation degree of 53%. Even though WPH has a hydrolyzation degree of only 16%, the better separation compared to GPH might be explained by the plant-based origin and the present additional metabolites.

A closer look at the loading plots from the PCA models for control samples and partial hydrolysate treated samples indicated that proteinogenic amino acids play a minor role for the variation between controls (and GPH) and WPH, both having a low hydrolyzation degree (15% and 16%, respectively). The variation of CPH from controls, which was to 53% hydrolyzed, was already dominated by proteinogenic amino acids. Especially the plant-based hydrolysate contained additional sugars such as maltose (Figure 4) and hexoses like glucose, detected with both technical approaches. In addition, higher levels of glycerol were detected in WPH. As mentioned above, future studies will have to elucidate to which extent different sugars are present in plant-based protein extracts used for hydrolysis. With the animal-based protein hydrolysates, GPH and CPH, the contents of ornithine (Figure 4) were increased as detected with ^{1}H-NMR and GC-MS. A low level of levulinic acid and 3-MCPD was observed in CPH treated breast muscle, which might be related to the CPH production process (CPH was commercially obtained). Using ^{1}H-NMR technique, higher amounts of acetate (GPH, WPH, CPH), butyrate (CPH), carnitine (WPH, CPH), citrate (WPH, CPH), glutathione (WPH, CPH), pantothenate (CPH) and putrescine (GPH, WPH, CPH), myo-inositol (GPH, WPH) were additionally detected (Figure 4 and Supplementary Table S4), whereas with GC-MS analysis we obtained increased levels of oxoproline (CPH), urea (CPH), hydroxylysine (GPH) and malic acid (GPH, WPH, CPH) among a few nonidentified compounds (Figure 4 and Supplementary Table S1). It can be concluded that the lower the hydrolyzation degree the more important are the additional compounds from the protein origins for the differentiation of nontreated samples and hydrolysate treated samples.

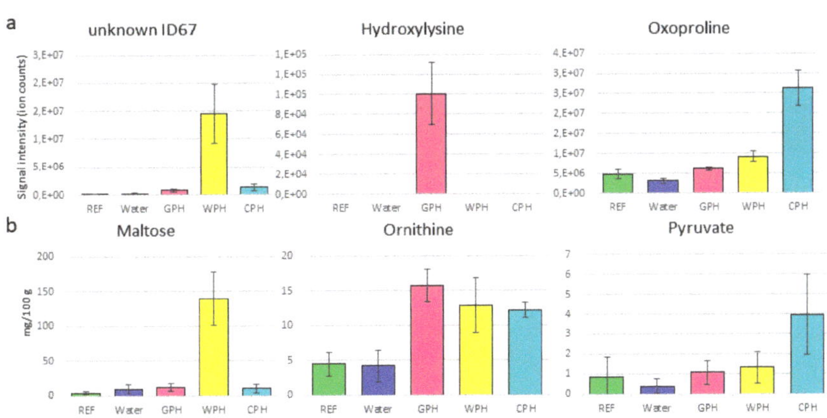

Figure 4. Selected signals that were present in partial hydrolysates and show significant differences between references and hydrolysate treated samples obtained via GC-MS (**a**) and ^{1}H-NMR (**b**).

In accordance with the reduced AA content in water treated samples, we observed for several endogenous metabolites of turkey breast muscle a similar reduction when samples were injected with the different kinds of hydrolysates. This effect was particularly obvious for highly water-soluble compounds such as creatinine and lactate, which were detected by GC-MS and ^{1}H-NMR. Additionally, reduced levels (not significant) of myo-inositol and the peptides glutathione and anserine were detected by ^{1}H-NMR. With GC-MS profiling we detected reduced levels of 4-hydroxybutanoic acid, myo-inositol, inosine and uracil among several nonidentified molecular features (Supplementary Table S1). In our approach, sample preparation was performed using ~2 g fresh turkey breast meat to minimize the effect of natural variation when comparing different hydrolysate types. Whether the observed wash-out effect can also be detected by using whole breast muscles has to be validated by further studies. It can be suspected that the natural variation has a greater impact than the detected small levels of a wash-out.

4. Conclusions

This study aimed at a comparison between different analytical methods and their possibility to detect adulteration of turkey breast meat with different hydrolysates. It showed that FAA profiling allows for the detection of protein hydrolysate treatments only above a certain threshold which is mainly related to the degree of hydrolyzation. The samples naturally strongly differ in their free amino acid contents as a result of feeding, genotype and meat age. Therefore, the FAA analyses under these conditions (e.g., determination of only ten FAA contents) were not suitable for the detection of food fraud in the case of partial hydrolysates. To overcome this limitation, the contents of more than ten FAA of the 20 proteinogenic AA should be analyzed. Further on, a much higher quantity of samples ought to be measured. The evaluation of these datasets enables the reduction of the variations and therefore more significant differences. The additions of hydrolysates with high amounts of AA to breast meat were easily proofed within this study.

The different profiling techniques revealed that protein sources contain different metabolites, which can be used as biomarkers for the detection of partial hydrolysates. Furthermore, byproducts formed during acidic hydrolysis provide additional evidence for the treatment of breast meat with protein hydrolysates. Therefore, a combination of FAA and metabolite (by-products) profiling makes it possible to identify and classify the addition of nondeclared hydrolysates to turkey breast meat. In addition, an advantage of this comprehensive analysis is that it might be possible to proof the addition of animal-based proteins or animal-based hydrolysates to vegetarian or vegan products. According to the advantages of NMR in terms of sample throughput and direct quantification of the identified compounds, ^{1}H-NMR is used in further/detailed studies analyzing food fraud of turkey breast meat.

Supplementary Materials: The following are available online at http://www.mdpi.com/2304-8158/9/8/1084/s1, Table S1: Molecular features used for GC-MS profiling. Annotation was performed with the NIST 14 library implemented in the GCMSsolution software (Shimadzu, Duisburg, Germany). The signal intensities of the quantifier ions were used for the relative quantitation and comparison between sample groups. According to the presented FAA contents the Dunnett's test was used for a comparison between control (REF) and the different treatments.; Table S2: Quantification of amino acids via GC-MS was performed using the following parameters.; Table S3: Assignment of ^{1}H-NMR signals which could be identified via the software ChenomX NMR Suite 8.4 library. 86 metabolites were identified and exemplary one NMR signal (ppm) was chosen which was the obvious signal for identification.; Table S4: Significantly different absolute concentrations of metabolites in turkey breast meat treated with different hydrolysates and analyzed via ^{1}H-NMR (mg/100 g).; Table S5: Significantly different absolute concentrations (mg/100 g) of amino acids in turkey breast meat treated with different hydrolysates and analyzed via HPLC-UV/VIS, GC-MS and ^{1}H-NMR.

Author Contributions: Conceptualization, L.W., M.P., B.K., N.G, S.A., D.A.B.; methodology, validation, data curation, formal analysis, investigation, B.K. (HPLC), M.P. (GC-MS), L.W. (NMR); investigation, N.G. (NMR); resources, U.B. (NMR), D.A.B.; writing original draft, B.K. (HPLC), M.P. (GC-MS), L.W. (NMR); writing- review and editing, all authors; L.W. coordinated the editing of the manuscript; visualization, B.K. (HPLC), M.P. (GC-MS), L.W. (NMR); supervision, D.A.B.; project administration, all authors. All authors have read and agreed to the published version of the manuscript.

Funding: This project was performed within the framework of the research project "Fremdeiweiß" (ProHydAdd), delegated by the Federal Ministry of Food and Agriculture, Germany.

Acknowledgments: The authors thank the technical staff Elke Gardill, Gabriele Schüßler and Katrin Weiß for their assistance in the laboratories.

Conflicts of Interest: The authors declare no conflict of interest.

References

1. Trivedi, D.K.; Hollywood, K.A.; Rattray, N.J.W.; Ward, H.; Trivedi, D.K.; Greenwood, J.; Ellis, D.I.; Goodacre, R. Meat, the metabolites: An integrated metabolite profiling and lipidomics approach for the detection of the adulteration of beef with pork. *Analyst* **2016**, *141*, 2155–2164. [CrossRef] [PubMed]
2. Campmajó, G.; Saez-Vigo, R.; Saurina, J.; Nunez, O. High-performance liquid chromatography with fluorescence detection fingerprinting combined with chemometrics for nut classification and the detection and quantitation of almond-based product adulterations. *Food Control* **2020**, *114*, 107265. [CrossRef]
3. Ballin, N.Z. Authentication of meat and meat products. *Meat Sci.* **2010**, *86*, 577–587. [CrossRef] [PubMed]
4. Kamruzzaman, M.; Makino, Y.; Oshita, S. Hyperspectral imaging in tandem with multivariate analysis and image processing for non-invasive detection and visualization of pork adulteration in minced beef. *Anal. Methods* **2015**, *7*, 7496–7502. [CrossRef]
5. Hamm, R. Colloid Chemistry of Meat; the Water-Binding Capacity of Muscle Protein in Theory and Practice. In *Kolloidchemie des Fleisches*; Parey: Singhofen, Germany, 1972.
6. Hahn, G.; Judas, M.; Berk, J. Forced locomotor activity improves walking ability of male turkeys and modifies carcass characteristics. *Br. Poult. Sci.* **2020**, *61*, 107–115. [CrossRef]
7. ISO (the International Organization for Standardization) 1442:1997(en). *Meat and Meat Products—Determination of Moisture Content (Reference Method)*; International Organization for Standardization: Geneva, Switzerland, 1997.
8. European Union. Commission regulation (EC) No 543/2008 of 16 June 2008 laying down detailed rules for the application of Council Regulation (EC) No 1234/2007 as regards the marketing standards for poultrymeat. *Off. J. Eur. Union* **2008**, *157*, 46–87.
9. BVL L 07.00-64 Technische Regel, 2014-08. *Untersuchung von Lebensmitteln—Bestimmung von Frei Vorliegenden Aminosäuren in Fleischerzeugnissen—Gaschromatographisches Verfahren (English: Analysing of Food—Determination of Free Amino Acids in Meat Products–Using GC)*; Beuth Verlag: Berlin, Germany, 2014.
10. Rodionova, O.Y.; Pomerantsev, A.L. Chemometric tools for food fraud detection: The role of target class in non-targeted analysis. *Food Chem.* **2020**, *317*, 7. [CrossRef]
11. Busch, D.U.; Krenz, O.; Schellenberg, D.A.; Huber, D.I.; Pavlovic, D.M. "Food fraud"—Analytische Herausforderungen von Lebensmittelverfälschungen. *Z. Das Gesamte Lebensm.* **2017**, *424*, 424–438.
12. Wishart, D.S. Metabolomics: Applications to food science and nutrition research. *Trends Food Sci. Amp. Technol.* **2008**, *19*, 482–493. [CrossRef]
13. Gallo, V.; Ragone, R.; Musio, B.; Todisco, S.; Rizzuti, A.; Mastrorilli, P.; Pontrelli, S.; Intini, N.; Scapicchio, P.; Triggiani, M.; et al. A Contribution to the Harmonization of Non-targeted NMR Methods for Data-Driven Food Authenticity Assessment. *Food Anal. Methods* **2019**, *13*, 530–541. [CrossRef]
14. Sentandreu, M.A.; Sentandreu, E. Authenticity of meat products: Tools against fraud. *Food Res. Int.* **2014**, *60*, 19–29. [CrossRef]
15. Böhme, K.; Calo-Mata, P.; Barros-Velázquez, J.; Ortea, I. Recent applications of omics-based technologies to main topics in food authentication. *Trac. Trends Anal. Chem.* **2019**, *110*, 221–232. [CrossRef]
16. Hahn, G.; Spindler, M. Method of dissection of turkey carcases. *Worlds Poult. Sci. J.* **2002**, *58*, 179–197. [CrossRef]
17. *Amino Acid Mix Solution*; Certificate (according to ISO Guide 31), Product 79248; GmbH Sigma-Aldrich Inc.: Darmstadt, Germany, 2020.
18. Wagner, L.; Trattner, S.; Pickova, J.; Gomez-Requeni, P.; Moazzami, A.A. ^{1}H NMR-based metabolomics studies on the effect of sesamin in Atlantic salmon (Salmo salar). *Food Chem.* **2014**, *147*, 98–105. [CrossRef] [PubMed]

19. Castejon, D.; Garcia-Segura, J.M.; Escudero, R.; Herrera, A.; Cambero, M.I. Metabolomics of meat exudate: Its potential to evaluate beef meat conservation and aging. *Anal. Chim. Acta* **2015**, *901*, 1–11. [CrossRef] [PubMed]
20. Bankefors, J.; Kaszowska, M.; Schlechtriem, C.; Pickova, J.; Brannas, E.; Edebo, L.; Kiessling, A.; Sandstrom, C. A comparison of the metabolic profile on intact tissue and extracts of muscle and liver of juvenile Atlantic salmon (*Salmo salar* L.)—Application to a short feeding study. *Food Chem.* **2011**, *129*, 1397–1405. [CrossRef]
21. Tatara, M.R.; Brodzki, A.; Pyz-Lukasik, R.; Pasternak, K.; Szpetnar, M. Sex-related Differences in Skeletal Muscle Amino Acid Concentrations in 20 Week Old Turkeys. *J. Poult. Sci.* **2012**, *49*, 219–223. [CrossRef]
22. Hamano, Y.; Kurimoto, Y. Effects of acetylated wood powder on growth performance, hepatic and muscular free amino acid profiles, and inosine 5′-monophosphate concentration of breast meat in broiler chickens. *Br. Poult. Sci.* **2016**, *57*, 643–654. [CrossRef]
23. Rikimaru, K.; Takahashi, H. Evaluation of the meat from Hinai-jidori chickens and broilers: Analysis of general biochemical components, free amino acids, inosine 5′-monophosphate, and fatty acids. *J. Appl. Poult. Res.* **2010**, *19*, 327–333. [CrossRef]
24. Girisuta, B.; Janssen, L.P.B.M.; Heeres, H.J. Green Chemicals: A Kinetic Study on the Conversion of Glucose to Levulinic Acid. *Chem. Eng. Res. Des.* **2006**, *84*, 339–349. [CrossRef]
25. Shen, J.; Wyman, C.E. Hydrochloric acid-catalyzed levulinic acid formation from cellulose: Data and kinetic model to maximize yields. *AIChE J.* **2012**, *58*, 236–246. [CrossRef]
26. Chen, Y.; Qiao, Y.; Xiao, Y.; Chen, H.; Zhao, L.; Huang, M.; Zhou, G. Differences in Physicochemical and Nutritional Properties of Breast and Thigh Meat from Crossbred Chickens, Commercial Broilers, and Spent Hens. *Asian Australas. J. Anim. Sci.* **2016**, *29*, 855–864. [CrossRef] [PubMed]
27. Gauza-Włodarczyk, M.; Kubisz, L.; Włodarczyk, D. Amino acid composition in determination of collagen origin and assessment of physical factors effects. *Int. J. Biol. Macromol.* **2017**, *104*, 987–991. [CrossRef] [PubMed]
28. Möller, J.; Sjödin, A. Determination of hydroxyproline in meat products by flow injection analysis. *Fresenius' Z. Anal. Chem.* **1988**, *329*, 732–734.
29. Fey, R. Über die Erfassung des Bindegewebsgehaltes von Fleischwaren mit der Hydroxyprolinmethode. *Z. Lebensm. Unters. Forsch.* **1977**, *164*, 233–238. [CrossRef] [PubMed]
30. Hofman, K.; Hall, B.; Cleaver, H.; Marshall, S. High-throughput quantification of hydroxyproline for determination of collagen. *Anal. Biochem.* **2011**, *417*, 289–291. [CrossRef]
31. Herbert, K.R.; Williams, G.M.; Cooper, G.J.; Brimble, M.A. Synthesis of glycosylated 5-hydroxylysine, an important amino acid present in collagen-like proteins such as adiponectin. *Org. Biomol. Chem.* **2012**, *10*, 1137–1144. [CrossRef]
32. Copley, S.D.; Frank, E.; Kirsch, W.M.; Koch, T.H. Detection and possible origins of aminomalonic acid in protein hydrolysates. *Anal. Biochem.* **1992**, *201*, 152–157. [CrossRef]
33. Van Buskirk, J.J.; Kirsch, W.M.; Kleyer, D.L.; Barkley, R.M.; Koch, T.H. Aminomalonic acid: Identification in Escherichia coli and atherosclerotic plaque. *Proc. Natl. Acad. Sci. USA* **1984**, *81*, 722. [CrossRef]
34. Baer, I.; de la Calle, B.; Taylor, P. 3-MCPD in food other than soy sauce or hydrolysed vegetable protein (HVP). *Anal. Bioanal. Chem.* **2010**, *396*, 443–456. [CrossRef]

foods

MDPI

Article

Comparison of Real-Time PCR Quantification Methods in the Identification of Poultry Species in Meat Products

Kerstin Dolch, Sabine Andrée * and Fredi Schwägele

Department of Safety and Quality of Meat, Max Rubner-Institute, E.-C.-Baumann-Str. 20, 95326 Kulmbach, Germany; kerstin.dolch@mri.bund.de (K.D.); fredischwaegele@gmx.de (F.S.)
* Correspondence: sabine.andree@mri.bund.de

Received: 26 June 2020; Accepted: 31 July 2020; Published: 3 August 2020

Abstract: Poultry meat is consumed worldwide and is prone to food fraud because of large price differences among meat from different poultry species. Precise and sensitive analytical methods are necessary to control poultry meat products. We chose species–specific sequences of the *cytochrome b* gene to develop two multiplex real-time polymerase chain reaction (real-time PCR) systems: one for chicken (*Gallus gallus*), guinea fowl (*Numida meleagris*), and pheasant (*Phasianus colchicus*), and one for quail (*Coturnix japonica*) and turkey (*Meleagris gallopavo*). For each species, added meat could be detected down to 0.5% *w/w*. No cross reactions were seen. For these two real-time PCR systems, we applied three different quantification methods: (A) with relative standard curves, (B) with matrix-specific multiplication factors, and (C) with an internal DNA reference sequence to normalize and to control inhibition. All three quantification methods had reasonable recovery rates from 43% to 173%. Method B had more accepted recovery rates, i.e., in the range 70–130%, namely 83% compared to 75% for method A or C.

Keywords: real-time PCR; quantification; chicken; guinea fowl; pheasant; quail; turkey

---

1. Introduction

Consumer awareness for food is growing. On one side, this may be due to health, religious, or ideological issues. On the other side, consumers are sensitized due to food fraud incidences like the horsemeat scandal [1]. Therefore, they want to know what they are getting for their money. For processed food products, the easiest information source is the label of ingredients. In the EU, regulation (EU) No 1169/2011 defines specifically what the label should contain and in which order [2]. For their control, if these regulations are complied, affordable and practical analytical methods are necessary. Hence, one of the main focuses of food authenticity testing is to have the right analytical methods in place and, if necessary, to develop new methods or to improve existing ones.

A change in detailedness of analytic results is one point where improvement is needed. Qualitative results are sufficient when many samples are screened to obtain a rough idea with respect to the contamination rate, and to find out suspected cases. However, in processed food with several ingredients, it is not always sufficient to detect a specific ingredient. Quite often, it is more important to know if the content of one ingredient is higher than another one [3], or if the concentration of an ingredient exceeds a certain threshold [4].

To check the correct declaration of ingredients of animal or plant origin, one strategy is to detect a specific sequence of the deoxyribonucleic acid (DNA) of the corresponding ingredient. This is possible as each species has a unique genome. One widely established method for this is the real-time polymerase chain reaction (real-time PCR). It has been used for a long time for different food products, and constantly new methods are published for meat, seafood, milk, and dairy products, as well as fruit juices [5]. The advantages of real-time PCR analysis are its easy handling and affordable laboratory equipment. The biggest issue, however, is to obtain valid results for legal purposes [6]. In literature, there exist several options.

The most straightforward idea is absolute quantification, where a serial dilution of the target DNA sequence is used as the standard curve [7]. Thereby, the template can be either directly isolated DNA of the pure target [8], a plasmid containing the cloned DNA sequence [9], or synthetically synthesized DNA [10]. This quantification method is well suited for raw food samples. However, for processed food, it is not practicable, as all three DNA sources for the standard curve have in common that they were not treated like the unknown sample. The measurement of DNA is an indirect quantification method for the added amount of animal or plant tissue. The detected concentration of DNA deriving from the target species should correlate directly with the amount of tissue added. But heat treatment may change the amount of DNA detectable due to heat degradation [11,12], which would lead to an underestimation of the actual amount of target tissue added.

The next possibility (method A) is to use a relative standard curve [13]. This step allows to perform quantification of processed and unprocessed food by co-analyzing the DNA of unknown samples together with the DNA of reference material. Therefore, reference material is produced under the same production conditions as the unknown sample. Before production, the target is added in quantities that comply with the measurement range, and DNA is isolated from these DNA standard samples as a reference. For each real-time PCR run, these DNA standard samples have to be applied and measured together with the DNA of the unknown samples [14].

Another possibility (method B) is to determine the matrix-specific multiplication factor of each species under each production condition. DNA is isolated from the reference material and from raw meat. These DNA samples are analyzed together to obtain the matrix-specific multiplication factor. In all further quantification experiments, the DNA of the reference material is not needed anymore. Instead, the DNA samples of the raw material are measured together with the DNA of the unknown samples, and corrected with the matrix-specific multiplication factors obtained earlier [3,15,16]. As the matrix-specific multiplication factors vary between laboratories [17], each laboratory has to determine their own multiplication factors for each animal species and each processing condition [17]. To overcome this time- and labor-consuming step, an internal DNA reference sequence is necessary to quantify via a normalized standard curve (method C). This can be a common DNA sequence like *myostatin* or a ribosome subunit [18], which detects the whole amount of eukaryotic DNA. In the subsequent analytical process, either the ratio is determined between target and reference sequence [19,20], or the difference between the detected amount of target and internal reference sequence (ΔCq) [21,22].

This leads to one of the most important decisions of real-time PCR: the choice of the target DNA sequence. While the usage of single-copy DNA sequences is preferred because of the more stable copy number per cell, the application of multi-copy DNA sequences is in favor of lower detection limits [23].

This study focused on the quantification of the relative meat content for five poultry species in meat products as poultry is the most consumed meat, and its consumption rate is still growing [24]. In addition, poultry products are ranked in the top-ten list of most susceptible product categories [25]. The main species for poultry meat are chicken (*Gallus gallus*) and turkey (*Meleagris gallopavo*), while guinea fowl (*Numida melegaris*), quail (*Coturnix japonica*), and pheasant (*Phasianus colchicus*) are less consumed in Germany [26]. For each bird species, a DNA sequence of the mitochondrial *cytochrome b* gene was chosen, which is often used for identifying animal species [27,28]. For the quantification method C, we chose a sequence of the *12S rRNA* gene as it is a mitochondrial DNA sequence as well. To determine if the processing temperature affects the possibility to detect meat from the five poultry species, sausages were prepared under two different temperatures and analyzed with the three different quantification methods.

2. Material and Methods

2.1. Material

2.1.1. Chemical Material

The following chemicals were used: Proteinase K (Machery-Nagel, Düren, Germany), hydrogen chloride, isopropanol, sodium chloride, and tris(hydroxymethyl)aminomethane (Merck, Darmstadt, Germany), dodecyl sulfate sodium salt (Serva, Heidelberg, Germany), ethylenediaminetetraacetic acid disodium salt (Riedel-de Haën, Seelze, Germany), guanidine hydrochloride, DNA-free water, and RNAse A (Sigma, St. Louis, USA), and ethanol (Th. Geyer, Renningen, Germany). The Wizard® Plus Minipreps DNA Purification System was from Promega (Mannheim, Germany), the QuantiTect Multiplex PCR NoRox from Qiagen (Hilden, Germany), and real-time PCR tubes from LTF-Labortechnik (Wasserburg, Germany).

Primers and probes were synthesized by Eurofins Genomics (Ebersberg, Germany).

2.1.2. Sample Material

All meat samples were pectoral muscle meat. Chicken and turkey meat were obtained from C + C, Kulmbach, Germany, and guinea fowl, pheasant, and quail as whole carcasses from a breeder in Bad Wörishofen, Germany (Josef Maier). All other meat and plant samples were bought in local stores.

Emulsified type sausages were produced twice as two independent batches A and B on separate days. If necessary, the carcasses were dissected, and the meat was minced in a Bizerba Ladenwolf (Baling, Germany). The basic formulation consisted of 50% meat, 25% sunflower oil, 23% ice, 1.7% nitrite salting mix, and 0.3% phosphate. All % values are (*w/w*) per sausage filling. For each species, the ingredients were added to a meat grinder (Food Machines Saarbrücken MK13, Germany) and mixed at 2600 rpm. The ground meat from all five poultry species were combined in various percentages (Table 1), and then the additional ingredients were added. Sausages were filled into cans (type 99/36 mm, Dosen-Zentrale Züchner GmbH, Cologne, Germany) and cooked at low (75 °C; batch A for 30 min, batch B for 4 min) or high temperatures (117 °C; with final F values of 5.68 and 6.01 for batches A and B, respectively, cf. [29]).

Table 1. Composition of standard (S1-5) and unknown emulsified type sausages (U1-5).

	Amount of Meat Added (%)									
Poultry Species	**S1**	**S2**	**S3**	**S4**	**S5**	**U1**	**U2**	**U3**	**U4**	**U5**
Chicken	1.0	0.0	69.0	25.0	5.0	2.0	0.5	57.5	32.0	8.0
Guinea fowl	25.0	5.0	1.0	0.0	69.0	32.0	8.0	2.0	0.5	57.5
Pheasant	0.0	69.0	25.0	5.0	1.0	0.5	57.5	32.0	8.0	2.0
Quail	5.0	1.0	0.0	69.0	25.0	8.0	2.0	0.5	57.5	32.0
Turkey	69.0	25.0	5.0	1.0	0.0	57.5	32.0	8.0	2.0	0.5

2.2. Methods

2.2.1. Bioinformatics

A DNA sequence of the mitochondrial genome was obtained from NCBI GenBank for chicken (NC_040970.1), guinea fowl (NC_034374.1), pheasant (NC_015526.1), quail (NC_003408.1), and turkey (NC_034374.1). These sequences were aligned with the software Molecular Evolutionary Genetics Analysis (MEGA) [30]. Regions with high similarity were chosen for primer and probe binding sites in the area coding for the *12S rRNA* gene.

The theoretical specificity of all primers was checked with the Primer-BLAST software (Basic Local Alignment Search Tool, NCBI) with the same parameters as described in [14].

2.2.2. DNA Isolation

The Wizard DNA isolation kit from Promega was used as the standard DNA isolation method [31]. All samples were prepared as duplicates according to the corresponding instruction. DNA-free water was used as negative control.

All DNA samples were quantified and qualified by measuring at 260, 280, and 340 nm with a spectrophotometer DU 7400 (Beckman Coulter, Brea, CA, USA).

2.2.3. Real-Time PCR

Reaction Set-Up

All real-time PCR assays were performed on a RotorGene 6000 (Qiagen, Hilden, Germany) according to the QuantiTect Multiplex PCR handbook (Qiagen, Hilden, Germany). The reaction was set up in 25 µL with primer and probe concentrations according to Table 2. The following cycler regime was used: 15 min at 95 °C, 35 cycles of 15 s at 95 °C, and 1 min at 60 °C, collecting the fluorescence signal at the end of each cycle.

Table 2. Sequences of primers and probes.

Multiplex Real-Time PCR	Animal Species	Gene	Code	DNA Sequence 5′–3′	Concentration (µM)	Reference
C-G-P	Chicken	*Cyt b*	C-for	AGC AAT TCC CTA CAT TGG ACA CA	0.20	[27]
			C-rev	GAT GAT AGT AAT ACC TGC GAT TGC A	0.20	
			C-probe	JOE-CAG TCG ACA ACC CAA CCC TTA CCC GAT TC-BHQ1	0.08	[32]
	Guinea fowl	*Cyt b*	G-for	GCA TAC GCC ATC CTC CGC TC	0.20	[33]
			G-rev	GCT GCC CAC TCA GGT TAG A	0.20	
			G-probe	DY682-TGG AGG CGT ACT AGC ACT AGC AGC CTC CG-BHQ2	0.08	[32]
	Pheasant	*Cyt b*	P-for	GAG ACA TGA AAC ACT GGA G	0.20	[33]
			P-rev	CAG GTC CAT TCT ACC AAG G	0.20	
			P-probe	ATTO633-CGT CCT ACT CCT CAC ACT CAT AGC AAC C-BHQ2	0.08	[32]
Q-T	Quail	*Cyt b*	Q-for	TGT ACC CTA CAT CGG CCA AAC C	0.20	[33]
			Q-rev	GTC AGA TGA GAT TCC TAA TGG G	0.20	
			Q-probe	FAM-CCT ACC CTA ACC CGA TTC TTC GCC CTC C-BHQ1	0.10	[32]
	Turkey	*Cyt b*	T-for	CAC TCT TGC ATT CTC TTC TGT GG	0.20	[33]
			T-rev	GGA GGT TAT GGA GGA GTC AAC	0.20	
			T-probe	ROX-CCT ACA CAT GCC GAA ACG TAC AAT ACG-BHQ2	0.08	[32]
ALL	Eukarya	*12S rRNA*	12S-for	AAA CTG GGA TTA GAT ACC CCA CTA TG	0.3	This work
			12S-rev	AGA ACA GGC TCC TCT AGG TGG	0.3	
			12S-probe	FAM-AGA ACT ACG AGC ACA AAC GCT TAA AAC TCT A-BHQ1	0.2	

Templates

After DNA isolation of the poultry sausages, the duplicates were combined. They were either adjusted to a DNA concentration of 20 ng/μL or diluted 1:10 with elution buffer. The other animal and plant samples were adjusted to a DNA concentration of 2 ng/μL. For the determination of efficiency, R^2, and limit of detection (LOD) values, the DNA samples of the poultry sausage and of the pure meat samples were diluted ten-fold with elution buffer.

All DNA samples were analyzed in triplicates (standard samples) or duplicates (unknown samples), or in sextets (influence of chicken DNA on detection of pheasant DNA and the LOD). Positive controls and no-template controls (water) were measured once.

2.2.4. Calculation

Method A: Quantification with Reference Material

All DNA samples were diluted 1:10 with elution buffer. The DNA samples from S1–5 (Table 1) were used as standard material, and the corresponding Cq values were plotted against the logarithmic starting quantity. This standard curve was used to quantify the amount of meat from each poultry species in the unknown samples U1–5 (Table 1) for both production temperatures.

Method B: Quantification with Matrix-Specific Multiplication Factors

For establishing the matrix-specific multiplication factors for each species, DNA was isolated from poultry meat and from the standard emulsified type sausages S1–5 (Table 1) which were adjusted to a DNA-concentration of 20 ng/μL with elution buffer. The DNA from the poultry meat was used as standard material to obtain standard curves. For the detection of chicken and quail meat, this was obtained by using 0.01, 0.1, 1.0, 10, and 100 ng DNA per real-time PCR reaction. For the detection of guinea fowl, pheasant, and turkey meat, 0.1, 1.0, 5.0, 10, and 100 ng DNA per real-time PCR reaction were used. The DNA from the emulsified type sausages S1–5 were used for calculating the respective multiplication factors according to Köppel et al. [13,15] for each meat from poultry species, separately for cooking at low or high temperatures. These multiplication factors were used to calculate the amount of meat from poultry species added in the unknown emulsified type sausages U1-5 (Table 1).

Method C: Quantification with Internal Reference Sequence

This method was performed according to Soares et al. [34] with Equation (1):

$$\Delta Cq = Cq_{target} - Cq_{reference} \quad (1)$$

The poultry-species-specific real-time PCR systems were used as the target, and the eukaryotic real-time PCR system was used as the reference.

All DNA samples were diluted 1:10 with elution buffer, and the ΔCq values from S1–5 (Table 1) were plotted against the logarithmic starting quantity for the standard curve. With this standard curve, we calculated the amount of DNA in the unknown samples U1–5 (Table 1). This was performed for the meat from each species and under both processing temperatures.

2.2.5. Statistical Analysis

Calculations were performed either with the Rotor-Gene Q Series Software (Qiagen, Hilden, Germany), Excel (Microsoft Office 2019, Redmond, WA, USA), or with JMP (SAS, Heidelberg, Germany). All factors were analyzed by multiple logistic regression, and the chi-squared values were recorded. The level of significance was set at 5%. Standard box plots were used to visualize the data. The box plots show the median, quantiles as boxes, and whiskers extend to 1.5 times the interquartile distance at most. Outliers were not omitted from the analysis.

3. Results

3.1. Bioinformatics

All primer pairs were checked theoretically for specificity against the ten most commonly eatable bird species (chicken, duck, emu, goose, guinea fowl, ostrich, partridge, pheasant, quail, and turkey). Additionally, the theoretical cross reactivity of the primers was checked for the triplex real-time PCR system (C-G-P) and for the duplex real-time PCR system (Q-T) against all eukaryotes. All false positive matches had several mismatches: the amplicons were either too short or too long, and/or the species were irrelevant as food. Consequently, there were no relevant false positive matches.

The primer pair for detecting all five species was checked theoretically against all entries for animal organisms in the NCBI GenBank database, and amplicons were obtained with a length of 143–146 bp. No mismatches were found for the five poultry species investigated.

3.2. Development of One Triplex and One Duplex Real-Time PCR System

A pentaplex real-time PCR system was proposed in a former publication. However, this system had a lack in precision and accuracy [32]. To overcome this problem, the pentaplex real-time PCR system was split into one triplex real-time PCR system for detecting meat of chicken, guinea fowl, or pheasant (C-G-P), and one duplex real-time PCR system for detecting meat of quail or turkey (Q-T).

DNA isolated from raw meat had concentrations of 333 ng/µL for chicken, 245 ng/µL for guinea fowl, 228 ng/µL for pheasant, 593 ng/µL for quail, and 209 ng/µL for turkey. Ten-fold dilution series of 10^{-1}–10^{-7} gave standard curves for detecting the meat of each species. For the triplex real-time PCR system, efficiency and R^2 values were 102% and 0.983 for chicken, 91% and 0.995 for guinea fowl, and 95% and 0.985 for pheasant; for the duplex system, the values were 94% and 0.998 for quail and 93% and 0.993 for turkey, respectively.

No signal was received with either real-time PCR system when DNA of the following animal species was used: bison, buffalo, camel, chamois, elk, fallow deer, goat, horse, llama, mouflon, pig, reindeer, roe deer, sheep, tuna, wild hare, zebra, or zebu, or DNA from the following plant species: bean, beetroot, black mustard, broccoli, Brussels sprouts, bunching onion, caraway, cardamom, carrot, cauliflower, celery, chili, Chinese cabbage, coriander, cress, cucumber, fennel, garden leek, garden radish, garlic, ginger, green cabbage, horseradish, Indian mustard, kohlrabi, lemon, marjoram, onion, parsley, pepper, pistachio, potato, pumpkin, radish, red cabbage, rutabaga, salsify, savoy cabbage, tomato, white mushroom, white mustard, white pepper, wood garlic, or zucchini.

However, signals were obtained for DNA of chicken, guinea fowl, pheasant, quail, or turkey, each with the respective real-time PCR system (Cq = 15–18) (Table 3). False positive signals were obtained for a few DNA samples with the earliest Cq value of 29. Additionally, all five real-time PCR systems had in common that the blank values gave signals with Cq values between 29 and 33. Therefore, a cut-off was set at Cq ≥ 29.

3.3. Quantification

The unknown emulsified type sausages were analyzed with three quantification methods. For each method, the predicted means were calculated for the unknown samples, together with standard deviations, coefficients of variation (CV), and bias. The CV represents the relative standard deviation of results obtained under repeatability conditions, and was accepted with CV ≤ 25 [13].

Bias was accepted in a range of ±25% relative to the mean. Additionally, recovery rates were calculated, and a range of ±30% was accepted, i.e., recovery rates of 70–130% [13].

3.3.1. Method A: Quantification with Reference Material

For this method, DNA of the reference material was used to obtain a standard curve. The reference material with known concentrations of meat from the five poultry species was produced under the same conditions as the unknown emulsified type sausages. Most of the CV and bias values were within the accepted range. Some of the values were out of range, especially when detecting small concentrations of poultry meat or a high concentration of quail meat (Table 4).

Table 3. Cq values for various animal and plant species tested with the triplex real-time polymerase chain reaction (real-time PCR) system (C-G-P) and the duplex real-time PCR system (Q-T) systems.

DNA	Triplex C-G-P						Duplex Q-T			
	Chicken		Guinea Fowl		Pheasant		Quail		Turkey	
Asparagus	32.28	32.96	-	-	-	-	-	-	-	-
Beef	32.20	-	-	-	-	-	-	-	-	-
Chicken	15.21	15.47	-	-	31.27	31.06	-	-	-	-
Deer	-	-	-	-	-	-	34.74	34.26	-	-
Duck	-	-	-	-	34.64	-	-	-	-	-
Goose	-	-	-	-	29.54	29.22	-	-	34.23	-
Guinea fowl	32.95	34.77	16.40	16.71	-	-	-	-	-	-
Kangaroo	-	-	-	-	-	-	32.33	31.46	-	-
Mace	32.13	-	-	-	29.53	-	-	-	-	-
Ostrich	-	-	31.01	30.59	30.49	30.12	30.94	31.20	-	-
Pheasant	-	-	-	-	14.70	14.73	-	-	-	-
Quail	-	-	-	-	-	-	17.69	17.73	-	-
Turkey	-	-	-	-	-	-	-	-	15.86	16.00
Wild boar	-	-	-	-	-	-	-	-	34.80	-
Blank value	32.66	-	32.36	32.33	29.13	29.73	32.23	31.35	31.71	32.38

All samples were measured in duplicates. - no Cq values were obtained until cycle 35.

Table 4. Predicted concentrations of meat from five poultry species in unknown emulsified type sausages under two temperature conditions, quantified with reference material.

Actual (% *w/w*)	Low Temperature				High Temperature			
	Mean Predicted (% *w/w*) [a]	SD [b]	CV (%) [c]	Bias (%) [d]	Mean Predicted (% *w/w*) [a]	SD [b]	CV (%) [c]	Bias (%) [d]
				Chicken				
0.5	0.28	0.04	14.41	−43.33	0.33	0.08	24.49	−33.33
2.0	1.88	0.26	14.01	−5.83	2.43	0.27	11.23	21.67
8.0	6.82	0.54	7.95	−14.79	7.22	1.64	22.71	−9.79
32.0	25.22	3.14	12.47	−21.20	36.45	7.15	19.61	13.91
57.5	50.53	10.56	20.90	−12.12	70.88	6.79	9.57	23.28
				Guinea fowl				
0.5	0.30	0.06	21.08	−40.00	0.37	0.08	22.27	−26.67
2.0	1.52	0.24	15.83	−24.17	2.48	0.69	27.95	24.17
8.0	8.45	0.73	8.62	5.63	10.30	2.42	23.48	28.75
32.0	26.63	3.93	14.74	−16.77	43.10	8.06	18.69	34.69
57.5	45.90	3.68	8.01	−20.17	57.42	17.53	30.53	−0.14
				Pheasant				
0.5	0.55	0.05	9.96	10.00	0.87	0.30	34.74	73.33
2.0	1.68	0.22	13.24	−15.83	1.78	0.15	8.25	−10.83
8.0	6.92	0.96	13.94	−13.54	10.45	1.35	12.96	30.63
32.0	24.85	3.93	15.81	−22.34	36.02	6.30	17.49	12.55
57.5	55.53	3.66	6.59	−3.42	57.60	13.94	24.19	0.17
				Quail				
0.5	0.47	0.16	34.99	−6.67	0.70	0.15	22.13	40.00
2.0	2.03	0.20	9.67	1.67	2.22	0.43	19.44	10.83
8.0	7.90	0.88	11.12	−1.25	9.68	2.41	24.91	21.04
32.0	28.83	7.56	26.23	−9.90	29.17	2.35	8.05	−8.85
57.5	50.32	10.41	20.69	−12.49	95.07	37.50	39.45	65.33
				Turkey				
0.5	0.52	0.16	31.01	3.33	0.45	0.05	12.17	−10.00
2.0	1.68	0.19	11.53	−15.83	2.03	0.30	14.81	1.67
8.0	4.95	0.23	4.56	−38.13	7.42	0.80	10.83	−7.29
32.0	29.05	2.81	9.68	−9.22	32.60	5.69	17.45	1.88
57.5	49.42	12.04	24.37	−14.06	54.68	5.37	9.82	−4.90

[a] Values are the means of replicate assays (n = 12); [b] SD—standard deviation; [c] CV—coefficient of variation; [d] Bias = 100 * ((mean value − actual value)/actual value).

Most recovery rates were within the accepted range of 70–130% (Figure 1). Lower recovery rates (<70%) were obtained for detecting small concentrations (0.5% meat) and higher recovery rates (>130%) for detecting 57.5% quail meat.

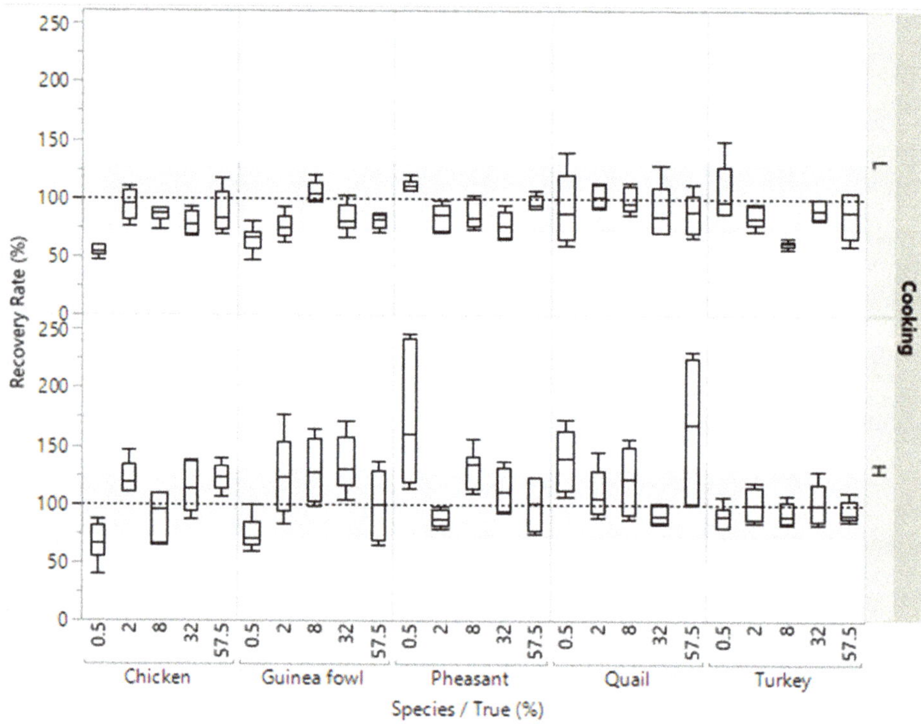

Figure 1. Recovery rates of meat from five poultry species in emulsified type sausages (with 0.5–57.5% meat) quantified with reference material from standard emulsified type sausages (with 0–69% meat). All concentration levels were cooked at low (L) or high temperature (H). DNA was isolated in duplicate from each sausage from both batches, and three independent real-time PCRs were performed, i.e., box plots are from twelve measurements. The grey areas represent the accepted range of 70–130%.

3.3.2. Method B: Quantification with Matrix-Specific Multiplication Factors

The DNA of the standard emulsified type sausages were used to calculate the matrix-specific multiplication factors, separately for low or high cooking temperatures, with the DNA of raw meat as standard material. The multiplication factors ranged from 0.90 (for pheasant meat) to 3.82 (for quail meat) at low cooking temperature, and from 0.09 (for pheasant meat) to 0.52 (for turkey meat) at high cooking temperature (Table 5).

All of the CV values were within the accepted range, only the CV value was slightly higher for detecting 0.5% of pheasant meat at low temperature (Table 6). Most of the bias values were as well within the given range, only for detecting chicken, pheasant, and turkey meat some values were out of the range.

Table 5. Matrix-specific multiplication factors to predict the concentration of meat from five poultry species in unknown emulsified type sausages under two temperature conditions.

Temperature	Batch	Chicken	Guinea Fowl	Pheasant	Quail	Turkey
Low	A	1.16	1.19	0.68	3.02	1.19
	B	1.44	1.49	1.12	4.61	1.17
	Mean	**1.30**	**1.34**	**0.90**	**3.82**	**1.18**
High	A	0.24	0.12	0.09	0.31	0.61
	B	0.23	0.09	0.10	0.29	0.42
	Mean	**0.24**	**0.11**	**0.09**	**0.30**	**0.52**

Table 6. Predicted concentrations of meat from five poultry species in unknown emulsified type sausages under two temperature conditions, quantified with matrix-specific multiplication factors.

Actual (% *w/w*)	Low Temperature				High Temperature			
	Mean Predicted (% *w/w*) [a]	SD [b]	CV (%) [c]	Bias (%) [d]	Mean Predicted (% *w/w*) [a]	SD [b]	CV (%) [c]	Bias (%) [d]
				Chicken				
0.5	0.40	0.05	11.70	−19.33	0.35	0.07	19.07	−29.67
2.0	2.85	0.20	6.94	42.58	2.71	0.41	15.24	35.42
8.0	10.23	0.55	5.39	27.83	10.33	0.74	7.13	29.10
32.0	34.77	1.90	5.47	8.65	31.76	2.37	7.47	−0.74
57.5	60.68	2.49	4.10	5.52	62.27	4.51	7.25	8.30
				Guinea fowl				
0.5	0.44	0.06	14.08	−13.00	0.37	0.08	21.66	−26.67
2.0	1.80	0.23	12.61	−10.25	1.81	0.09	4.71	−9.58
8.0	7.75	0.73	9.45	−3.10	7.73	1.11	14.42	−3.35
32.0	29.86	2.96	9.91	−6.68	32.29	2.84	8.81	0.90
57.5	51.62	4.15	8.04	−10.22	51.76	2.72	5.26	−9.98
				Pheasant				
0.5	0.74	0.19	25.55	47.33	0.69	0.17	24.82	38.00
2.0	2.50	0.33	13.21	24.75	2.00	0.28	14.08	0.00
8.0	9.25	1.60	17.33	15.65	8.50	1.64	19.33	6.19
32.0	31.41	3.44	10.96	−1.83	31.11	4.83	15.53	−2.79
57.5	63.42	5.17	8.16	10.30	63.70	5.12	8.03	10.79
				Quail				
0.5	0.43	0.10	22.58	−13.67	0.45	0.06	13.19	−10.33
2.0	1.81	0.22	11.95	−9.75	1.90	0.19	9.94	−5.17
8.0	8.90	0.70	7.83	11.21	9.54	1.06	11.10	19.23
32.0	35.00	4.17	11.93	9.36	35.42	2.81	7.95	10.68
57.5	53.81	3.19	5.93	−6.42	57.99	4.04	6.96	0.86
				Turkey				
0.5	0.66	0.13	19.06	32.33	0.50	0.06	12.78	−0.67
2.0	1.74	0.17	9.89	−13.00	1.38	0.16	11.85	−31.00
8.0	5.68	0.94	16.46	−28.98	4.37	0.42	9.65	−45.42
32.0	26.62	4.58	17.19	−16.81	26.32	4.03	15.32	−17.76
57.5	57.65	2.79	4.84	0.26	54.78	2.57	4.69	−4.74

[a] Values are the means of replicate assay (n = 12); [b] SD—standard deviation; [c] CV—coefficient of variation; [d] Bias = 100 * ((mean value—actual value)/actual value).

Most recovery rates were within the accepted range of 70–130% (Figure 2). Lower recovery rates were obtained for detecting small concentrations of chicken and higher recovery rates were obtained for detecting small concentrations of pheasant meat.

3.3.3. Method C: Quantification with an Internal Reference Sequence

A mitochondrial reference sequence was chosen because the specific target sequences were mitochondrial. This additional step did not only normalize the results, it also worked well as an amplification and PCR inhibition control, which is recommended for processed food products [13]. For detecting the meat of guinea fowl, quail, and turkey, most of these values are either close to the limit of the range or above (Table 7). On the contrary, the CV and bias values are mostly within the range for detecting chicken or pheasant meat.

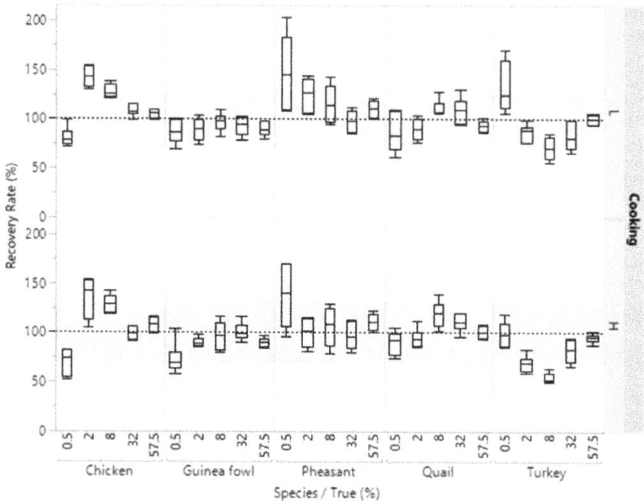

Figure 2. Recovery rates of meat from five poultry species in emulsified type sausages (with 0.5–57.5% meat) quantified with pre-defined multiplication factors. All concentration levels were cooked at low (L) or high temperature (H). DNA was isolated in duplicate from each sausage from both batches, and three independent real-time PCRs were performed, i.e., box plots are from twelve measurements. The grey areas represent the accepted range of 70–130%.

Table 7. Predicted concentrations of meat from five poultry species in unknown emulsified type sausages under two temperature conditions, quantified with an internal reference sequence.

Actual (% *w/w*)	Low Temperature				High Temperature			
	Mean Predicted (% *w/w*) [a]	SD [b]	CV (%) [c]	Bias (%) [d]	Mean Predicted (% *w/w*) [a]	SD [b]	CV (%) [c]	Bias (%) [d]
				Chicken				
0.5	0.69	0.11	16.37	37.67	0.57	0.21	37.23	14.00
2.0	2.06	0.20	9.65	3.08	2.03	0.32	15.90	1.58
8.0	7.38	2.64	35.76	−7.79	7.08	0.68	9.56	−11.56
32.0	26.22	2.48	9.47	−18.05	26.61	4.78	17.97	−16.84
57.5	72.45	9.83	13.56	25.99	73.04	9.12	12.49	27.02
				Guinea fowl				
0.5	0.21	0.04	21.00	−57.67	0.25	0.07	30.08	−50.67
2.0	1.87	0.42	22.56	−6.50	2.09	0.50	23.74	4.75
8.0	10.85	3.17	29.24	35.56	9.36	1.99	21.25	17.02
32.0	33.86	7.46	22.03	5.82	32.24	5.27	16.36	0.73
57.5	57.65	17.42	30.21	0.26	60.87	13.56	22.28	5.85
				Pheasant				
0.5	0.67	0.15	21.82	34.67	0.67	0.05	7.32	34.67
2.0	1.85	0.76	41.35	−7.50	1.75	0.25	14.53	−12.58
8.0	6.99	0.66	9.46	−12.62	7.22	0.85	11.76	−9.79
32.0	31.17	3.02	9.68	−2.59	34.09	2.49	7.31	6.52
57.5	53.74	6.71	12.49	−6.54	52.72	8.00	15.18	−8.31
				Quail				
0.5	0.35	0.16	47.44	−31.00	0.41	0.11	26.76	−18.00
2.0	1.92	0.57	29.44	−3.92	1.65	0.30	18.42	−17.50
8.0	7.33	1.64	22.44	−8.40	8.21	1.02	12.38	2.67
32.0	31.10	5.97	19.20	−2.83	32.02	6.33	19.76	0.06
57.5	57.61	8.82	15.31	0.19	66.62	10.82	16.24	15.87
				Turkey				
0.5	0.52	0.14	26.07	4.33	0.50	0.09	17.35	0.33
2.0	1.62	0.51	31.38	−18.92	1.45	0.39	27.19	−27.67
8.0	7.33	1.34	18.28	−8.42	6.59	0.51	7.73	−17.63
32.0	34.91	8.85	25.36	9.10	26.89	7.77	28.90	−15.97
57.5	42.73	9.91	23.19	−25.68	47.90	14.35	29.96	−16.70

[a] Values are the means of replicate assay ($n = 12$); [b] SD—standard deviation; [c] CV—coefficient of variation; [d] Bias = 100 * ((mean value—actual value)/actual value).

The median of most of the recovery rates were within the accepted range of 70–130% (Figure 3). However, the scattering of the values for the recovery rates were wide for detecting the five poultry meat species. Lower recovery rates were obtained for detecting small concentrations of guinea fowl meat (0.5% meat).

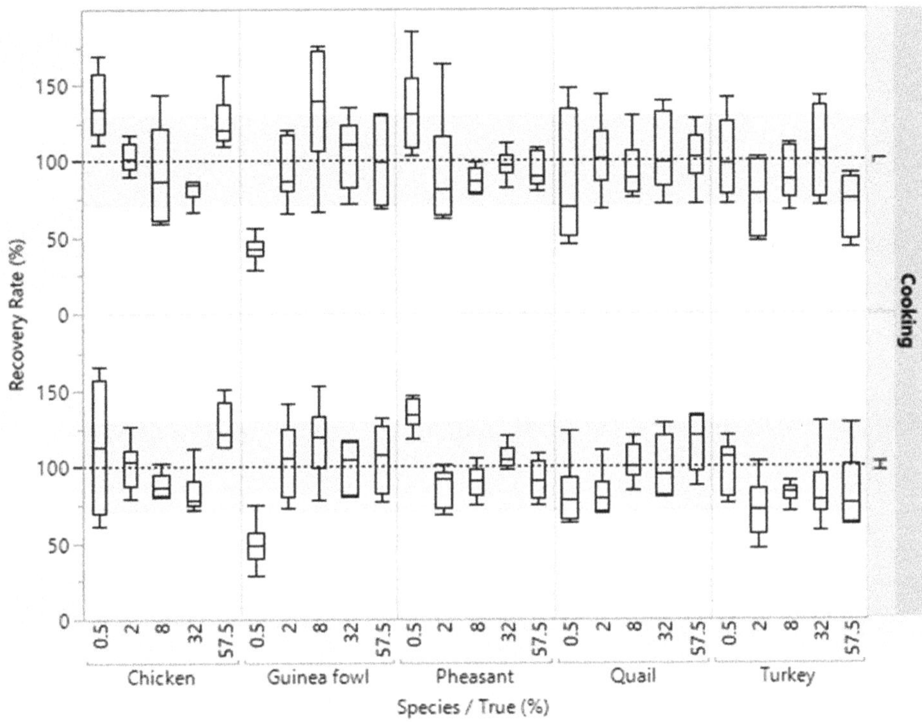

Figure 3. Recovery rates of meat from five poultry species in emulsified type sausages (with 0.5–57.5% meat) quantified with an internal reference sequence. All concentration levels were cooked at low (L) or high temperature (H). DNA was isolated in duplicate from each sausage from both batches, and three independent real-time PCRs were performed, i.e., box plots are from twelve measurements. The grey areas represent the accepted range of 70–130%.

3.3.4. Comparison

Repeatability (CV) and bias were used to compare the three quantification methods. With low cooking temperature, the CV values differed ($\chi^2 = 0.0079$). For method A, 88% of the CV values were within the limits, 96% for method B, and 67% for method C. For the bias, no obvious difference was seen between method A, B, or C ($\chi^2 = 0.3679$). With high cooking temperature, the percentage of accepted CV values differed between the three methods ($\chi^2 = 0.0395$). For method A, 84% of the CV values were within the limits, 100% for method B, and 76% for method C. No differences were seen for the bias values ($\chi^2 = 0.4000$) (data not shown).

Another criterion is to compare the recovery rates, where the limits for acceptance were set to ±30%. A multiple logistic regression was performed, and all predictors which did not significantly contribute to the whole model (p-value > 5%) were removed from analysis. Thus, cooking temperature and batches were omitted from the model. The three quantification methods differed in the percentage of accepted recovery rates ($p = 0.0110$), with 75% for method A, 83% for method B, and 75% for method C. There was no obvious pattern for under- or for overestimation. Recovery rates varied between poultry species ($p = 0.0129$) as well as between concentration levels of poultry meat ($p < 0.0001$). For detecting chicken meat, all three quantification methods had low accepted recovery rates (Table 8). For detecting guinea fowl meat, the quantification method B showed high accepted recovery rates. For detecting pheasant and quail meat, all three quantification methods are similar and well suited. For detecting turkey meat, method A showed the highest accepted recovery rates.

Table 8. Overview of the applied quantification methods, with percentages of accepted bias, coefficients of variation (CV), and recovery rate.

Method	Technical Summary	Bias	CV	Recovery Rate	Species
		(% within Accepted Range) [d]			
		Quantification with reference material			
A [a]	- Fast - Low costs	80	100	68	Chicken
		60	80	65	Guinea fowl
		80	90	80	Pheasant
		80	70	77	Quail
		90	90	85	Turkey
		Quantification with matrix-specific multiplication factors			
B [b]	- More time and more costs for establishment of multiplication-factors - Suited for repeated use	50	100	70	Chicken
		90	100	93	Guinea fowl
		80	90	82	Pheasant
		100	100	97	Quail
		60	100	73	Turkey
		Quantification with an internal reference sequence			
C [c]	- More time and more costs due to second real-time PCR assay - With inhibition control	70	80	77	Chicken
		70	70	62	Guinea fowl
		80	90	80	Pheasant
		90	70	80	Quail
		80	40	77	Turkey

[a] Isolation of DNA from reference material and unknown sample; one real-time PCR for calculation of unknown sample. [b] Isolation of DNA from unknown sample, reference material, and raw meat; one real-time PCR for calculation of multiplication factor and another one for calculation of unknown sample. [c] Isolation of DNA from reference material and unknown sample; two real-time PCR assays (one for target and one for reference sequence) for calculation of unknown sample. [d] for explanation see Section 3.3: Quantification.

4. Discussion

In this study, we compared three different methods to quantify the amount of meat from chicken, guinea fowl, pheasant, quail, or turkey in meat products, cooked at low or high temperatures [24–26].

For the detection of the two main poultry meat species—chicken and turkey—there is a large variety of real-time PCR systems. Most of them are single real-time PCR systems to detect a mitochondrial gene like *cytochrome b* [4,23,27,28,35,36]. For the detection of chicken meat, a few chromosomal genes are used like *interleukin-2* gene [37] or *β-actin* gene [38]. Fewer real-time PCR systems have been published for the detection of meat from guinea fowl [39], pheasant [35,39], or quail [35,39]. There is only one multiplex real-time PCR system for the combination of chicken and turkey meat [16]. To our knowledge, less prominent poultry meat species like guinea fowl, pheasant, or quail have not been considered so far. However, these species are also relevant as they are a delicacy and high-priced. Therefore, the focus was set on the combined detection of the two main poultry meat species, chicken and turkey, together with the high-priced poultry meat species guinea fowl, pheasant, and quail.

The bioinformatic testing of the primers resulted in no false positive matches as single systems, as well as within their combinations (C-G-P and Q-T). However, as the DNA databases are not complete, there is always the chance to miss a species [14]. Therefore, different animal and plant DNAs were tested with both multiplex systems. False positive signals were obtained with a few of these DNA samples. However, each of these Cq values appeared later than the Cq values of the blank samples, and each was below the cut-off value. Furthermore, no influence of chicken DNA on the Cq value of detecting pheasant DNA was shown and it was not possible to establish values for the LOD as all dilutions were to 100% detectable until the cut-off. This, together with high efficiency and R^2 values, indicates that both multiplex real-time PCR systems are precise, specific, sensitive, and suitable to differentiate meat from these five poultry species in meat products.

The two real-time PCR systems were established with DNA that was isolated from 300 mg fresh meat from each species. The DNA content for quail meat was almost twice as high than for each of the other four species. One explanation for this observation is the small size of quails, which is the smallest of the five poultry species investigated. A positive correlation between cell size and body mass among birds [40] implies higher DNA content per body weight in smaller than in larger bird species.

In the literature, there is a large number of methods to quantify material of animal or plant origin in food products [13,41]. Some of these methods are not suited for processed food products, and were therefore not considered in this study. All other methods have in common that standard reference material is required which should be prepared under identical conditions, with similar content, and in similar concentrations [42,43]. Therefore, standard and unknown emulsified type sausages were prepared under comparable conditions.

The three quantification methods compared in this study are in wide use. Quantification method A used DNA from reference material to establish a standard curve that was applied to quantify the amount of meat of each bird species in the unknown samples. At low cooking temperature, the recovery rates were between 70% (chicken or turkey meat) and 90% (pheasant meat) within the accepted limits. At high cooking temperature, the recovery rates were lower than at low temperature for most species, but 100% for turkey meat. Combined with the high bias values for the detection of a low concentration of 0.5% pheasant meat, and 57.5% of quail meat, it can be concluded that quantification with reference material at high cooking conditions is not suited for the whole concentration range for all poultry species. Overall, the idea of this quantification method is quite straight forward, but the main problem is to have the right standard material in stock. For research purposes, this is feasible, and this method was successfully applied to our unknown emulsified type sausages using the standard emulsified type sausages.

Quantification method B applied multiplication factors. This method was first published by Köppel and colleagues in 2011 to detect cow, pig, horse, and sheep [15]. It has been applied to many different animal species since [3,17]. In this study, the multiplication factors were established separately for each bird species and each cooking temperature. The multiplication factors were smaller for high than for low cooking temperatures. This might be due to the higher degradation rate of the DNA due to the higher processing temperature [12]. At low cooking temperature, the percentage of recovery rate values within the limits of ±30% reached from 73% for chicken or pheasant meat to 97% for guinea fowl or quail meat. At high cooking temperature, the percentage of recovery rate values within the limits ranged from 67% (chicken or turkey meat) to 97% (quail meat). Only for the detection of a concentration of 0.5% of pheasant meat, the CV and the bias values were out of range for both cooking temperatures. This implies that, with such a low concentration of pheasant meat, the quantification is not accurate. Overall, the quantification via multiplication factors was effectively applied to the unknown emulsified type sausages. However, as this quantification method normalizes the concentration determined for each species, this method is only practicable when all species added are both known and analyzed together. Therefore, both real-time PCR systems should be expanded if e.g., pork or beef meat were additional ingredients.

For quantification method C, an additional real-time PCR system was necessary. This system should amplify a specific sequence from all eukaryote species. Therefore, this system is a way to measure the total amount of eukaryotic DNA in a sample. In the literature, many different universal systems have been published. The most common system for quantification of mammal or poultry DNA is the *myostatin* gene. There are several systems which differ in their amplicon length [44–46]. However, none of these systems are suited for both of our multiplex real-time PCR systems which detect sequences of the multicopy and mitochondrial *cytochrome b* gene, while *myostatin* is single-copy and nuclear. Another gene which is used quite often is the *18S rRNA* sequence [34,47–49]. This gene is multicopy, however, it is also nuclear. Therefore, it was necessary to develop a new real-time PCR system for eukaryotes which amplifies a multicopy and mitochondrial sequence: the *12S rRNA* gene. Because the calculation of the ΔCq is widely used [13], this method was applied in our study.

At low cooking temperature, the percentage of the recovery rate within the limits ranged from 50% (guinea fowl meat) to 80% (pheasant meat), and at high cooking temperature, the values were between 73% (guinea fowl meat) and 87% (quail meat). Under both conditions, the CV and bias values were especially large for the lower concentrations. This method allowed the detection of meat from all five poultry species. As an advantage of this quantification system, the amplification of a reference sequence serves also as an inhibition control. However, the addition of another real-time PCR system duplicates the number of samples necessary, and consequently also the costs. Moreover, this quantification is not always precise.

In summary, each quantification method was successfully applied to detect meat from the five species in poultry meat products. While method A had a simple and easy line of action (just a standard curve from standard emulsified sausages), the other two methods were more labor-intensive. For method B, the multiplication factors had to be determined additionally, and for method C, an additional real-time PCR system had to be established and performed. For highly processed food products, an inhibition control is recommended and already included in method C. If the detection system is to be used more often, quantification method B was the easiest to operate: in future experiments, the standard emulsified type sausages are not needed anymore, and the DNA from raw meat can be used for preparing standard curves. In addition, with some minor exceptions, the percentage of acceptable values for CV and recovery rate were the highest for method B.

5. Conclusions

Overall, splitting the pentaplex real-time PCR system into one triplex and one duplex real-time PCR system led to a stable, precise, and specific detection method to identify chicken, guinea fowl, pheasant, quail, and turkey meat. All three quantification methods were successfully applied, although mitochondrial gene sequences were chosen. While each quantification method had its pros and cons, a final choice of the quantification method depends on the purpose of its application and the expected concentration of poultry meat species in the meat product.

Author Contributions: Conceptualization, K.D., F.S., and S.A.; methodology, K.D., F.S., and S.A.; validation, K.D., F.S., and S.A.; formal analysis, K.D.; investigation, K.D., F.S., and S.A.; resources, F.S.; writing—original draft preparation, K.D.; writing—review and editing, F.S. and S.A.; visualization, K.D.; supervision, F.S.; project administration, K.D., F.S., and S.A. All authors have read and agreed to the published version of the manuscript.

Funding: This research received no external funding.

Acknowledgments: We thank K. Fischer and E. Müller for their technical assistance in the lab, E. Schlimp, J. Haida, M. Spindler, and M. Zäh for producing the emulsified type sausages, and M. Judas for proof-reading.

Conflicts of Interest: The authors declare no conflict of interest.

References

1. European Commission. *The Rapid Alert System for Food and Feed 2013 Annual Report*; European Union: Luxembourg, 2014. Available online: https://ec.europa.eu/food/sites/food/files/safety/docs/rasff_annual_report_2013.pdf (accessed on 27 July 2020).
2. Regulation 1169/2011. Commission regulation (EU) No 1169/2011 of the European parliament and of the council of 25 October 2011 on the provision of food information to consumers. *Off. J. Eur. Union L 304* **2011**, *20*, 18–63.
3. Köppel, R.; Ganeshan, A.; van Velsen, F.; Weber, S.; Schmid, J.; Graf, C.; Hochegger, R. Digital duplex versus real-time PCR for the determination of meat proportions from sausages containing pork and beef. *Eur. Food Res. Technol.* **2018**, *245*, 151–157. [CrossRef]
4. Kesmen, Z.; Yetiman, A.E.; Sahin, F.; Yetim, H. Detection of chicken and turkey meat in meat mixtures by using real-time PCR assays. *J. Food Sci.* **2012**, *77*, C167–C173. [CrossRef] [PubMed]
5. Böhme, K.; Calo-Mata, P.; Barros-Velázquez, J.; Ortea, I. Review of recent DNA-based methods for main food-authentication topics. *J. Agric. Food Chem.* **2019**, *67*, 3854–3864. [CrossRef]

6. Ballin, N.Z.; Vogensen, F.K.; Karlsson, A.H. Species determination—Can we detect and quantify meat adulteration? *Meat Sci.* **2009**, *83*, 165–174. [CrossRef]
7. Kang, T.S.; Tanaka, T. Comparison of quantitative methods based on SYBR Green real-time qPCR to estimate pork meat adulteration in processed beef products. *Food Chem.* **2018**, *269*, 549–558. [CrossRef]
8. Waiblinger, H.-U.; Boernsen, B.; Geppert, C.; Demmel, A.; Peterseil, V.; Koeppel, R. Ring trial validation of single and multiplex real-time PCR methods for the detection and quantification of the allergenic food ingredients mustard, celery, soy, wheat and rye. *J. Verbrauch. Lebensm.* **2017**, *12*, 55–72. [CrossRef]
9. Sakai, Y.; Kotoura, S.; Yano, T.; Kurihara, T.; Uchida, K.; Miake, K.; Akiyama, H.; Tanabe, S. Quantification of pork, chicken and beef by using a novel reference molecule. *Biosci. Biotechnol. Biochem.* **2011**, *75*, 1639–1643. [CrossRef]
10. Lick, S.; Max Rubner-Institut, Kulmbach, Germany. Personal Communication, 2020.
11. Gryson, N. Effect of food processing on plant DNA degradation and PCR-based GMO analysis: A review. *Anal. Bioanal. Chem.* **2010**, *396*, 2003–2022. [CrossRef]
12. Mano, J.; Nishitsuji, Y.; Kikuchi, Y.; Fukudome, S.-I.; Hayashida, T.; Kawakami, H.; Kurimoto, Y.; Noguchi, A.; Kondo, K.; Teshima, R.; et al. Quantification of DNA fragmentation in processed foods using real-time PCR. *Food Chem.* **2017**, *226*, 149–155. [CrossRef]
13. Kang, T.S. Basic principles for developing real-time PCR methods used in food analysis: A review. *Trends Food Sci. Technol.* **2019**, *91*, 574–585. [CrossRef]
14. Dolch, K.; Judas, M.; Schwägele, F.; Brüggemann, D.A. Development and validation of two triplex real-time PCR systems for the simultaneous detection of six cereal species in processed meat products. *Food Control* **2019**, *101*, 180–188. [CrossRef]
15. Köppel, R.; Ruf, J.; Rentsch, J. Multiplex real-time PCR for the detection and quantification of DNA from beef, pork, horse and sheep. *Eur. Food Res. Technol.* **2011**, *232*, 151–155. [CrossRef]
16. Köppel, R.; Ruf, J.; Zimmerli, F.; Breitenmoser, A. Multiplex real-time PCR for the detection and quantification of DNA from beef, pork, chicken and turkey. *Eur. Food Res. Technol.* **2008**, *227*, 1199–1203. [CrossRef]
17. Köppel, R.; Ganeshan, A.; Weber, S.; Pietsch, K.; Graf, C.; Hochegger, R.; Griffiths, K.; Burkhardt, S. Duplex digital PCR for the determination of meat proportions of sausages containing meat from chicken, turkey, horse, cow, pig and sheep. *Eur. Food Res. Technol.* **2019**, *245*, 853–862. [CrossRef]
18. Rojas, M.; González, I.; Pávon, M.A.; Pegels, N.; Lago, A.; Hernández, P.E.; García, T.; Martin, R. Novel TaqMan real-time polymerase chain reaction assay for verifying the authenticity of meat and commercial meat products from game birds. *Food Addit. Contam. Part A-Chem. Anal. Control Expo. Risk Assess.* **2010**, *27*, 749–763. [CrossRef]
19. Druml, B.; Grandits, S.; Mayer, W.; Hochegger, R.; Cichna-Markl, M. Authenticity control of game meat products—A single method to detect and quantify adulteration of fallow deer (*Dama dama*), red deer (*Cervus elaphus*) and sika deer (*Cervus nippon*) by real-time PCR. *Food Chem.* **2015**, *170*, 508–517. [CrossRef]
20. Iwobi, A.; Sebah, D.; Spielmann, G.; Maggipinto, M.; Schrempp, M.; Kraemer, I.; Gerdes, L.; Busch, U.; Huber, I. A multiplex real-time PCR method for the quantitative determination of equine (horse) fractions in meat products. *Food Control* **2017**, *74*, 89–97. [CrossRef]
21. Costa, J.; Amaral, J.S.; Grazina, L.; Oliveira, M.B.P.P.; Mafra, I. Matrix-normalised real-time PCR approach to quantify soybean as a potential food allergen as affected by thermal processing. *Food Chem.* **2017**, *221*, 1843–1850. [CrossRef]
22. Soares, S.; Amaral, J.S.; Oliveira, M.B.P.P.; Mafra, I. Quantitative detection of soybean in meat products by a TaqMan real-time PCR assay. *Meat Sci.* **2014**, *98*, 41–46. [CrossRef]
23. López-Andreo, M.; Lugo, L.; Garrido-Pertierra, A.; Prieto, M.I.; Puyet, A. Identification and quantitation of species in complex DNA mixtures by real-time polymerase chain reaction. *Anal. Biochem.* **2005**, *339*, 73–82. [CrossRef] [PubMed]
24. Produktion von Fleisch weltweit nach Art in den Jahren 2014 bis 2019 (in Millionen Tonnen). Available online: https://de.statista.com/statistik/daten/studie/28778/umfrage/erzeugung-von-fleisch-weltweit-seit-1990/ (accessed on 2 June 2020).
25. Wisniewski, A.; Buschulte, A. How to tackle food fraud in official food control authorities in Germany. *J. Consum. Prot. Food Saf.* **2019**, *14*, 319–328. [CrossRef]
26. Statistisches Bundesamt. Geflügelschlachtereien, Geschlachtete Tiere und Schlachtmenge in Deutschland im Jahr 2018. Available online: https://de.statista.com/statistik/daten/studie/232533/umfrage/gefluegelstatistik-erhebung-in-gefluegelschlachtereien-in-deutschland/ (accessed on 2 June 2020).

27. Dooley, J.J.; Paine, K.E.; Garrett, S.D.; Brown, H.M. Detection of meat species using TaqMan real-time PCR assays. *Meat Sci.* **2004**, *68*, 431–438. [CrossRef] [PubMed]
28. Jonker, K.M.; Tilburg, J.J.; Hägele, G.H.; de Boer, E. Species identification in meat products using real-time PCR. *Food Addit. Contam. Part A Chem. Anal. Control Expo. Risk Assess.* **2008**, *25*, 527–533. [CrossRef]
29. Laube, I.; Zagon, J.; Broll, H. Quantitative determination of commercially relevant species in foods by real-time PCR. *Int. J. Food Sci. Technol.* **2007**, *42*, 336–341. [CrossRef]
30. Kumar, S.; Stecher, G.; Li, M.; Knyaz, C.; Tamura, K. MEGA X: Molecular Evolutionary Genetics Analysis across Computing Platforms. *Mol. Biol. Evol.* **2018**, *35*, 1547–1549. [CrossRef]
31. Hübner, P. Isolation genomischer DNA aus Lebensmitteln (Wizard-Extraktion). *Kanton. Labor Zürich* **2007**, *Z6521/9902*, 1–6.
32. Andrée, S.; Schwägele, F. Authentifizierung ausgewählter Geflügelspezies mittels Pentaplex-Real-Time-PCR. *Mitt. Fleischforsch. Kulmb.* **2011**, *50*, 163–171.
33. Stirtzel, S.; Andrée, S.; Schwägele, F. Authentifizierung von Geflügel mit tierartspezifischen Primersystemen. *Mitt. Fleischforsch. Kulmb.* **2006**, *45*, 251–258.
34. Soares, S.; Amaral, J.S.; Oliveira, M.B.P.P.; Mafra, I. A SYBR Green real-time PCR assay to detect and quantify pork meat in processed poultry meat products. *Meat Sci.* **2013**, *94*, 115–120. [CrossRef]
35. Chisholm, J.; Sanchez, A.; Brown, J.; Hird, H. The development of species-specific real-time PCR assays for the detection of pheasant and quail in food. *Food Anal. Methods* **2008**, *1*, 190–194. [CrossRef]
36. Tanabe, S.; Hase, M.; Yano, T.; Sato, M.; Fujimura, T.; Akiyama, H. A real-time quantitative PCR detection method for pork, chicken, beef, mutton, and horseflesh in foods. *Biosci. Biotechnol. Biochem.* **2007**, *71*, 3131–3135. [CrossRef] [PubMed]
37. Zdeňková, K.; Akhatova, D.; Fialová, E.; Krupa, O.; Kubica, L.; Lencová, S.; Demnerová, K. Detection of meat adulteration: Use of efficient and routine-suited multiplex polymerase chain reaction-based methods for species authentication and quantification in meat products. *J. Food Nutr. Res.* **2018**, *57*, 351–362.
38. Xiang, W.J.; Shang, Y.; Wang, Q.; Xu, Y.C.; Zhu, P.Y.; Huang, K.L.; Xu, W.T. Identification of a chicken (*Gallus gallus*) endogenous reference gene (*Actb*) and its application in meat adulteration. *Food Chem.* **2017**, *234*, 472–478. [CrossRef] [PubMed]
39. Rojas, M.; González, I.; Pavón, M.A.; Pegels, N.; Hernández, P.E.; García, T.; Martín, R. Polymerase chain reaction assay for verifying the labeling of meat and commercial meat products from game birds targeting specific sequences from the mitochondrial D-loop region. *Poult. Sci.* **2010**, *89*, 1021–1032. [CrossRef]
40. Gregory, T.R. A bird's-eye view of the C-value enigma: Genome size, cell size, and metabolic rate in the class aves. *Evolution* **2002**, *56*, 121–130. [CrossRef]
41. Sánchez, A.; Quinteiro, J.; Vázquez, J.A.; Perez-Martín, R.I.; Sotelo, C.G. Comparison of real-time PCR methods for quantification of European hake (*Merluccius merluccius*) in processed food samples. *Food Chem.* **2019**, *272*, 279–285. [CrossRef]
42. Eugster, A.; Ruf, J.; Rentsch, J.; Köppel, R. Quantification of beef, pork, chicken and turkey proportions in sausages: Use of matrix-adapted standards and comparison of single versus multiplex PCR in an interlaboratory trial. *Eur. Food Res. Technol.* **2009**, *230*, 55–61. [CrossRef]
43. Kaltenbrunner, M.; Hochegger, R.; Cichna-Markl, M. Red deer (*Cervus elaphus*)-specific real-time PCR assay for the detection of food adulteration. *Food Control.* **2018**, *89*, 157–166. [CrossRef]
44. Druml, B.; Kaltenbrunner, M.; Hochegger, R.; Cichna-Markl, M. A novel reference real-time PCR assay for the relative quantification of (game) meat species in raw and heat-processed food. *Food Control.* **2016**, *70*, 392–400. [CrossRef]
45. Laube, I.; Zagon, J.; Spiegelberg, A.; Butschke, A.; Kroh, L.W.; Broll, H. Development and design of a 'ready-to-use' reaction plate for a PCR-based simultaneous detection of animal species used in foods. *Int. J. Food Sci. Technol.* **2007**, *42*, 9–17. [CrossRef]
46. Ren, Y.F.; Li, X.; Liu, Y.M.; Yang, L.T.; Cai, Y.C.; Quan, S.; Pan, L.W.; Chen, S.S. A novel quantitative real-time PCR method for identification and quantification of mammalian and poultry species in foods. *Food Control* **2017**, *76*, 42–51. [CrossRef]
47. Fajardo, V.; González, I.; Martín, I.; Rojas, M.; Hernández, P.E.; García, T.; Martín, R. Real-time PCR for detection and quantification of red deer (*Cervus elaphus*), fallow deer (*Dama dama*), and roe deer (*Capreolus capreolus*) in meat mixtures. *Meat Sci.* **2008**, *79*, 289–298. [CrossRef] [PubMed]

48. Meira, L.; Costa, J.; Villa, C.; Ramos, F.; Oliveira, M.; Mafra, I. EvaGreen real-time PCR to determine horse meat adulteration in processed foods. *LWT-Food Sci. Technol.* **2017**, *75*, 408–416. [CrossRef]
49. Pegels, N.; González, I.; Martín, I.; Rojas, M.; García, T.; Martín, R. Applicability assessment of a real-time PCR assay for the specific detection of bovine, ovine and caprine material in feedstuffs. *Food Control* **2011**, *22*, 1189–1196. [CrossRef]

foods

MDPI

Communication

Species Identification of Red Deer (*Cervus elaphus*), Roe Deer (*Capreolus capreolus*), and Water Deer (*Hydropotes inermis*) Using Capillary Electrophoresis-Based Multiplex PCR

Mi-Ju Kim, Yu-Min Lee, Seung-Man Suh and Hae-Yeong Kim *

Institute of Life Sciences & Resources and Department of Food Science & Biotechnology, Kyung Hee University, Yongin 17104, Korea; mijukim79@gmail.com (M.-J.K.); lym5373@naver.com (Y.-M.L.); teri2gogo@naver.com (S.-M.S.)

* Correspondence: hykim@khu.ac.kr

Received: 26 June 2020; Accepted: 22 July 2020; Published: 23 July 2020

Abstract: To provide consumers correct information on meat species, specific and sensitive detection methods are needed. Thus, we developed a capillary electrophoresis-based multiplex PCR assay to simultaneously detect red deer (*Cervus elaphus*), roe deer (*Capreolus capreolus*), and water deer (*Hydropotes inermis*). Specific primer sets for these three species were newly designed. Each primer set only amplified target species without any reactivity against non-target species. To identify multiple targets in a single reaction, multiplex PCR was optimized and combined with capillary electrophoresis to increase resolution and accuracy for the detection of multiple targets. The detection levels of this assay were 0.1 pg for red deer and roe deer and 1 pg for water deer. In addition, its applicability was demonstrated using various concentrations of meat DNA mixtures. Consequently, as low as 0.1% of the target species was detectable using the developed method. This capillary electrophoresis-based multiplex PCR assay for simultaneous detection of three types of deer meat could authenticate deer species labeled on products, thus protecting consumers from meat adulteration.

Keywords: red deer; roe deer; water deer; multiplex PCR; capillary electrophoresis

1. Introduction

The inaccurate information on the meat species in meat products has been globally concerned by consumers and regulatory agencies [1,2]. Since it is illegal to substitute meat species undeclared on the label of meat products, food manufactures must authenticate correct ingredients declared on their products [3,4]. In the meat industry, game meat consumed commercially is more expensive than meat from domesticated animals. This is because game meat has high nutritional value, such as higher protein and lower fat levels. In addition, it does not contain residues of antibiotics or growth hormones [3,5,6]. Accordingly, replacing game meat with relatively cheaper domesticated meat has taken place for the economic benefit [5]. For game meat products containing deer species, red deer (*Cervus elaphus*) and roe deer (*Capreolus capreolus*) are commonly used, meaning that these species are particularly susceptible to fraudulent labeling [7,8]. Several European countries traditionally permit game hunting [7]. Meanwhile, in Korea, wild animals, such as water deer (*Hydropotes inermis*), that damage crops can be temporarily hunted. However, their distribution and sale are limited, according to the Ministry of Environment guideline. In addition, water deer cannot be used as raw meat or processed food in Korea. To prevent food adulteration, an authentication method for differentiating red deer, roe deer, and water deer is essential.

Methods for detecting meat species have been developed based on DNA molecules and proteins [1,9]. Protein-based methods for deer species authentication have been used by enzyme-linked immunosorbent assay (ELISA), high-performance liquid chromatography (HPLC), and liquid chromatography-mass spectrometry (LC-MS) [10–12]. However, the thermal stability of nucleic acids compared to proteins can increase the amplification efficiency of target species in processed foods [13,14]. PCR, a representative DNA-based detection method, has been utilized for species identification in various fields [15–18]. For deer species, PCR-based detection methods, such as conventional PCR and real-time PCR, have been developed [3,8,19]. To differentiate closely related animal species, the development of specific primers for a target species is very crucial. Mitochondrial DNAs, such as cytochrome b, 12 S rRNA, and D-loop, are commonly used as target genes due to their sequence variations [2,20–22]. Furthermore, to increase the sensitivity of the DNA-based detection method in processed foods, a short fragment of PCR amplification is required because of DNA degradation during the manufacturing process [22,23]. Meanwhile, a multiplex PCR can simultaneously detect several species in a single reaction tube, resulting in effective detection [15,24,25]. Recently, to clearly separate similar sizes of amplicons of short PCR products, multiplex PCR methods combined with capillary electrophoresis have been developed and applied to simultaneously identify various target species [15,26].

The aim of this study was to develop a capillary electrophoresis-based multiplex PCR (CE-mPCR) method to verify the presence of wild animal species, such as red deer, roe deer, and water deer, in processed foods. The developed assay not only saves time and labor because it can simultaneously detect three target species but also can be utilized as a specific and sensitive method for a clear separation of these three species.

2. Materials and Methods

2.1. Sample Preparation

Raw tissue samples of 10 animal species (red deer: *Cervus elaphus*, water deer: *Hydropotes inermis*, roe deer: *Capreolus capreolus*, beef: *Bos taurus*, pork: *Sus scrofa domestica*, lamb: *Ovis aries*, goat: *Capra hircus*, horse: *Equus caballus*, chicken: *Gallus gallus*, and duck: *Anas platyrhynchos*) were collected from the Conservation Genome Resource Bank (CGRB, Seoul, Korea) or purchased from online and local markets of Korea. All samples were cut into small pieces and immediately stored at −20 °C until analysis.

2.2. DNA Extraction

DNAs were extracted from meat samples of animal species and processed products using a DNeasy Blood and Tissue Kit (Qiagen, Hilden, Germany), according to the manufacturer's instructions with slight modifications. For good quality of DNA, 25 mg of meat was ground, and all buffers for extraction were used at double quantity. The purity and concentration of extracted DNAs were measured with a Maestro spectrophotometer (Maestro, Las Vegas, NV, USA). DNAs with a 260/280 nm ratio between 1.8 and 2.0 were used as templates for PCR.

2.3. Primer Design

To select species-specific regions for red deer, roe deer, and water deer, nucleotide sequences of target genes of 19 various animals were downloaded from the GenBank database (Table S1) and aligned using Clustal Omega program (http://www.ebi.ac.uk/Tools/msa/clustalo/) (Figure 1). Species-specific primer sets were newly designed using Primer Designer, version 3.0 (Scientific and Educational Software, Durham, NC, USA). Primers used in this study are listed in Table 1. They were synthesized by Bionics (Seoul, Korea).

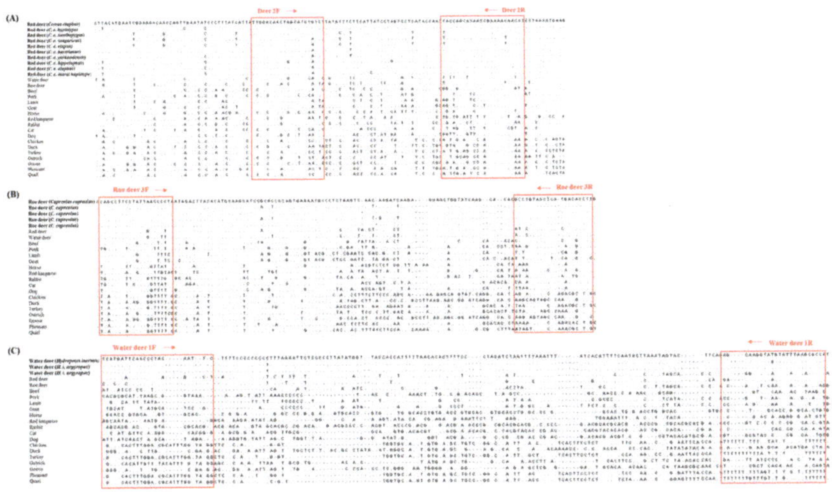

Figure 1. The sequence alignment of red deer (**A**), roe deer (**B**), and water deer (**C**) specific primers in the mitochondrial cytochrome b, 12 S rRNA, and D-loop regions against various animal species.

Table 1. Primers used in this study.

Target Species	Primer Name	Sequences (5′ → 3′)	Target Genes	Amplicon Size (bp)	Accession No.
Red deer	Deer 2 F Deer 2 R	TGGACAACTAGCATCTGTCT GAGGTTGTTTTCGATTGTGCTGGTG	cyt b	79	JF489133.1
Roe deer	Roe deer 3 F Roe deer 3 R	CAGCCTTCCTATTAACCCCT AGGTGTCATGAGCTACAGGC	12 S rRNA	126	KJ681490.1
Water deer	Water deer 1 F Water deer 1 R	CATGATTCAACCCTACAATTC GGCGCTTAAATACATACCTTGCT	D-loop	160	NC011821

2.4. Single and Multiplex PCR Conditions

Single PCR was performed in a 25 μL final volume containing 10 × Buffer (Bioneer, Daejeon, Korea), 10 mM of dNTPs (Bioneer), 5 units of Hot Start *Taq* DNA polymerase (Bioneer), 0.4 μM of each primer, and 10 ng of DNA template. PCR reaction was carried out in a thermal cycler (Model PC 808, ASTEC, Fukuoka, Japan) as follows: pre-denaturation at 95 °C for 5 min, followed by 35 cycles of 95 °C for 30 s, 60 °C for 30 s, and 72 °C for 30 s, with a final extension step at 72 °C for 5 min.

PCR mixture for multiplex PCR was similar to single PCR except that it used 10 units of Hot Start *Taq* DNA polymerase (Bioneer) and optimized concentrations of primers. Annealing temperature concentrations of primers were optimized, considering specificity between three deer species. The annealing temperatures were estimated at 58, 59, 60, and 61 °C, and the red deer/roe deer/water deer primers combinations were 0.2/0.4/0.4, 0.2/0.4/0.5, and 0.4/0.4/0.4 μM. Finally, 0.2 μM of primers for red deer and 0.4 μM of primers for roe deer and water deer were used for multiplex PCR. Multiplex PCR reactions were carried out under the same conditions as single PCR. All PCR amplicons were electrophoresed on 3% agarose gels stained with ethidium bromide at 150 V for 25 min and confirmed by capillary electrophoresis using an Agilent 2100 Bioanalyzer (Agilent Technologies, Santa Clara, CA, USA) with DNA 1000 Lab Chip kit (Agilent Technologies). Briefly, 1 μL of PCR product and 5 μL of markers were loaded into each of the 12 wells and applied with a gel-dye mix in the chip, which was run in the bioanalyzer.

2.5. Specificity and Sensitivity of Multiplex PCR

The specificity of each primer set was performed using DNAs (10 ng each) isolated from 10 animal samples, including red deer, roe deer, and water deer. The specificity of the developed multiplex

PCR was conducted using DNAs of the three target species to determine whether there was any cross-reactivity between closely related species.

The sensitivity of multiplex PCR was estimated using serially diluted DNAs (from 10 ng to 0.01 pg per reaction) of the three target species. Detection limits were tested using meat DNA mixtures. The ratio of DNA used in the mixture is shown in Table 2. This test was validated independently using different PCR instruments by different operators. All PCR reactions included a positive control (target DNA) and negative control (no-template).

Table 2. The ratio of meat DNA mixtures used in this study.

	The Ratio of Meat DNA Mixtures (%)		
	Red Deer	**Roe Deer**	**Water Deer**
Roe deer and water deer in red deer	80	10	10
	90	5	5
	98	1	1
	99	0.5	0.5
	99.8	0.1	0.1
Red deer and roe deer in water deer	10	10	80
	5	5	90
	1	1	98
	0.5	0.5	99
	0.1	0.1	99.8

3. Results and Discussion

3.1. The Specificity of Newly Designed Species-Specific Primers

In this study, the species-specific primer sets targeting mitochondrial genes of cytochrome b, 12 S rRNA, and D-loop for red deer, roe deer, and water, respectively, were newly designed. As shown in Figure 1, sequences of each target species were compared with two closely related species and 16 other animal species. Considering the intraspecific variation of target species, each primer was selected to have specific sequences of target species (Figure 1). Primer design is very important in the development of multiplex PCR because the primer has to selectively amplify the target in a single reaction containing several primer sets [12]. For multiplex PCR, the sizes of PCR products amplified by each primer set were different for the three target species (79, 126, and 160 bp for red deer, roe deer, and water deer, respectively, Table 1). Each set of species-specific primers amplified only the target species without showing cross-reactivity with nine other species (Figure 2), demonstrating high primer specificity for the target species.

3.2. Specificity and Sensitivity of Capillary Electrophoresis-Based Multiplex PCR

Using these newly designed primers for the identification of red deer, roe deer, and water deer, a multiplex PCR was first optimized by adjusting the concentration of each primer and annealing temperature of PCR condition. The specificity of this optimized assay was then evaluated using DNAs isolated from 10 animal species. As shown in Figure 3, each primer set for red deer, roe deer, and water deer in CE-mPCR specifically amplified target species, showing a high resolution between target species. These results indicated that these red deer-, roe deer-, and water deer-specific primers were sufficient to differentiate these closely related three species by multiplex PCR without causing any cross-amplification against non-target species.

The sensitivity of this CE-mPCR developed in this study was evaluated using DNA at different amounts ranging from 10 ng to 0.01 pg. The results are shown in Figure 4. In lane 6 of Figure 4, peaks of electropherogram were detected for red deer and water deer, but the peak of roe deer was not detected in lane 6 and shown in lane 5. Therefore, sensitivities for red deer, roe deer, and water deer were 0.1, 1, and 0.1 pg, respectively. Such high sensitivity of this assay might lead to accurate and reliable detection and differentiation of meat from three target deer species.

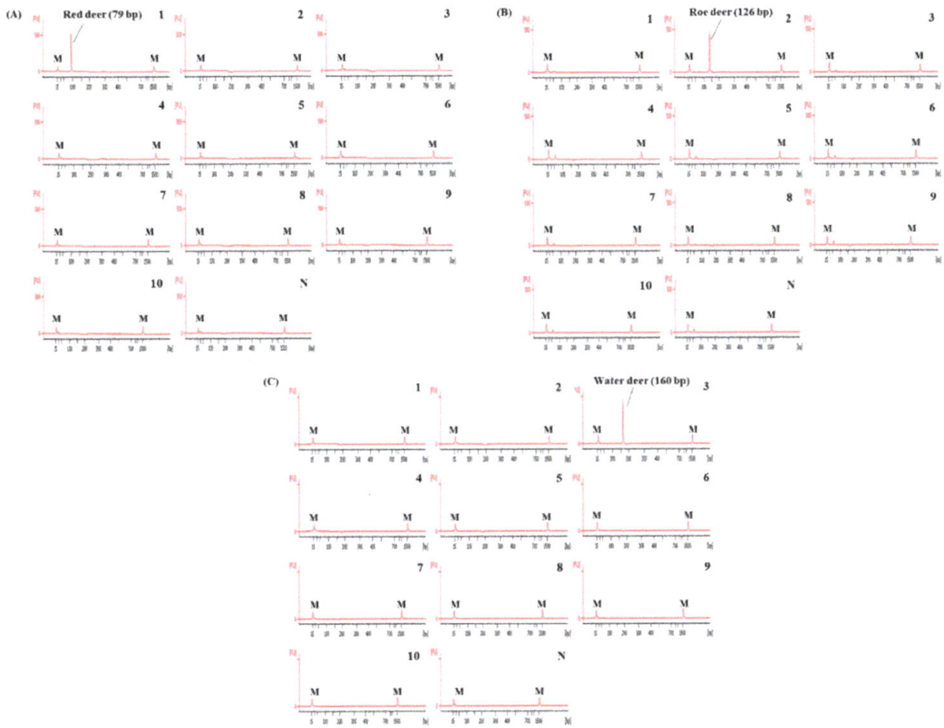

Figure 2. Electropherograms of specificity results of the single PCRs using newly designed primer sets for red deer (**A**), roe deer (**B**), and water deer (**C**). FU: fluorescence, M: alignment marker, lane 1: red deer, 2: roe deer, 3: water deer, 4: beef, 5: pork, 6: lamb, 7: goat, 8: horse, 9: chicken, 10: duck, and N: non-template.

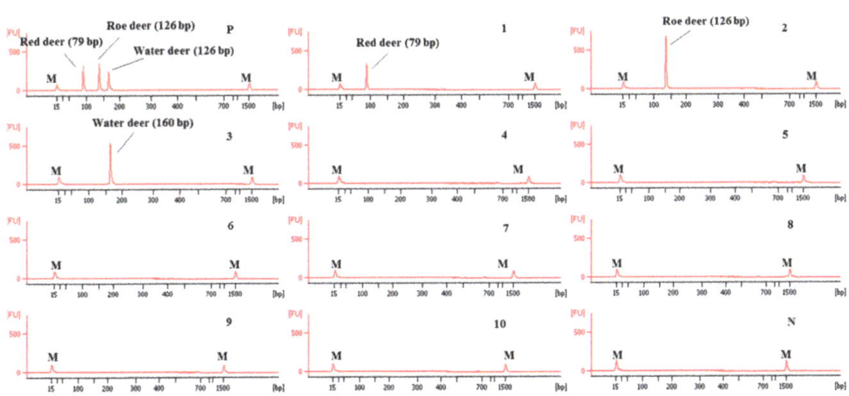

Figure 3. Electropherograms of specificity result of the multiplex PCR assay for red deer, roe deer, and water deer. FU: fluorescence, M: alignment marker, lane P: positive control (red deer, roe deer, and water deer), lane 1: red deer, 2: roe deer, 3: water deer, 4: beef, 5: pork, 6: lamb, 7: goat, 8: horse, 9: chicken, 10: duck, and N: non-template.

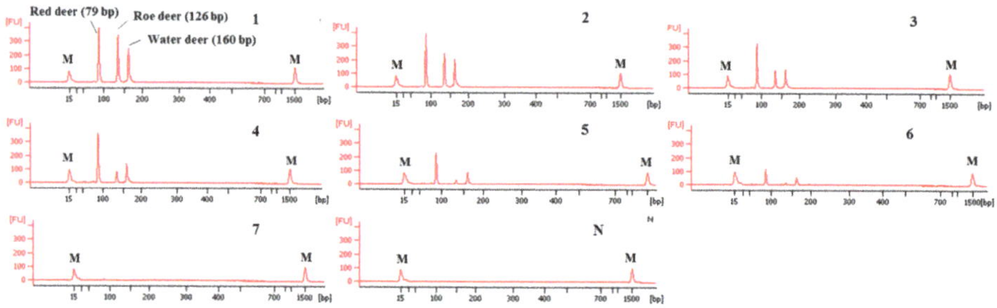

Figure 4. Electropherograms of sensitivity results of the multiplex PCR assay. FU: fluorescence, M: alignment marker, lanes 1–7: 1.0×10^{1}, 10^{0}, 10^{-1}, 10^{-2}, 10^{-3}, 10^{-4}, and 10^{-5} ng of three target species, and lane N: non-template.

3.3. *Application and Validation of Capillary Electrophoresis-Based Multiplex PCR Using Meat DNA Mixtures*

To determine detection limits of CE-mPCR and confirm its applicability to a real sample, two sets of meat DNA mixtures were prepared as follows: (1) roe deer and red deer commonly used as game meat were added in water deer to authenticate game meat species present in commercial deer meats, and (2) red deer was contaminated with roe deer and water deer to detect wild animal species not permitted commercially in several countries. As shown in Figure 5, the detection limit of this assay was at least 0.1% for roe deer and red deer in meat DNA mixtures. In another meat DNA mixture, as low as 0.1% of roe deer and water deer could be detected (Figure 6). Microchip-based capillary electrophoresis technology used in this study is known to provide better accuracy and resolution in multiple target detection [15]. In the present study, at a low concentration of 0.1% for water deer (lane 6 in Figure 6), the result obtained by capillary electrophoresis was clearer than a PCR band visualized on agarose gel (Figure S1). This can help overcome the drawback, such as a false-negative result. In addition, the detection limit was validated independently in duplicate. All results obtained through intra-laboratory validation analysis were similar. The 0.1% of roe deer and red deer mixed in water deer and 0.1% of roe deer and water deer mixed in red deer were detected in two independent PCR reactions using the developed primer sets. Thus, this CE-mPCR assay developed in this study was able to simultaneously detect up to 0.1% of red deer, roe deer, and water deer in meat DNA mixtures. Compared to the limit of detection of 0.1% for roe deer and red deer [4] and 0.5% for red deer [3], our method showed higher or similar sensitivity. Meanwhile, since this was the first study to apply a detection method for water deer to meat DNA mixtures, the detection limit of 0.1% of our developed method could not be compared to previous reports. However, this method might be sufficient to be utilized as a specific and sensitive molecular tool for monitoring these three types of deer meat.

(A)

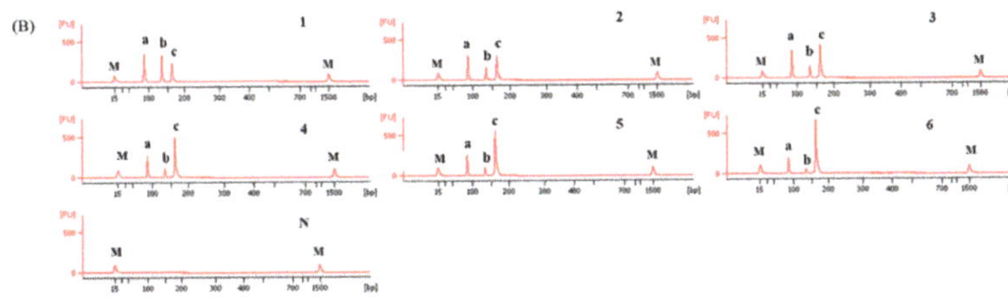

Figure 5. Detection limits of red deer and roe deer in water deer by the multiplex PCR assay. Gel image (**A**) and electropherograms (**B**). FU: fluorescence, M: alignment marker, lane L: 100 bp DNA ladder, lane 1: positive control (10 ng of DNA from target species), lanes 2–6: 10, 5, 1, 0.5, and 0.1% red deer and roe deer in water deer, and lane N: non-template. a, b, and c indicate red deer, roe deer, and water deer, respectively.

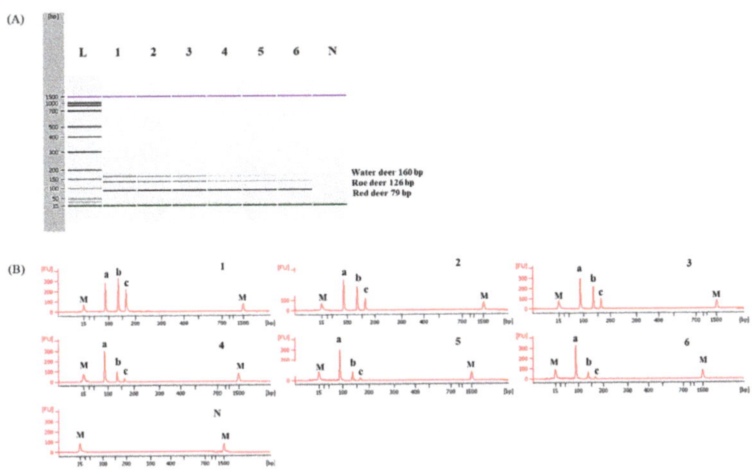

Figure 6. Detection limits of roe deer and water deer in red deer by the multiplex PCR assay. Gel image (**A**) and electropherograms (**B**). FU: fluorescence, M: alignment marker, lane L: 100 bp DNA ladder, lane 1: positive control (10 ng of each DNA from target species), lanes 2–6: 10, 5, 1, 0.5, and 0.1% roe deer and water deer in red deer, and lane N: non-template. a, b, and c indicate red deer, roe deer, and water deer, respectively.

4. Conclusions

The CE-mPCR assay developed in this study could successfully detect three types of deer meat. Its applicability for authentication of meat species was verified using various ratios of meat DNA mixtures. This method is simple and user-friendly. It has high specificity and sensitivity for the simultaneous detection of red deer, roe deer, and water deer. However, despite several advantages of this method developed, since it is utilized for only qualitative detection, further study is required to the application of real-time PCR to quantify meat of target deer species in processed game meat.

Supplementary Materials: The following are available online at http://www.mdpi.com/2304-8158/9/8/982/s1. Figure S1: Detection limits of the multiplex PCR assay. Lane M: 100 bp DNA ladder, lanes 1: positive control (10 ng of each DNA from target species), lanes 2-6: 10, 5, 1, 0.5, and 0.1% roe deer and water deer in red deer, and lane N: non-template. Table S1: Mitochondrial gene sequences of various animals used for the sequence alignment.

Author Contributions: Formal analysis, M.-J.K., Y.-M.L., and S.-M.S.; Funding acquisition, H.-Y.K.; Methodology, M.-J.K.; Project administration, H.-Y.K.; Supervision, H.-Y.K.; Validation, Y.-M.L.; Writing-original draft, M.-J.K.; Writing-review and editing, M.-J.K., Y.-M.L., S.-M.S., and H.-Y.K. All authors have read and agreed to the published version of the manuscript.

Funding: This research was funded by the Ministry of Food and Drug Safety in Korea, grant number 17162MFDS065.

Acknowledgments: This research was supported by a grant (17162MFDS065) from the Ministry of Food and Drug Safety in Korea.

Conflicts of Interest: The authors declare no conflict of interest.

References

1. Rahmati, S.; Julkapli, N.M.; Yehye, W.A.; Basirun, W.J. Identification of meat origin in food products-A review. *Food Control* **2016**, *68*, 379–390. [CrossRef]
2. Kim, M.J.; Suh, S.M.; Kim, S.Y.; Qin, P.; Kim, H.R.; Kim, H.Y. Development of a real-time PCR assay for the detection of donkey (*Equus asinus*) meat in meat processing conditions. *Foods* **2020**, *9*, 130. [CrossRef]
3. Druml, B.; Grandits, S.; Mayer, W.; Hochegger, R.; Cichna-Markl, M. Authenticity control of game meat products—A single method to detect and quantify adulteration of fallow deer (*Dama dama*), red deer (*Cervus elaphus*) and sika deer (*Cervus Nippon*) by real-time PCR. *Food Chem.* **2015**, *170*, 506–517. [CrossRef]
4. Kaltenbrunner, M.; Hochegger, R.; Cichna-Markl, M. Tetraplex real-time PCR assay for the simultaneous identification and quantification of roe deer, red deer, fallow deer and sika deer for deer meat authentication. *Food Chem.* **2018**, *269*, 486–494. [CrossRef]
5. Fajardo, V.; Conzalez, I.; Martin, I.; Rojas, M.; Hernandez, P.E.; Garcia, T.; Martin, R. Real-time PCR for detection and quantification of red deer (*Cervus elaphus*), fallow deer (*Dama dama*), and roe deer (*Capreolus capreolus*) in meat mixtures. *Meat Sci.* **2008**, *79*, 289–298. [CrossRef]
6. Kaltenbrunner, M.; Hochegger, R.; Cichna-Markl, M. Development and validation of a fallow deer (*Dama dama*)-specific TaqMan real-time PCR assay for the detection of food adulteration. *Food Chem.* **2018**, *243*, 82–90. [CrossRef]
7. Druml, B.; Hochegger, R.; Cichna-Markl, M. Duplex real-time PCR assay for the simultaneous determination of the roe deer (*Capreolus capreolus*) and deer (sum of fallow deer, red deer and sika deer) content in game meat products. *Food Control* **2015**, *57*, 370–376. [CrossRef]
8. Koppel, R.; Van Velsen, F.; Ganeshan, A.; Pietsch, K.; Weber, S.; Graf, C.; Murmann, P.; Hochegger, R.; Licina, A. Multiplex real-time PCR for the detection and quantification of DNA from chamois, roe, deer, pork and beef. *Eur. Food Res. Technol.* **2020**, *246*, 1007–1015. [CrossRef]
9. Kim, M.J.; Yoo, I.; Lee, S.Y.; Hong, Y.; Kim, H.Y. Quantitative detection of pork in commercial meat products by TaqMan real-time PCR assay targeting the mitochondrial D-loop region. *Food Chem.* **2016**, *210*, 102–106. [CrossRef]
10. Jiang, X.; Rao, Q.; Mittl, K.; Hsieh, Y.H.P. Monoclonal antibody-based sandwich ELISA for the detection of mammalian meats. *Food Control* **2020**, *110*, 107045. [CrossRef]
11. Chou, C.C.; Lin, S.P.; Lee, K.M.; Hsu, C.T.; Vickroy, T.W.; Zen, J.M. Fast differentiation of meats from fifteen animal species by liquid chromatography with electrochemical detection using copper nanoparticle plated electrodes. *J. Chromatogr. B.* **2007**, *846*, 230–239. [CrossRef]
12. Varani, M.; Faccini, A.; Lambertini, F.; Suman, M.; Laporati, A.; Tedeschi, T.; Sforza, S. Species specific marker peptides for meat authenticity assessment: A multispecies quantitative approach applied to Bolognese sauce. *Food Control* **2019**, *97*, 15–24.
13. Soares, S.; Amaral, J.S.; Oliveira, M.B.; Mafra, I. A SYBR Green real-time PCR assay to detect and quantify pork meat in processed poultry meat products. *Meat Sci.* **2013**, *94*, 115–120. [CrossRef]
14. Kim, M.J.; Kim, H.Y. A fast multiplex real-time PCR assay for simultaneous detection of pork, chicken, and beef in commercial processed meat products. *LWT-Food Sci. Technol.* **2019**, *114*, 108390. [CrossRef]
15. Ali, M.E.; Razzak, M.A.; Hamid, S.B.; Rahman, M.M.; Amin, M.A.; Rashid, N.R. Multiplex PCR assay for the detection of five meat species forbidden in Islamic foods. *Food Chem.* **2015**, *177*, 214–224. [CrossRef]
16. Kim, M.J.; Yoo, I.; Yang, S.M.; Suh, S.M.; Kim, H.Y. Development and validation of a multiplex PCR assay for simultaneous detection of chicken, turkey, and duck in processed meat products. *Int. J. Food Sci. Technol.* **2018**, *53*, 2673–2679. [CrossRef]
17. Kim, M.J.; Kim, S.Y.; Jung, S.K.; Kim, M.Y.; Kim, H.Y. Development and validation of ultrafast PCR assays to detect six species of edible insects. *Food Control* **2019**, *103*, 21–26. [CrossRef]

18. Park, S.H.; Ricke, S.C. Development of multiplex PCR assays for simultaneous detection of *Salmonella* genus, *Salmonella* subspecies I., *Salm.* Eteritidis, *Salm.* Heidelberg and *Salm.* Typhimurium. *J. Appl. Micobiol.* **2015**, *118*, 152–160. [CrossRef]
19. Rak, L.; Knapik, K.; Bania, J.; Sukowski, J.; Gadzinowski, A. Detection of roe deer, red deer, and hare meat in raw materials and processed products available in Poland. *Eur. Food Res. Technol.* **2014**, *239*, 189–194. [CrossRef]
20. Hou, B.; Meng, X.; Zhang, L.; Guo, J.; Li, S.; Jin, H. Development of a sensitive and specific multiplex PCR method for the simultaneous detection of chicken, duck and goose DNA in meat products. *Meat Sci.* **2015**, *101*, 90–94. [CrossRef] [PubMed]
21. Kim, M.J.; Kim, H.Y. Species identification of commercial jerky products in food and feed using direct pentaplex PCR assay. *Food Control* **2017**, *78*, 1–6. [CrossRef]
22. Kitpipit, T.; Sittichan, K.; Thanakiatkrai, P. Direct-multiplex PCR assay for meat species identification in food products. *Food Chem.* **2014**, *163*, 77–82. [CrossRef] [PubMed]
23. Lin, W.F.; Hwang, D.F. A multiplex PCR assay for species identification of raw and cooked bonito. *Food Control* **2008**, *19*, 879–885. [CrossRef]
24. He, H.; Hong, X.; Feng, Y.; Wang, Y.; Ying, J.; Liu, Q.; Qian, Y.; Zhou, X.; Wang, D. Application of Quadruple Multiplex PCR Detection for Beef, Duck, Mutton and Pork in Mixed Meat. *J. Food Nutr. Res.* **2015**, *3*, 392–398. [CrossRef]
25. Hossain, M.M.; Ali, M.E.; Hamid, S.B.A.; Mustafa, S.; Desa, M.N.M.; Zaidul, I.S.M. Targeting double genes in multiplex PCR for discriminating bovine, buffalo and porcine materials in food chain. *Food Control* **2017**, *73*, 175–184. [CrossRef]
26. Suh, S.M.; Kim, M.J.; Kim, H.I.; Kim, H.J.; Kim, H.Y. A multiplex PCR assay combined with capillary electrophoresis for the simultaneous detection of tropomyosin allergens from oyster, mussel, abalone, and clam mollusk species. *Food Chem.* **2020**, *317*, 126451. [CrossRef]

foods

Article

Development of a Real-Time PCR Assay for the Detection of Donkey (*Equus asinus*) Meat in Meat Mixtures Treated under Different Processing Conditions

Mi-Ju Kim †, Seung-Man Suh †, Sung-Yeon Kim, Pei Qin, Hong-Rae Kim and Hae-Yeong Kim *

Department of Food Science and Biotechnology, Institute of Life Sciences & Resources, Kyung Hee University, Yongin 17104, Korea; mijukim79@gmail.com (M.-J.K.); teri2gogo@naver.com (S.-M.S.); sungyeon94@naver.com (S.-Y.K.); qinp77@163.com (P.Q.); star6783204@nate.com (H.-R.K.)

* Correspondence: hykim@khu.ac.kr; Tel.: +82-31-201-2660; Fax: +82-31-204-8116

† These authors contributed equally to the work.

Received: 25 December 2019; Accepted: 23 January 2020; Published: 26 January 2020

Abstract: In this study, a donkey-specific primer pair and probe were designed from mitochondrial *cytochrome b* gene for the detection of raw donkey meat and different processed meat mixtures. The PCR product size for donkey DNA was 99 bp, and primer specificity was verified using 20 animal species. The limit of detection (LOD) was examined by serially diluting donkey DNA. Using real-time PCR, 0.001 ng of donkey DNA could be detected. In addition, binary meat mixtures with various percentages of donkey meat (0.001%, 0.01%, 0.1%, 1%, 10%, and 100%) in beef were analyzed to determine the sensitivity of this real-time PCR assay. At least 0.001% of donkey meat was detected in raw, boiled, roasted, dried, grinded, fried, and autoclaved meat mixtures. The developed real-time PCR method showed sufficient specificity and sensitivity in identification of donkey meat and could be a useful tool for the identification of donkey meat in processed products.

Keywords: food adulteration; food fraud; donkey; *cytochrome b*; real-time PCR; meat products

1. Introduction

Identification of animal species in meat products is important for preventing food adulteration and providing accurate information regarding meat species to consumers. Donkey meat products are highly nutritious; moreover, in many countries, including Korea, it is considerably more expensive than other meats owing to its low supply [1]. In Islamic countries, donkey meat consumption is prohibited on religious grounds [2]. Due to donkey meat being expensive, it is likely to be mixed with other cheaper meats for economic benefits, and there is a need to avoid donkey meat entering the food chain in Islamic countries. Therefore, it is necessary to develop reliable and specific detection methods that can accurately identify animal species from meat products to prevent cases of disguising meat from one species as another [3,4].

To date, many protein- and DNA-based detection methods have been developed to determine animal species in food products. In particular, DNA-based methods have been used to detect target species in processed foods because DNA is stable at high temperatures and pressures [5–7]. Real-time PCR is an effective tool that accurately amplifies target DNA. Several real-time PCR methods, particularly based on detection via TaqMan probes, have been developed with high sensitivity and accuracy to distinguish common meat species such as pork, lamb, and beef [8–12].

Mitochondrial DNA (mtDNA) has been mainly used to detect target species in meat products, and mtDNA sequences from related species have been phylogenetically studied [13]. In addition, because mtDNA evolves faster than nuclear DNA, mtDNA has been used to discriminate target species from similar species. Further, mitochondria are present in high copy numbers in cells. Thus, real-time PCR based on specific mtDNA sequences can amplify target DNA degraded by food processing or mixed with other species [14,15]. In many studies, mitochondrial *cytochrome b* has been used to develop species-specific real-time PCR detection methods [12,16,17].

Here, we designed a donkey-specific primer and probe from mitochondrial *cytochrome b* and developed a real-time PCR method to accurately identify donkey meat. Although there have been several previous studies for donkey meat detection [1,9], no study has applied donkey meats treated under a variety of processing conditions. Thus, in this study, we evaluated the applicability of the developed method for the detection of donkey meat using raw, boiled, roasted, dried, grinded, fried, and autoclaved meat mixtures.

2. Materials and Methods

2.1. Preparation of Samples and Binary Meat Mixtures

A total of 20 raw meat samples were obtained from the Conservation Genome Resource Bank for Korean Wildlife (CGRB), the National Institute of Animal Science (NIAS), and local markets in South Korea. All samples were homogenized in small pieces and stored at −20 °C until analysis.

Binary meat mixtures were prepared to determine the detection limit of donkey-specific real-time PCR assay. For binary raw meat mixtures, 10 g of each of donkey and beef was lyophilized for 24 h using a freeze dryer (Ilsin Biobase, Dongduchon, Korea) to remove moisture of raw meats without DNA degradation, and then ground. In addition, to evaluate the applicability of the developed method in processed meat products, two meats were treated under six different processing conditions as follows: (1) boiled at 100 °C for 15 min in water bath (MONO-TECH, Daegu, Korea), (2) roasted at 200 °C for 5 min in hot plate (Corning Co., New York, NY, USA), (3) dried at 65 °C for 12 h in dry oven (HANKUK S&I, Hwaseong, Korea), (4) grinded for 5 min in commercial grinder (Buwon Electronics, Daegu, Korea), (5) fried at 180 °C for 5 min in cooking oil, and (6) autoclaved at 121 °C 150 kPa for 30 min. Each meat for the six different mixtures was prepared in triplicate, which was made on different days and from meats of different origins. After treatments, each of binary meat mixtures containing six different percentages (0.001%, 0.01%, 0.1%, 1%, 10%, and 100% (*w/w*)) of donkey meats in beef was prepared. Samples (final weight, 100 mg) of various meat mixtures were used for analysis.

2.2. DNA Extraction

Genomic DNA was extracted from raw and autoclaved meat mixtures using the DNeasy Blood and Tissue kit (Qiagen, Hilden, Germany), following the manufacturer's instructions with minor modifications. Briefly, 100 mg of each sample was lysed with 3600 μL of ATL buffer and 400 μL of proteinase K (20 mg/mL) in a water bath at 56 °C for 1 h. After adding 40 μL of RNase A (100 mg/mL), the mixture was incubated at room temperature for 2 min. AL buffer (4000 μL) and 100% ethanol (4000 μL) were mixed with the DNA mixture, and the sample was centrifuged through a spin column. After washing with AW1 and AW2 buffers, the column-bound DNA was eluted with purified water. The purity and concentration of the extracted DNA were confirmed using a Maestro Nano spectrophotometer (Maestrogen, Las Vegas, NV, USA).

2.3. Primer and Probe Design

A donkey-specific primer pair and probe for the detection of donkey were designed to amplify the specific target DNA (Table 1). To design a donkey-specific primer pair, mitochondrial *cytochrome b* sequences from 20 animal species, including donkey, beef, and horse (Accession No.: FJ428510.1, D34635.1, and MH594485.1, respectively) were obtained from GenBank. All sequences were aligned using Clustal Omega program (http://www.ebi.ac.uk/Tools/msa/clustalo/). The primer pair and probe were designed using the Primer Designer program version 3.0 (Scientific and Education Software, Durham, NC, USA) and synthesized by Bionics (Seoul, Korea) and Bioneer (Daejeon, Korea).

Table 1. Sequences of primers and probes used in this study.

Primer Name	Sequences (5′→3′)	Target Genes	Amplicon Size (bp)	Reference
Don3 F	CGCTCCATTCCCAACAAACTAGGTGGT	*Cytochrome b*	99	This study
Don3 R	GCTTCGTTGTTTTGACATGTGTAGGGTA			
Don3 P	FAM-GCCCTTATCCTTTCCATCTTAATCC-TAMRA			
18SpEU-DIR	GGTAGTGACGAAAAATAACAATACAGGAC	18S rRNA	141	[18]
18SpEU-INV	ATACGCTATTGGAGCTGGAATTACC			
18S probe	FAM-AAGTGGACTCATTCCAATTACAGGGCCT-TAMRA			

2.4. Conventional PCR Reaction

Conventional PCR was performed using a thermal cycler (PC808, ASTEC, Kyoto, Japan) under the following conditions: pre-incubation at 94 °C for 5 min, 30 cycles of denaturation for 30 s at 94 °C, annealing for 30 s at 60 °C, extension for 30 s at 72 °C, and final extension for 5 min at 72 °C. The PCR reaction mixture comprised 400 nM of each primer, 0.5 U of Ampli-Gold Taq polymerase (Applied Biosystems, Foster City, CA, USA), 10× PCR buffer (Applied Biosystems), 2.5 mM of each dNTP (Applied Biosystems), 1.5 mM of $MgCl_2$ (Applied Biosystems), and 10 ng of DNA template isolated from each of animal species for the specificity test in a total reaction volume of 25 μL. All PCR products were electrophoresed on a 2% agarose gel and then visualized under UV irradiation.

2.5. Real-Time PCR Reaction

Real-time PCR amplification was performed using an ABI 7500 Real-time PCR instrument (Applied Biosystems). The PCR reaction was performed in a final volume of 25 μL, containing 2× TaqMan Universal Master mix (Applied Biosystems), 400 nM of primer pairs, 200 nM of the probe, and 10 ng of the DNA template. Real-time PCR was performed with a holding stage at 95 °C for 10 min, followed by 40 cycles at 95 °C for 15 s and 60 °C for 1 min. All real-time PCR reactions were performed in triplicates; no-template control (NTC) was used as a negative control. Data were analyzed using 7500 Software V.2.3 (Applied Biosystems).

2.6. Specificity and Sensitivity of Real-Time PCR

The specificity of the donkey-specific primer pair and probe was tested using 10 ng of genomic DNA extracted from 20 animal species. To confirm the presence of DNA, endogenous primer pair and probe targeting the 18S rRNA gene were also used [18]. The sensitivity of the real-time PCR was measured using 10-fold serially diluted DNA (from 10 to 0.001 ng) extracted from donkey. The detection limit of real-time PCR in 6 processed binary mixtures containing donkey meat (ranging in concentration from 10% to 0.001%) mixed with beef meat was used.

3. Results and Discussions

3.1. Specificity

The donkey-specific primer and probe were designed to get a small product size of 99 bp from mitochondrial *cytochrome b*. The specificity of donkey primer set was confirmed using the DNA from 20 animal species as templates for this assay. Only the DNA fragment specific for donkey was amplified by conventional PCR, and there was no amplification in 19 nontarget species (Table 2). The PCR product amplified by the donkey-specific primer was sequenced to verify the donkey species (Figure S1). The specificity of the real-time PCR method was additionally verified, and the donkey DNA was specifically amplified without any cross-reactivity against the 19 other animal species tested (Table 2). To confirm the presence of DNA, eukaryotic PCR targeting the 18S rRNA gene was performed. As shown in Table 2, positive signals were observed in all PCR reactions. Thus, our results proved that the donkey-specific primer and probe were accurately amplified the target DNA.

Table 2. Specificity results using conventional and real-time PCR assays.

Common Name	Scientific Name	Conventional PCR		Real-Time PCR	
		Donkey-Specific PCR	Eukaryotic PCR	Donkey-Specific PCR	Eukaryotic PCR
Donkey	*Equus asinus*	+	+	+	+
Horse	*Equus caballus*	−	+	−	+
Beef	*Bos taurus*	−	+	−	+
Lamb	*Ovis aries*	−	+	−	+
Goat	*Capra hircus*	−	+	−	+
Deer	*Cervus elaphus*	−	+	−	+
Pork	*Sus scrofa domestica*	−	+	−	+
Rabbit	*Oryctolagus cuniculus*	−	+	−	+
Raccoon dog	*Nyctereutes procyonoides*	−	+	−	+
Dog	*Canis lupus familiaris*	−	+	−	+
Cat	*Felis catus*	−	+	−	+
Siberian chipmunk	*Tamias sibiricus*	−	+	−	+
Turkey	*Meleagris gallopavo*	−	+	−	+
Ostrich	*Struthio camelus*	−	+	−	+
Chicken	*Gallus gallus*	−	+	−	+
Pheasant	*Phasianus colchicus*	−	+	−	+
Duck	*Anas platyrhynchos*	−	+	−	+
Goose	*Anser anser*	−	+	−	+
Pigeon	*Columba livia domestica*	−	+	−	+
Japanese quail	*Coturnix japonica*	−	+	−	+

3.2. Sensitivity of the Donkey-Specific Real-Time PCR Assay

The sensitivity of the donkey-specific real-time PCR targeting *cytochrome b* gene was determined using 10-fold serially diluted donkey DNA from 10 to 0.001 ng. Ct values were plotted against logarithmic DNA concentrations to construct the standard curve for the donkey DNA. The slope and correlation coefficient (R^2) of the standard curve were −3.79 and 0.997, respectively. PCR efficiency was calculated using the equation "$E = (10^{(-1/slope)} - 1)$" and was determined to be 83.47% (Figure 1). Each PCR reaction was performed thrice, and 0.001 ng of the donkey DNA was detected in all the reactions. The absolute detection limit of the donkey-specific real-time PCR assay was as low as 0.001 ng. These results demonstrated that the real-time PCR method developed in this study has good linearity and sensitivity.

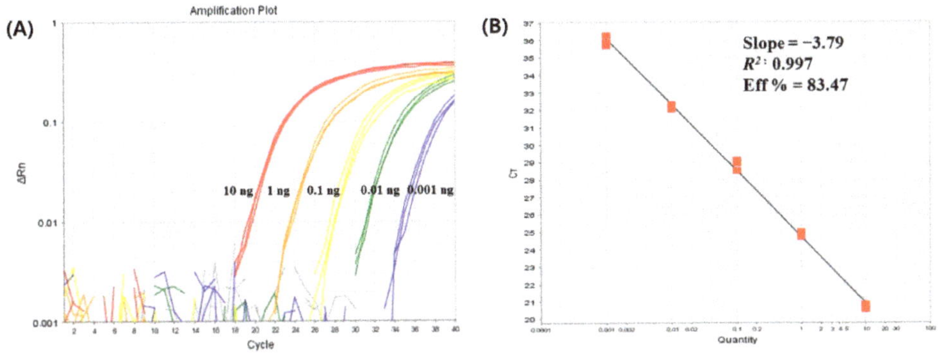

Figure 1. Amplification plot (**A**) and standard curve (**B**) for the detection of donkey DNA using 10-fold serial dilutions (from 10 to 0.001 ng).

3.3. Application of the Real-Time PCR Assay to Meat Mixtures Treated under Different Processing Conditions

The meat mixtures treated under six conditions were used to confirm the applicability of the developed method using highly processed meat products as well as raw meat, and various concentrations of binary meat mixtures were prepared for the limit of detection (LOD) test of this method. As shown in Table 3, 0.001% of donkey meat was successfully detected in all processed meat mixtures, despite high heat and pressure treatments of donkey meat. The average Ct values of three replicates using donkey DNA were 18.45 ± 0.7, 20.24 ± 0.97, 18.74 ± 0.06, 18.59 ± 0.31, 19.17 ± 0.6, 21.17 ± 0.55, and 20.86 ± 0.26 for raw, boiled, roasted, dried, ground, fried, and autoclaved meats, respectively. Ct values of the target species in the boiled, fried, and autoclaved meat mixtures were relatively higher than in other meat mixtures; this may be attributable to the fact that the DNA was degraded under the high pressure and temperature treatments [7,19].

The lowest percentage of donkey meat adulteration that could be detected by the real-time PCR method developed in this study was 0.001%, which was lower than 1% of detection limit reported by Chen et al. [1] and same or lower than 0.001% and 0.01% of detection limits reported by Kesmen et al. [19]. Therefore, this real-time PCR method can help to confirm the presence of donkey meat in highly processed meat products and provide accurate information on target meat species. For a more efficient detection method tool, a further study can be performed the development of multiplex real-time PCR for the detection of two genes, including the endogenous 18S rRNA gene.

Table 3. Results of the developed real time-PCR method when applied to six meat mixtures processed following different technologies.

Target Species	Ratio of Donkey Meat in the Binary Meat Mixture (%)	Ct Values						
		Raw	Boiled	Roasted	Dried	Grinded	Fried	Autoclaved
Donkey	100	18.45 ± 0.70 [a]	20.24 ± 0.97	18.74 ± 0.06	18.59 ± 0.31	19.17 ± 0.60	21.17 ± 0.55	20.86 ± 0.26
	10	21.55 ± 0.52	23.34 ± 0.64	21.79 ± 0.08	21.27 ± 0.25	22.30 ± 0.57	24.01 ± 0.33	23.86 ± 0.21
	1	24.58 ± 0.47	26.4 ± 0.63	24.97 ± 0.0.8	24.35 ± 0.43	25.71 ± 0.10	27.31 ± 0.42	26.68 ± 0.30
	0.1	27.37 ± 1.03	29.15 ± 1.06	28.19 ± 0.07	27.47 ± 0.43	28.84 ± 0.10	30.61 ± 0.38	29.53 ± 0.19
	0.01	30.48 ± 1.03	32.11 ± 1.20	31.32 ± 0.05	30.63 ± 0.56	31.83 ± 0.43	33.71 ± 0.27	32.67 ± 0.08
	0.001	33.59 ± 1.08	34.89 ± 0.96	34.35 ± 0.07	33.57 ± 0.51	34.13 ± 0.51	36.72 ± 0.33	35.69 ± 0.17

[a] Average Ct value ± standard deviation obtained from triplicate reactions.

4. Conclusions

This study described the development of a real-time PCR method to identify donkey DNA. By targeting a 99 bp fragment of mitochondrial *cytochrome b*, the designed primer pair and probe specifically amplified the donkey DNA. The standard curve of the developed real-time PCR method has good linearity and sensitivity, which is adequate to successfully amplify the target DNA. Raw and highly processed meat mixtures were analyzed with a sensitivity of 0.001% to demonstrate the applicability of the method developed in the present study for detecting donkey meat in processed meat products. The applicability of this method was verified with six processing conditions that can be used for meat processing, and the applicability was confirmed under all processing conditions. Therefore, the real-time PCR method developed in this study could be a useful tool for the detection of donkey and determination of intentional adulterations or food fraud in highly processed meat products.

Supplementary Materials: The following are available online at http://www.mdpi.com/2304-8158/9/2/130/s1, Figure S1: The identity result of sequences of the PCR products for donkey-specific primer sets.

Author Contributions: H.-Y.K. conceived of the overall study. M.-J.K., S.-Y.K., S.-M.S., P.Q., and H.-R.K. carried out the experiment. M.-J.K. wrote the first draft of the manuscript. SMS reviewed the manuscripts. All authors have read and agreed to the published version of the manuscript.

Funding: This research was supported by grant 17162MFDS065 from the Ministry of Food and Drug Safety in Korea.

Conflicts of Interest: The authors declare no conflict of interest.

References

1. Chen, A.; Wei, C.; Chen, G.; Zhao, Y.; Yang, S. Duplex PCR approach for the detection and quantification of donkey, horse and mule in raw and heat-processed meat products. *Int. J. Food Sci. Technol.* **2015**, *50*, 834–839. [CrossRef]
2. Mousavi, S.M.; Khaniki, G.J.; Eskandari, S.; Rabiei, M.; Samiee, S.M.; Mehdizadeh, M. Applicability of species-specific polymerase chain reaction for fraud identification in raw ground meat commercially sold in Iran. *J. Food Compos. Anal.* **2015**, *40*, 47–51. [CrossRef]
3. Kim, K.H.; Lee, H.Y.; Kim, Y.S.; Kim, M.R.; Jung, Y.K.; Lee, J.H.; Chang, H.S.; Park, Y.C.; Kim, S.Y.; Choi, J.D.; et al. Development of species-specific PCR to determine the animal raw Material. *J. Food Hyg. Saf.* **2014**, *29*, 347–355. [CrossRef]
4. Sul, S.Y.; Kim, M.J.; Kim, H.Y. Development of a direct loop-mediated isothermal amplification (LAMP) assay for rapid and simple on-site detection of chicken in processed meat products. *Food Control* **2019**, *98*, 194–199. [CrossRef]
5. Herrero, B.; Vieites, J.M.; Espiñeira, M. Development of an in-house fast real-time PCR method for detection of fish allergen in foods and comparison with a commercial kit. *Food Chem.* **2014**, *151*, 415–420. [CrossRef] [PubMed]
6. Pascoal, A.; Barros-Velazques, J.; Ortea, I.; Cepeda, A.; Gallardo, J.M.; Calo-Mata, P. Molecular identification of the black tiger shrimp (*Penaeus monodon*), the white leg shrimp (*Litopenaeus vannamei*) and the Indian white shrimp (*Fenneropenaeus indicus*) by PCR targeted to the 16S rRNA mtDNA. *Food Chem.* **2011**, *125*, 1457–1461. [CrossRef]
7. Kim, M.J.; Kim, H.Y. Development of a fast duplex real-time PCR assay for simultaneous detection of chicken and pigeon in raw and heat-treated meats. *Food Control* **2018**, *85*, 1–5. [CrossRef]
8. Eaqub Ali, M.; Hashim, U.; Sabar Dhahi, T.; Mustafa, S.; Che, Y.B.; Abdul Latif, M.M. Analysis of pork adulteration in commercial burgers targeting porcine-specific mitochondrial cytochrome B gene by TaqMan probe real-time polymerase chain reaction. *Food Anal. Methods* **2012**, *5*, 784–794.
9. Kesmen, Z.; Yetiman, A.E.; Sain, F.; Yetim, H. Detection of chicken and turkey meat in meat mixtures by using real-time PCR assay. *J. Food Sci.* **2012**, *77*, 167–173. [CrossRef] [PubMed]
10. Rodriguez, M.A.; García, T.; González, I.; Hernández, P.E.; Martín, R. TaqMan real-time PCR for detection and quantitation of pork in meat mixtures. *Meat Sci.* **2005**, *70*, 113–120. [CrossRef] [PubMed]

11. Tanabe, S.; Hase, M.; Yano, T.; Sato, M.; Fujimura, T.; Akiyama, H. A real-time quantitative PCR detection method for pork, chicken, beef, mutton, and horseflesh in foods. *Biosci. Biotechnol. Biochem.* **2007**, *71*, 3131–3135. [CrossRef] [PubMed]
12. Fang, X.; Zhang, C. Detection of adulterated murine components in meat products by TaqMan real-time PCR. *Food Chem.* **2016**, *192*, 485–490. [CrossRef] [PubMed]
13. Kumar, A.; Kumar, R.R.; Sharma, B.D.; Gokulakrishnan, P.; Mendiratta, S.K.; Sharma, D. Identification of species origin of meat and meat products on the DNA basis: A review. *Crit. Rev. Food Sci. Nutr.* **2015**, *55*, 1340–1351. [CrossRef] [PubMed]
14. Girish, P.S.; Anjaneyulu, A.S.R.; Viswas, K.N.; Anand, M.; Rajkumar, N.; Shivakumar, B.M.; Bhaskar, S. Sequence analysis of mitochondrial 12S rRNA gene can identify meat species. *Meat Sci.* **2004**, *66*, 551–556. [CrossRef]
15. Kim, S.Y.; Kim, M.J.; Jung, S.K.; Kim, H.Y. Development of a fast real-time PCR assay based on TaqMan probe for identification of edible rice grasshopper (*Oxya chinensis*) in processed food products. *Food Res. Int.* **2019**, *116*, 441–449. [CrossRef] [PubMed]
16. Fajardo, V.; Gonzalez, I.; Rojas, M.; Garcia, T.; Martin, R. A review of current PCR-based methodologies for the authentication of meats from game animal species. *Trends Food Sci. Technol.* **2010**, *21*, 408–421. [CrossRef]
17. Ali, M.E.; Amin, M.A.; Razzak, M.A.; Hamid, S.B.A.; Rahman, M.M.; Rashid, N.R.A.; Asing. Short Amplicon-Length PCR Assay Targeting Mitochondrial Cytochrome b Gene for the Detection of Feline Meats in Burger Formulation. *Food Anal. Methods* **2016**, *9*, 571–581. [CrossRef]
18. Fajardo, V.; Gonzalez, I.; Martin, I.; Rojas, M.; Hernandez, P.E.; Garca, T.; Martin, R. Real-time PCR for quantitative detection of chamois (Rupicapra rupicapra) and Pyrenean ibex (Capra pyrenaica) in mea tmixtures. *J. AOAC Int.* **2008**, *97*, 103–111. [CrossRef]
19. Kesmen, Z.; Güllüce, A.; Yilmaz, M.T.; Yetiman, A.E.; Yetim, H. TaqMan-based duplex real-time polymerase chain reaction approach for the detection and quantification of donkey and pork adulterations in raw and heat-processed meats. *Int. J. Food Prop.* **2014**, *17*, 629–638. [CrossRef]

foods

MDPI

Article

Metabolite Profiling and Chemometric Study for the Discrimination Analyses of Geographic Origin of Perilla (*Perilla frutescens*) and Sesame (*Sesamum indicum*) Seeds

Tae Jin Kim [1,†], Jeong Gon Park [1,†], Hyun Young Kim [2], Sun-Hwa Ha [3], Bumkyu Lee [4], Sang Un Park [5], Woo Duck Seo [2,*] and Jae Kwang Kim [1,*]

1 Division of Life Sciences, College of Life Sciences and Bioengineering, Incheon National University, Incheon 22012, Korea; f91gd@inu.ac.kr (T.J.K.); parkjk1132@naver.com (J.G.P.)
2 Division of Crop Foundation, National Institute of Crop Science, Rural Development Administration, Wanju, Jeonbuk 55365, Korea; hkkim84@korea.kr
3 Department of Genetic Engineering and Graduate School of Biotechnology, Kyung Hee University, Yongin 17104, Korea; sunhwa@khu.ac.kr
4 Department of Environment Science & Biotechnology, Jeonju University, Jeonju 55069, Korea; leebk@jj.ac.kr
5 Department of Crop Science, Chungnam National University, 99 Daehak-ro, Yuseong-gu, Daejeon 34134, Korea; supark@cnu.ac.kr
* Correspondence: swd2002@korea.kr (W.D.S.); kjkpj@inu.ac.kr (J.K.K.); Tel.: +82-63-238-5305 (W.D.S.); +82-32-835-8241 (J.K.K.)
† These authors contributed equally to this work.

Received: 26 June 2020; Accepted: 21 July 2020; Published: 24 July 2020

Abstract: Perilla and sesame are traditional sources of edible oils in Asian and African countries. In addition, perilla and sesame seeds are rich sources of health-promoting compounds, such as fatty acids, tocopherols, phytosterols and policosanols. Thus, developing a method to determine the geographic origin of these seeds is important for ensuring authenticity, safety and traceability and to prevent cheating. We aimed to develop a discriminatory predictive model for determining the geographic origin of perilla and sesame seeds using comprehensive metabolite profiling coupled with chemometrics. The orthogonal partial least squares-discriminant analysis models were well established with good validation values (Q^2 = 0.761 to 0.799). Perilla and sesame seed samples used in this study showed a clear separation between Korea and China as geographic origins in our predictive models. We found that glycolic acid could be a potential biomarker for perilla seeds and proline and glycine for sesame seeds. Our findings provide a comprehensive quality assessment of perilla and sesame seeds. We believe that our models can be used for regional authentication of perilla and sesame seeds cultivated in diverse geographic regions.

Keywords: perilla; sesame; geographic origin; metabolomics; multivariate analysis; metabolite profiling

---

1. Introduction

Perilla (*Perilla frutescens*) seed is a rich source of health-promoting compounds, such as tocopherols, phytosterols, policosanols and fatty acids, which have various bioactivities [1]. Tocopherols have an antioxidant effect and are known as vitamin E. Phytosterols show reduction of total cholesterols in the serum. They increase high-density lipoprotein cholesterol levels and reduce low-density lipoprotein cholesterol levels in the blood. Policosanols also have a serum lipid- and cholesterol-lowering effect and other beneficial effects, such as cytoprotection, antiaging, liver protection, antioxidant and anti-parkinsonian effects [2]. In addition, perilla seeds contain high levels of octacosanol (C28-ol) [1,2].

The fatty acid α-linolenic acid (C18:3n3) is found in high levels in perilla seeds, which is essential to human health; moreover, perilla seeds contain omega-3 fatty acid, which lowers inflammation and risk of cancer and cardiovascular and atopic diseases [3]. Sesame (*Sesamum indicum* L.) seeds also contain the abovementioned health-promoting compounds, and they are a good source of proteins rich in sulfur-containing amino acids [4,5]. Linoleic acid (C18:2n6), which is an essential fatty acid for humans, is the main fatty acid found in sesame seeds; in addition, oleic acid (C18:1n9) is the second most abundant fatty acid in sesame seeds [6]. In addition, γ-tocopherol is the main tocopherol in sesame seeds [6,7]. Sesame seeds reportedly contain high levels of phytosterols [5]. Although the composition and contents of various health beneficial compounds in the perilla and sesame seeds have been reported, to the best of our knowledge, a comprehensive comparative-analysis involving hydrophilic and lipophilic compounds has not been reported.

Metabolomics has been widely used to distinguish food products on the basis of differences in their chemical composition and metabolite contents [8,9]. Food metabolomics comprises analytical techniques and multivariate discriminant analysis (MVDA) techniques used for food substances. The analytical techniques usually used in food metabolomics are mass spectrometry (MS) coupled with separation techniques such as liquid chromatography (LC) and gas chromatography (GC) and nuclear magnetic resonance (NMR) [10]. For MVDA, the most commonly used methods are principal component analysis (PCA), partial least squares-discriminant analysis (PLS-DA) and orthogonal partial least squares discriminant analysis (OPLS-DA), which are useful tools for describing correlations and diagnosing differences among the studied samples and their metabolites. Therefore, food metabolomics strategies are suitable for analyzing food safety, authenticity, traceability and quality assessment and these strategies have been used to assess various foods and beverages, such as adzuki bean, olive oil, cabbage, wine, rice, coffee and tomato [11–16].

Perilla and sesame seeds are traditionally used as sources of edible oils in Korea, China, India and other Asian countries. Perilla is cultivated in Korea, China, Japan, India, Nepal and Thailand [17,18]. In Korea, the production of perilla seeds was average 40,448 tons per year over the last decade, and approximately 24,411 tons were imported per year [19]. Out of the imported perilla seeds, almost of 99% are Chinese perilla seeds [20]. Sesame is mainly produced in China, Myanmar, India and African countries such as Sudan, Nigeria and Tanzania. In Korea, the average production of sesame was 12,168 tons over the last decade, whereas approximately 76,812 tons were imported; the self-sufficiency rate in sesame production was 14% [19,20]. In particular, more than 90% of sesame seeds were imported from China (50%) and India (40%) [19]. The price of perilla and sesame seeds is influenced by their places of origin; therefore, identification of the geographic origin of these seeds is important [21]. Forging or mislabeling domestic seeds as imported seeds to gain economic benefits has increasingly become a crucial issue for both producers and consumers, and it affects food quality assurance and safety [22]. To prevent this problem, developing a precise and accurate method to identify the geographic origin of perilla and sesame seeds is needed. Recently, genomic and analytical approaches have been developed for such identification [4,6,15,23–25]. The genomics method is considerably accurate; however, it cannot determine the geographic origins of the same plant variety [14]. On the contrary, the analytical methods can accurately determine the different geographic origins of the same variety based on the differences in chemical composition. Previous studies have used multivariate analysis for discriminating between geographic origins of perilla and sesame seeds using genomics and analytical methods [4,22]. In the case of perilla, however, genomic methods have been reported to determine geographic origin, but analytical methods have not been developed [23].

We aimed to develop a method to discriminate the geographic origin of perilla and sesame seeds and to assess their nutritional quality. To discriminate the geographic origin, MVDA was performed with targeted metabolite profiling using gas chromatography-mass spectrometry (GC-MS). The hydrophilic and lipophilic metabolite profiling (including amino acids, organic acids, sugars, sugar alcohols, tocopherols, sterols, policosanols and fatty acids) of perilla and sesame seeds originated in the Korea and China was performed. Using this, a discrimination model was established for the determination of geographic origins of perilla and sesame seeds. This is the first attempt to construct a discrimination model for perilla seeds using metabolomics. Further, potential biomarkers for distinguishing the geographic origins of perilla and sesame seeds were proposed. A comprehensive food quality assessment was also performed. Our findings can offer reliable information about food authenticity and traceability of perilla and sesame seeds.

2. Materials and Methods

2.1. Sample and Chemicals

Korean perilla and sesame cultivars were grown at the National Institute of Crop Science, Rural Development Administration, Wanju-gun, Korea, during the 2018 growing season (June to November). Chinese perilla and sesame samples were procured from a local market in Xinzhou and JiangXia district (Wuhan city), China. The Chinese samples including perilla and sesame were from the recent harvests of November 2017 and 2016, respectively. Three biologic replicates were prepared for each sample. 5α-Cholestane, ribitol, pentadecanoic acid, fatty acid methyl ester (FAME) mixture, *N*-methyl-*N*-trimethylsilyl trifluoroacetamide (MSTFA) and pyridine were purchased from Sigma-Aldrich (St. Louis, Mo, USA). All other chemicals used in this study were reagent grade unless stated otherwise.

2.2. Extraction and Analysis of Hydrophilic Compounds

The extraction and analysis of hydrophilic compounds was performed as described previously [26]. A finely ground sample (10 mg) was mixed with 1 mL of a mixture of methanol, water and chloroform in the ratio 2.5:1:1 (*v/v*/v). Sixty microliters of ribitol (200 μg/mL) was added to the mixture as an internal standard (IS) and the mixture was incubated using a Thermomixer Comfort (model 5355, Eppendorf AG, Hamburg, Germany) at 37 °C for 30 min at a mixing frequency of 1200 rpm. The mixture was centrifuged at 16,000× *g* for 3 min. The upper layer (methanol/water phase) of 800 μL was pipetted into a fresh tube and mixed with 400 μL of water. The methanol/water fraction was centrifuged at 16,000× *g* for 3 min and 900 μL of the supernatant was collected into a fresh tube. The aliquots were evaporated for 2 h in a centrifugal concentrator (CC-105; TOMY, Tokyo, Japan) and freeze-dried for over 16 h. For derivatization, 80 μL of 2% methoxyamine hydrochloride (MOX) in pyridine (*w/v*) was added in freeze-dried samples and the mixture was incubated at 30 °C and 1200 rpm for 90 min using a Thermomixer Comfort (Eppendorf AG). Subsequently, 80 μL of MSTFA was added and the mixture was further incubated at 37 °C and 1200 rpm for 30 min. The hydrophilic compounds were separated on the GCMS-QP2010 Ultra system equipped with autosampler AOC-20i (Shimadzu, Kyoto, Japan) and a DB-5 column (30 m length, 0.25-mm diameter and 1.00 μm thickness). The temperatures for injection, interface and ion source were set at 280, 280 and 200 °C, respectively. The carrier gas was helium and the column flow rate was 1.1 mL/min. The temperature was held for 4 min at 100 °C, after which it was increased at a rate of 10 °C/min up to 320 °C and held for 11 min. The runtime was 4.00 to 37.00 min and the scan mode was used with a mass range of 45 to 600 *m/z*. The compounds were confirmed using standards and the Wiley9, NIST11 and OA TMS DB5 (Shimadzu) libraries (Table S1). For relative quantification, we used ribitol as an IS and the calculated the integrated peak area of all the analyte ratios relative to the peak area of the IS.

2.3. Extraction and Analysis of Lipophilic Compounds

Extraction and analysis of lipophilic compounds (policosanols, phytosterols, tocopherols and other terpenoids) was performed as described previously [27]. Finely ground samples weighing 10 mg were collected in 15-mL conical tubes, and 3 mL of ethanol containing 0.1% ascorbic acid (*w/v*) was added to the tubes. Fifty microliters of 5α-cholestane (10 μg/mL) was added to the mixture as an IS. Next, the samples were vortexed for 20 s and placed in a water bath at 85 °C for 5 min. Subsequently, 120 μL of potassium hydroxide (80%, *w/v*) was added for saponification, and the mixture was vortexed for 20 s. The mixture was returned to the water bath at 85 °C for 10 min. The samples were then cooled on ice for 5 min, and 1.5 mL each of deionized water and hexane was added to each sample and vortexed for 20 s. The mixture was centrifuged at 1200× *g* for 5 min at 4 °C and the upper layer was pipetted into afresh tube. In order to re-extract the remaining compounds, 1.5 mL of hexane was added again into the remaining pellets. The hexane fraction was collected in fresh tubes and evaporated under a stream of N_2 gas in a centrifugal concentrator (TOMY). For the derivatization step, 30 μL of MSTFA and 30 μL of pyridine were added and incubated at 60 °C and 1200 rpm for 30 min using a Thermomixer Comfort (model 5355, Eppendorf AG, Hamburg, Germany). The GCMS-QP2010 Ultra system, equipped with the autosampler AOC-20i (Shimadzu), was installed with a Rtx-5MS column (30 m length, 0.25-mm-diameter and 0.25-μm-thickness) and used for the separation of lipophilic compounds. In total 1.0 μL of each sample was injected with split mode (10:1 ratio) and the injection temperature was set at 290 °C. Helium was used as a carrier gas and the column flow rate was 1.0 mL/min. The oven temperature was held for 2 min at 150 °C, increased at the rate of 15 °C/min up to 320 °C and finally held for 10 min. The chromatography runtime was 2.00–23.33 min. The MS interface and ion source temperatures were 280 and 230 °C, respectively. The Labsolutions GCMSsolution software version 4.20 (Shimadzu Kyoto, Japan) was used for the analysis of chromatograms and mass spectra. The calibration curve range of each lipophilic compound was 0.025–5.00 μg, and a fixed concentration (0.50 μg each) of the internal standard was used. Qualitative and quantitative analyses were conducted using standards (Table S2).

Extraction of fatty acids was performed according to a method described previously, but with slight modifications [28,29]. Briefly, 10-mg of sample was mixed with 2.5 mL of chloroform/methanol (2:1, *v/v*) and 10 μL of pentadecanoic acid (100 μg/mL) as an IS. The mixture was sonicated for 15 min. Next, 2.5 mL of 0.58% (*w/v*) sodium chloride (NaCl) in water was added to separate the extract into two phases (methanol-water and chloroform) and to remove proteinaceous matter from the chloroform fraction. The mixture was briefly vortexed and then centrifuged at 13,000× *g* for 5 min at 4 °C. Thereafter, the chloroform phase (bottom layer) was pipetted into a new tube and evaporated using a centrifugal concentrator (TOMY). Toluene (100 μL), 5 M sodium hydroxide (NaOH, 20 μL) and methanol (180 μL) were added to the dried sample, and the tube was incubated at 85 °C for 5 min. Next, 300 μL of 14% (*w/v*) boron trifluoride (BF_3) in methanol was added for methylation, and the reaction was performed at 85 °C for 5 min. Afterward, 800 μL of pentane and 400 μL of distilled water were added to the tube, and the tube was centrifuged at 750 ×*g* for 15 min at 4 °C. The supernatant was collected into a new 2-mL tube and concentrated using the centrifugal concentrator. The concentrated sample was finally dissolved in 300 μL of hexane, filtered through a 0.5-μm syringe filter and analyzed by gas chromatography–quadrupole mass spectrometry (GC-qMS) (Shimadzu). The methylated fatty acids (1 μL) were separated in a DB-5 column (30 m × 0.25 mm × 1.00 μm; Agilent, Palo Alto, CA, USA) using a GCMS-QP2010 Ultra system with autosampler AOC-20i (Shimadzu). Injection volume of the samples was 1.0 μL and split mode was set at 10:1 ratio. Injection, ion source and interface temperatures were set at 280 °C, 200 °C and 280 °C, respectively. The column temperature conditions were as follows. The initial temperature was maintained at 40 °C for 2 min and raised to 320 °C at a rate of 6 °C/min. Helium was used as a carrier gas at a flow rate of 1.42 mL/min. Runtime was 2.86 to 49.00 min and scan mode was used with a mass range of 45 to 500 *m/z*. Qualitative and quantitative analyses of fatty acids were conducted using standards and a FAME Mix (C8–C24) (Table S3).

2.4. Statistical Analysis

All analyses were performed no fewer than three times. Data obtained from GC-qMS were analyzed using PCA and OPLS-DA (SIMCA-P version 13.0; Umetrics, Umea, Sweden) to discriminate the geographic origin of perilla and sesame seeds. To determine the optimal OPLS-DA model, all the data were normalized with unit variance (UV)-scaling and pareto-scaling. PCA and OPLS–DA were based on the calculated eigenvectors and eigenvalues. The external validation test, permutation test and analysis of variance of the cross-validated residuals (CV-ANOVA) were conducted using SIMCA-P version 13.0 (Umetrics). The receiver operating characteristic (ROC) analysis and student's *t*-test were performed using MetaboAnalyst 4.0 (https://www.metaboanalyst.ca).

3. Results

3.1. Metabolite Profiling of Perilla and Sesame Seeds

To discriminate the geographic origin of perilla and sesame seeds, we analyzed hydrophilic and lipophilic compounds using GC-qMS. We detected 35 hydrophilic compounds in 19 samples of perilla seeds and 31 hydrophilic compounds in 25 samples of sesame seeds (Tables S4 and S5). The lipophilic compounds, such as fatty acids, sterols, policosanols and tocopherols, were detected and quantified in all seed samples (Tables S6–S11). In total, 28 lipophilic compounds, including 11 fatty acids, 9 policosanols, 3 tocopherols, 3 sterols and 2 amyrins, were identified in perilla seeds (Tables S6, S8 and S10). In addition, 23 lipophilic compounds, including 10 fatty acids, 9 policosanols, 1 tocopherol and 3 sterols were detected in sesame seeds (Tables S7, S9 and S11). Unlike perilla seeds, α- and β-tocopherols, α- and β-amyrins and C18:3n3 were not detected in sesame seeds.

3.2. PCA and OPLS-DA for Geographic Discrimination of Perilla and Sesame Seeds

To discriminate the geographic origins of perilla and sesame seeds, the metabolite profiling data were processed using multivariate statistical analysis (PCA and OPLS-DA), which is an important tool for identifying the features of samples in complex data matrices. PCA uses an orthogonal linear transformation to transform the original data into a new set of variables, the principal component (PC). The scores and loading of PCs are represented in a bi-dimensional plot, which can formulate a dataset pattern from the raw data. The data were normalized with UV-scaling. In the PCA score plots, the two seeds did not show any variance according to geographic origins (Figures S1 and S2).

To improve the geographic discrimination of perilla and sesame seeds, we used OPLS-DA to determine the differences in metabolites arising due to differences in the geographic origin. OPLS-DA is a supervised classification method that features (X variables: metabolites) divides into two parts to separate the systematic variation: one that models the correlation between X and Y (prediction) and another that models the orthogonal components [30]. Thus, OPLS-DA has maximum separation by geographic origins based on their metabolites. The geographic origins (Y-variables) were set to 0 for Korea and 1 for China. Internal validation method was used to validate the model. The quality of the predictive model was measured by R^2 and Q^2 values of the validation results. The R^2 value indicates how much the proportion of variation in the data is explained by the model and the goodness of fit. The Q^2 value indicates how much proportion of variation in the data is predictable by the model and the goodness of prediction. The parameters R^2 and Q^2 were calculated minimum zero to maximum one; the R^2 value closer to 1 indicates a good value, $Q^2 > 0.5$ is regarded as a good prediction model and $Q^2 > 0.9$ is regarded as excellent prediction model. To develop a better discrimination model, the data were normalized by UV and pareto scaling. The optimal OPLS-DA model was established using UV-scaling, which showed higher R^2Y (perilla; 0.822, sesame; 0.844) and Q^2 (perilla; 0.761, sesame; 0.799) values than pareto-scaling (R^2Y: perilla; 0.575, sesame; 0.744/ Q^2: perilla; 0.480, sesame; 0.715) (Table 1). The OPLS-DA models of both perilla and sesame seeds showed the Q^2 values to be above 0.5, indicating a good prediction model.

Table 1. Model validation results from orthogonal partial least squares discriminant analysis (OPLS–DA) with various scaling methods for discriminating the geographic origin of perilla and sesame seeds.

Sample	X Variables Number	Scaling Method	R^2X	R^2Y	Q^2
Perilla	57	UV	0.316	0.822	0.761
		Par	0.473	0.575	0.480
Sesame	78	UV	0.303	0.844	0.799
		Par	0.526	0.744	0.715

UV—unit variance; Par—pareto.

The OPLS-DA analysis was performed with UV-scaling data. The OPLS score plot of perilla seeds showed good separation on the basis of geographic origins (Korea and China) (Figure 1A). To identify the potential biomarkers for the geographic discrimination of perilla seeds, variable importance in projection (VIP) plots were used to explain the contribution of metabolites to the prediction models wherein VIP values greater than 1.00 indicate the significant influence on the model. In total, 29 metabolites had greater than 1.00 VIP values (Table S12). Glycolic acid, α-tocopherol and C20:0 were top-ranked metabolites in the VIP plots. The OPLS score plot of sesame seeds also showed good separation by region (Korea and China) (Figure 1B). In total, 26 metabolites showed a VIP cut off value of over 1.00 (Table S13). Proline, glycine and alanine were top-ranked in VIP plots.

The established OPLS-DA model for the discrimination of perilla and sesame seeds on the basis of geographic origin was subjected to an external validation test to determine its accuracy. In the case of perilla seeds, 57 samples were divided into 49 training samples and 8 test samples. The Y-variables were set to 0 for Korea and 1 for China. The OPLS projection model was established using 49 training samples, and then the 8 test samples were projected on the established OPLS projection model. The results of external validation test showed good discrimination of geographic origin of perilla seeds in the OPLS prediction model with $R^2X = 0.298$, $R^2Y = 0.788$ and $Q^2 = 0.674$. In addition, this OPLS model showed a root mean square error of prediction (RMSEP) = 0.229, which indicates the accuracy of prediction. The RMSEP value, being close to zero, indicated a good value. Furthermore, perilla seeds cultivated in Korea and China did not fall on the borderline of 0.5, which was a threshold level in the external validation test. Additionally, a permutation test and CV-ANOVA were conducted to test the risk of over-fitting the OPLS model. The permutation test was performed with 200 permuted models, which was constructed using randomized Y-variables. The reference distribution of the Q^2 value for random data from permuted models was compared with the Q^2 value of the real (unpermuted) OPLS model. When the Q^2 value from the permuted model is smaller than the Q^2 value of the original OPLS model, the model is considered as a predictable model. The results of the permutation test showed the Q^2 value of −0.496, which was lower than the Q^2 value of the original OPLS model (Figure 2A). The CV-ANOVA test was performed to testify the validity of the model. When the p-value was smaller than 0.05, the model was regarded as a validated model. The p-value of perilla seeds from the CV-ANOVA test was 3.05×10^{-10}.

To perform the external validation test for the OPLS-DA model of sesame seeds, the 78 samples were divided into 68 training samples and 10 test samples. The 68 training samples were used for the construction of the OPLS prediction model, and the 10 test samples were projected on the OPLS model. The external validation test results displayed good separation of sesame seeds samples on the basis of geographic origin in the OPLS projection model, which showed validation values with $R^2X = 0.320$, $R^2Y = 0.812$, $Q^2 = 0.754$ and RMSEP = 0.208. The results of the permutation test for the OPLS predictive model for sesame seeds showed the Q^2 value of −0.383, which was smaller than the Q^2 value of the real OPLS model. The CV-ANOVA test results of sesame seeds showed the p-value of 1.61×10^{-18}. Therefore, the OPLS-DA model for geographic discrimination of both of perilla and sesame seeds were successfully established and validated.

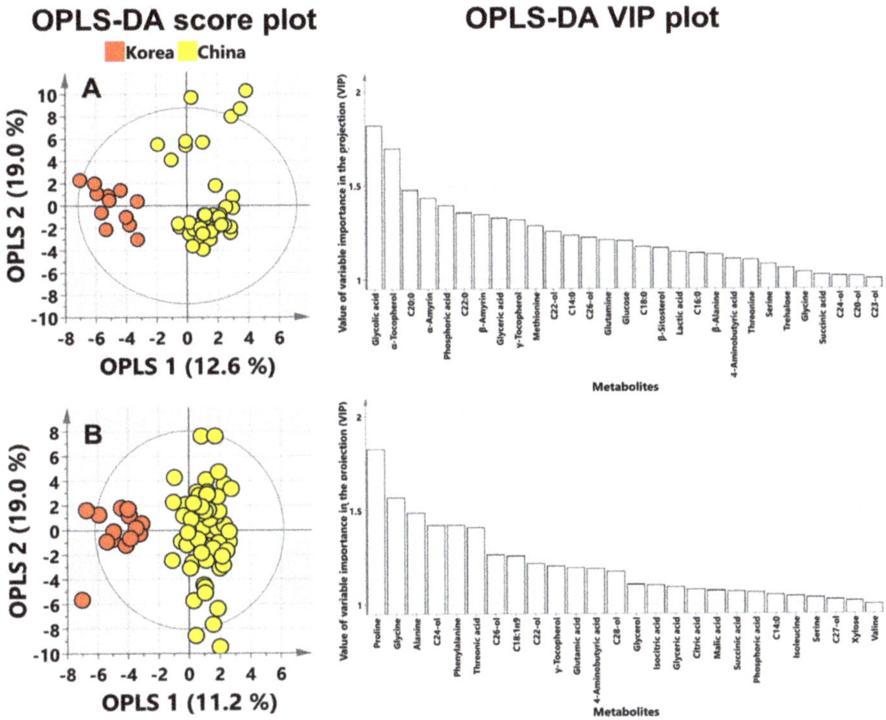

Figure 1. OPLS–DA score plots and VIP (variable importance in the projection) plots of (**A**) perilla and (**B**) sesame seeds from Korea and China. C20-ol—eicosanol; C21-ol—heneicosanol; C22-ol—docosanol; C23-ol—tricosanol; C24-ol—tetracosanol; C26-ol—hexacosanol; C27-ol—heptacosanol; C28-ol—octacosanol; C30-ol—triacontanol; C12:0—lauric acid; C14:0—myristic acid; C16:1n7—palmitoleic acid; C16:0—palmitic acid; C18:2n6—linoleic acid; C18:3n3—α-linolenic acid; C18:1n9—oleic acid; C18:0—stearic acid; C20:0—arachidic acid; C22:0—behenic acid; C24:0—lignoceric acid.

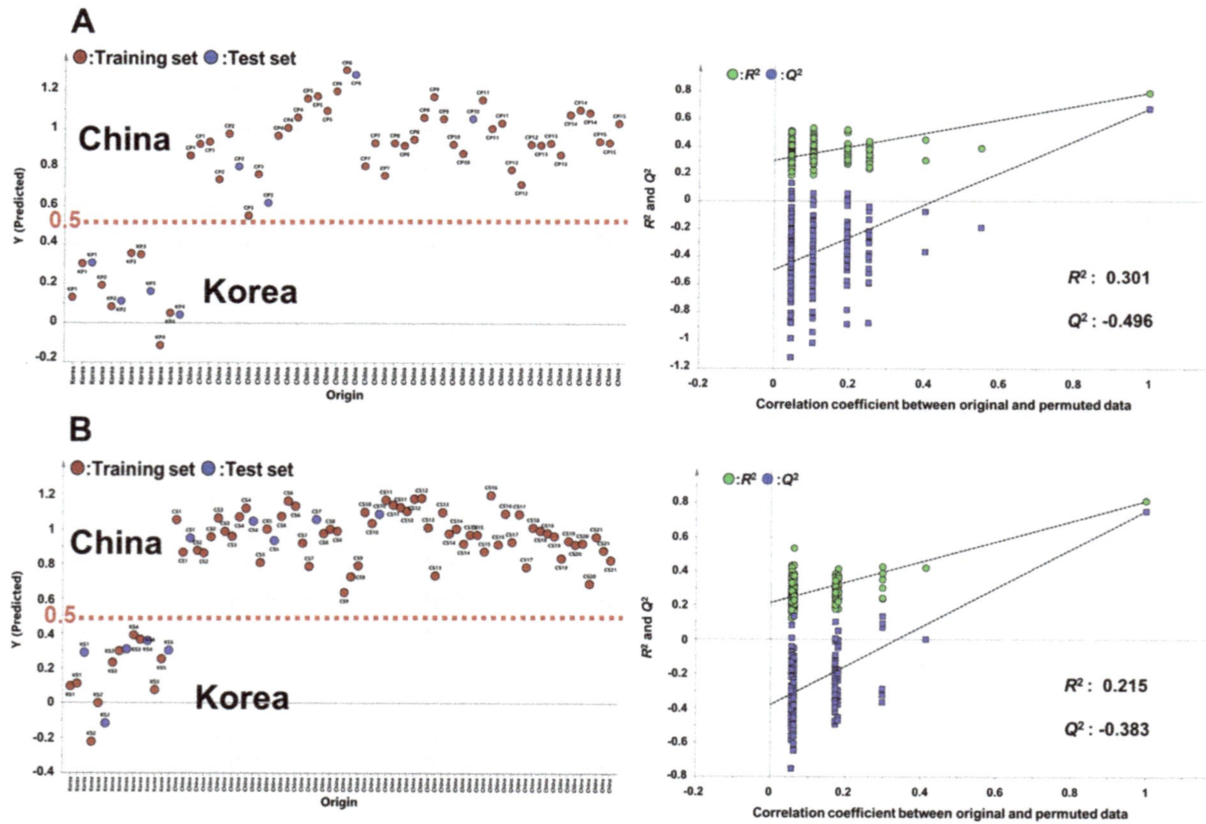

Figure 2. External validation test and permutation test by OPLS-DA for discriminating the geographic origin of (**A**) perilla and (**B**) sesame seeds from Korea and China. The number of permutations for the permutation test was 200. (A: $R^2X = 0.298$, $R^2Y = 0.788$, $Q^2 = 0.674$, RMSEP = 0.229; B: $R^2X = 0.320$, $R^2Y = 0.812$, $Q^2 = 0.754$, RMSEP = 0.208).

3.3. Potential Biomarkers for the Discrimination of Perilla and Sesame Seeds Based on Their Geographic Origins

The OPLS-biplot displayed a combination of observations (samples), X-variables (metabolites) and Y-variables (geographic origin) in a bi-dimensional space. This could easily explain the correlation of variables and the clustering of samples. The three ellipses—inner (0.50), middle (0.75) and outer (1.00)—indicate that the explained variances are 50%, 75% and 100%, respectively. If the variables are located close to the observations, the sample group has high levels of metabolites, whereas if they are opposite, the levels of metabolites are low. If the variables are closer to the outer circle (1.00) of the OPLS-biplot, the metabolites have more significantly contributed to the model.

In the OPLS-biplot of perilla seeds, glycolic acid, α-tocopherol and C20:0 were significant contributors, which were notably positioned the closest to the outer (1.00) circle and Y-variables (Figure 3A). In particular, only glycolic acid was located within middle (0.75) and outer (1.00) circles among these metabolites. In addition, these metabolites had top-ranked VIP values (glycolic acid, 1.82; α-tocopherol, 1.70; and C20:0, 1.48) in VIP plot. Therefore, to evaluate the predictive performance of these metabolites as potential biomarkers, ROC analysis was conducted. When the area under curve (AUC) values, which were a result of the ROC analysis, are to be closer to 1.00, the outcome is desirable [4]. Glycolic acid showed the AUC value of 1.000, indicating the excellent accuracy of discriminating Korean and Chinese perilla seeds (Figure 4A). In addition, α-tocopherol (AUC: 0.900) and C20:0 (AUC: 0.856) showed good accuracy to be considered as potential biomarkers. Therefore, glycolic acid was proposed as a potential biomarker for Chinese perilla seeds.

As shown in Figure 4B, proline, glycine and alanine, which were top-ranked (proline, 1.82; glycine, 1.57; and alanine, 1.49) in the VIP plot of sesame seeds, were located the closest to the outer circle and Y-variables. These metabolites showed AUC values in the range of 0.915–0.944, indicating their excellent accuracy as potential biomarkers for discriminating Korean and Chinese sesame seeds. Thus, proline, glycine and alanine were proposed as potential biomarkers for discriminating sesame seeds on the basis of geographic origin.

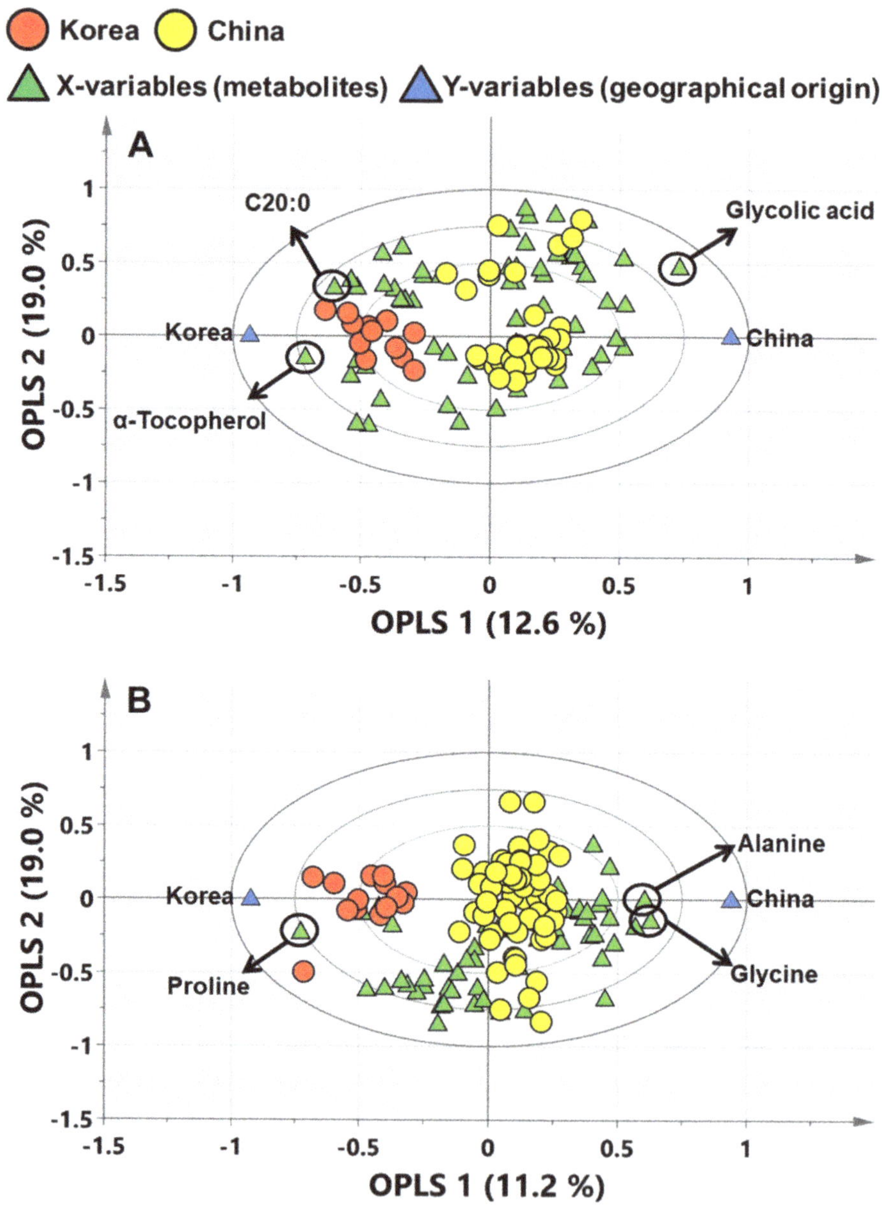

Figure 3. The OPLS-biplot for discriminating the geographic origin of (**A**) perilla and (**B**) sesame seeds using metabolite profiling data. The OPLS-biplot showed correlation of all metabolites (X-variables), sample clusters (observations) and geographic origins (Y-variables). C20:0; arachidic acid.

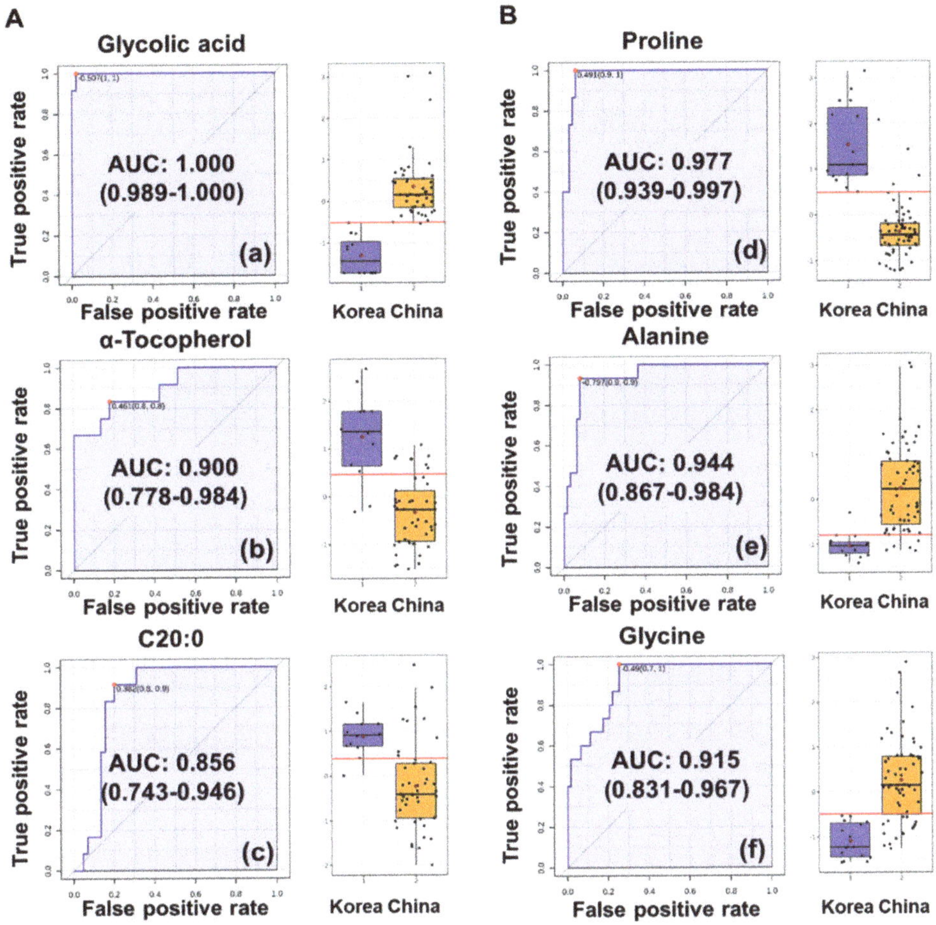

Figure 4. Receiver operating characteristic (ROC) curves for discriminating the geographic origins of (**A**) perilla and (**B**) sesame seeds using metabolite profiling data. ROC curves for (a) glycolic acid, (b) α-tocopherol and (c) C20:0 (arachidic acid) on discriminating (**A**) perilla seeds from Korea and China. ROC curves for (d) proline, (e) alanine and (f) glycine on discriminating (**B**) sesame seeds from Korea and China.

4. Discussion

The quality of perilla and sesame seeds and oils based on various health-related compounds such as fatty acids, tocopherols and sterols has been assessed previously [1,5]. However, to the best of our knowledge, a comprehensive metabolite profiling, which combines primary and secondary metabolites, has not been reported for perilla and sesame seeds. Therefore, we analyzed the primary metabolites and health-promoting compounds, which are abundantly found in perilla and sesame seeds, using GC-qMS. Perilla and sesame seeds are important oil crops, and they contain high levels of lipophilic compounds. In our analysis, perilla seeds showed high levels of α-linolenic acid (C18:3n3) and linoleic acid (C18:2n6), which are essential omega-3 and -6 fatty acids, respectively (Tables S10 and S11). On the contrary, α-linolenic acid (C18:3n3) was not detected in sesame seeds. However, linoleic acid (C18:2n6) and oleic acid (C18:1n9) were detected in higher levels in sesame seeds than in perilla seeds. Among tocopherols, γ-tocopherol was found in the highest amount in both perilla and sesame seeds; however, α- and β-tocopherols were not detected in sesame seeds. Phytosterols were found in high amounts in perilla and sesame seeds (Tables S8 and S9). The levels of phytosterols in sesame seeds were approximately three times higher than those in perilla seeds. The above results were consistent with those of the previous studies [1]. Perilla seeds showed high levels of policosanols (Table S6). In particular, C28-ol was found in the highest level among policosanols in perilla seeds. However, sesame seeds showed low levels of policosanols (Table S7). These results agreed with those of the previous studies, which showed that perilla seeds and oils contain the highest levels of policosanols among other oil crops, while sesame seeds and oils contain negligible amounts of policosanols [31,32]. The hydrophilic metabolites, such as amino acids, organic acids and sugars, were detected in both perilla and sesame seeds (Tables S4 and S5). Almost all amino acids were found at higher levels in sesame seeds than in perilla seeds, except methionine and β-alanine. Sesame seeds are known as a good source of proteins rich in high sulfur-containing amino acids [4,5]. Therefore, sesame seeds may be consumed methionine for generating protein, which including high sulfur-containing amino acids. For the synthesis of high amount methionine, aspartic acid metabolism is activated. As a result, aspartic acid levels were higher in sesame seeds than in perilla seeds. In addition, sesame seeds have high levels of phenylalanine. Sesame seeds are also known to contain high amounts of lignans such as sesamin, sesamolin and sesamol [6,7]. Therefore, sesame seeds may have an activated phenylpropanoid pathway for the synthesis of lignans, resulting in the upregulated levels of phenylalanine.

To compare the compositional differences in seeds according to their origins, student's *t*-test was performed with metabolite profile data of perilla and sesame seeds. The *t*-test results of perilla seeds showed that 22 metabolites were considered statistically significant ($0.05 \geq p$-value) between Korean and Chinese perilla seeds. In addition, these metabolites were shown to have compositional differences with geographic origins of perilla seeds. In the OPLS-DA loading plots of perilla seeds, the Korean perilla seeds had higher amounts of five terpenoids (α-, γ-tocopherols, β-sitosterol and α-, β-amyrin), five fatty acids (C14:0, C16:0, C18:0, C20:0 and C22:0) and methionine than Chinese seeds (Figure S3B). On the other hand, four policosanols (C20-ol, C22-ol, C24-ol and C26-ol), five organic acids (glycolic acid, phosphoric acid, nicotinic acid, lactic acid, glyceric acid), 4-aminobutyric acid and sucrose were shown to be present in higher levels in Chinese perilla seeds. In the case of sesame seeds, 25 metabolites were considered statistically significant between Korean and Chinese seeds. In the OPLS-DA loading plots of sesame seeds, three fatty acids (C14:0, C18:1n-9 and C24:0), four organic acids (citric acid, isocitric acid, malic acid and threonic acid), threonine and C22-ol were higher in concentration in Korean sesame seeds than in Chinese sesame seeds (Figure S4B). Whereas, the Chinese sesame seeds contained higher amounts of four amino acids (glycine, alanine, phenylalanine and 4-aminobutyric acid), two organic acids (succinic acid and glyceric acid), four policosanols (C24-ol, C28-ol, C26-ol and C30-ol), γ-tocopherol, glycerol, phosphoric acid, inositol and fructose than the Korean sesame seeds.

We determined and predicted the geographic origins of perilla and sesame seeds cultivated in China and Korea using OPLS-DA (Figure 1). The score plot of OPLS-DA showed good separation of both perilla and sesame seeds using appropriate data pretreatment. The optimal data preprocessing method for the OPLS-DA model was the UV-scaling method with the highest Q^2 and R^2Y values in both of perilla and sesame seeds (Table 1). The selection of normalization methods is particularly important to reduce the unwanted instrumental errors of peak intensity measurements for relevant biologic differences. Thus, data normalization and scaling strategies should be chosen in such a way that the model shows optimal predictive ability of MVDA and retains meaningful biologic information [33].

The OPLS-biplots and VIP plots were generated to identify the biomarkers for discriminating perilla and sesame seeds on the basis of their geographic origins. Glycolic acid, α-tocopherol and C20:0 were identified as potential biomarkers for perilla seeds discrimination. Furthermore, proline, alanine and glycine were found to be potential biomarkers for sesame seeds discrimination. These potential biomarkers were further validated using ROC curve analysis. All AUC values of potential biomarkers were higher than 0.85, indicating that these metabolites significantly contribute to discriminating the seeds on the basis of their geographic origins. Kim et al. have reported that the VIP values of proline and glycine derived from the OPLS-DA model for discriminating the geographic origin of sesame seeds were higher than 1.0, indicating that these metabolites can be potential biomarkers for determining the regional origins of sesame seeds [4]. Thus, our results were consistent with those of a previous study. Glycolic acid is generated during photorespiration. Under low atmospheric CO_2 condition, C3 photosynthetic metabolism fixes the competing substrate O_2 instead of CO_2. The oxygen fixation generates one molecule of 3-phosphoglycerate (3-PGA) and one molecule of 2-phosphoglycolate (2-PG) instead of two molecules of 3-PGA. Glycolic acid is generated from the dephosphorylation of 2-PG, and it can inhibit the rate of photosynthesis in the chloroplast. As a result, photorespiration under current atmospheric CO_2 concentrations reduces the efficiency of C3 photosynthesis by ~15% to 50%, depending upon the temperature in the growing season at that particular geographic location [34]. Therefore, this study suggests that glycolic acid could be a potential biomarker for geographic discrimination of perilla seeds and proline and glycine could be the same for sesame seeds.

Outlier detection is an important issue in chemometrics analysis. The outliers are observations that are extreme or that do not fit the PCA model. Furthermore, outliers can be both serious and interesting observations in the data. To discover the outliers in the PCA model, we used the Hotelling's T^2. The Hotelling's T^2 is a multivariate generalization of student's *t*-test and provides a check for observations adhering to multivariate normality. In the PCA score plots, the ellipse of Hotelling's T^2 indicates 95% confidence. When observations fall outside the confidence ellipse, they are termed as strong outliers. Observations suggested as outliers were removed from the entire data set. This process was repeated until no outliers were displayed on the PCA score plot. Figures S5 and S6 show the outlier removal process. A total of 11 samples were identified as outliers, and 46 samples remained in the data set of perilla seeds. In the OPLS-DA score plot of perilla seeds (Figure 1), Chinese perilla seeds were more dispersed than Korean perilla seeds because the outliers were clustered in the upper right of the score plot (Figure S3A). In addition, the data set of sesame seeds retained 69 samples and eliminated 9 samples. These pretreated data sets of perilla and sesame seeds were subjected to OPLS-DA. Figure S7 shows OPLS-DA scores and VIP plots of the outlier removal data sets. The OPLS-DA model was established using UV-scaling, which showed higher R^2Y (perilla; 0.928, sesame; 0.876) and Q^2 (perilla; 0.874, sesame; 0.842) values than the original data set R^2Y (perilla; 0.822, sesame; 0.844) and Q^2 (perilla; 0.761, sesame; 0.799) values. The OPLS-DA score plots for the outlier removal data sets showed good separation of both perilla and sesame seeds. In particular, the OPLS-DA score plots of the outlier removal data set of perilla seeds showed clearer clustering of the Chinese samples than that of the original data set. Furthermore, the VIP plots of the outlier removal data sets of perilla and sesame seeds showed results that were almost same as those of the original data sets. Although the number of samples was reduced by more than 10% due to the outlier removal, the potential biomarker candidates were the same as those from the original data sets. These results demonstrated that the established OPLS-DA discrimination models for perilla and sesame seeds were reliable predictive models.

In conclusion, we performed comprehensive metabolite profiling, which included primary metabolites and health-promoting secondary metabolites, for perilla and sesame seeds cultivated in Korea and China. In addition, we established the OPLS-DA discriminative model for perilla and sesame seeds and validated it with good test results. The OPLS-DA results showed a clear separation of perilla and sesame seeds sourced from Korea and China on the basis of their geographic origins. The OPLS-biplot and VIP plot showed that glycolic acid was a notable metabolite for discrimination of perilla seeds based on geographic origin; therefore, we propose it as a potential biomarker for such discrimination. Furthermore, proline and glycine most significantly contributed for determining the geographic origins of sesame seed, and thus, they could be potential biomarkers for discrimination of sesame seeds based on the geographic origin. This study provides a reliable discriminatory predictive model to determine the geographic origins of perilla and sesame seeds cultivated in Korea and China. In addition, to the best of our knowledge, this is the first attempt to construct a discrimination model for perilla seeds using metabolomics. We believe that this model will be helpful in dealing with issues of selling domestic perilla and sesame seeds in the name of imported ones. In this study, the number of samples and their source countries was limited. A future work should involve a larger sample size from more cultivated regions in various countries and evaluate the predictive ability of this model.

Supplementary Materials: The following are available online at http://www.mdpi.com/2304-8158/9/8/989/s1, Figure S1: PCA score (A) and loading (B) plots of perilla (*Perilla frutescens*) seeds from Korea and China, Figure S2: PCA score (A) and loading (B) plots of sesame (*Sesamum indicum*) seeds from Korea and China, Figure S3: OPLS-DA score (A) and loading (B) plots of perilla (*Perilla frutescens*) seeds from Korea and China, Figure S4: OPLS-DA score (A) and loading (B) plots of sesame (*Sesamum indicum*) seeds from Korea and China, Figure S5: PCA score plots and Hotelling's T^2 range column plots of perilla (*Perilla frutescens*) seeds from Korea and China for outlier removal process, Figure S6: PCA score plots and Hotelling's T^2 range column plots of sesame (*Sesamum indicum*) seeds from Korea and China for outlier removal process, Figure S7: OPLS–DA score plots and VIP (variable importance in the projection) plots of perilla (A) and sesame (B) seeds from Korea and China outlier removal data sets, Table S1: Relative retention times (RRT) and mass spectral data of hydrophilic compounds as trimethylsilyl derivatives, Table S2: Relative retention times (RRT) and mass spectral data of lipophilic compounds as trimethylsilyl derivatives, Table S3: Relative retention times (RRT) and concentration of fatty acid methyl esters (FAME) mixture and fatty acids, Table S4: Composition and content (ratio/g) of hydrophilic compounds in perilla (*Perilla frutescens*) cultivars, Table S5: Composition and content (ratio/g) of hydrophilic compounds in sesame (*Sesamum indicum*) cultivars, Table S6: Composition and content (μg/g) of policosanol compounds in perilla (*Perilla frutescens*) cultivars, Table S7: Composition and content (μg/g) of policosanol compounds in sesame (*Sesamum indicum*) cultivars, Table S8: Composition and content (μg/g) of sterol and terpenoid compounds in perilla (*Perilla frutescens*) cultivars, Table S9: Composition and content (μg/g) of sterol and terpenoid compounds in sesame (*Sesamum indicum*) cultivars, Table S10: Composition and content (mg/g) of fatty acids in perilla (*Perilla frutescens*) cultivars, Table S11: Composition and content (mg/g) of fatty acids in sesame (*Sesamum indicum*) cultivars, Table S12: OPLS-DA loading plots and VIP values of variables of perilla (*Perilla frutescens*) cultivars, Table S13: OPLS-DA loading plots and VIP values of variables of sesame (*Sesamum indicum*) cultivars.

Author Contributions: Conceptualization, methodology: J.K.K., W.D.S. and T.J.K.; formal analysis: J.G.P.; resources: H.Y.K. and W.D.S.; data curation: J.G.P. and T.J.K.; writing—original draft preparation: T.J.K. and J.G.P.; writing—review and editing: S.-H.H., B.L. and J.K.K.; project administration: S.U.P., J.K.K. and W.D.S. All authors have read and agreed to the published version of the manuscript.

Funding: This research received no external funding.

Acknowledgments: This work was supported "Cooperative Research Program for Agriculture Science & Technology Development (Project No. PJ013483042020)" funded by the Rural Development Administration (RDA), Republic of Korea and by Research Assistance Program (2019) in the Incheon National University, Republic of Korea.

Conflicts of Interest: The authors declare no conflict of interest.

References

1. Kim, J.K.; Park, S.-Y.; Na, J.-K.; Seong, E.S.; Yu, C.Y. Metabolite profiling based on lipophilic compounds for quality assessment of perilla (*Perilla frutescens*) cultivars. *J. Agric. Food Chem.* **2012**, *60*, 2257–2263. [CrossRef]
2. Shen, J.; Luo, F.; Lin, Q. Policosanol: Extraction and biological functions. *J. Funct. Foods* **2019**, *57*, 351–360. [CrossRef]
3. Lee, K.-R.; Kim, K.-H.; Kim, J.B.; Hong, S.-B.; Jeon, I.; Kim, H.U.; Lee, M.H.; Kim, J.K. High accumulation of γ-linolenic acid and Stearidonic acid in transgenic perilla (*Perilla frutescens* var. frutescens) seeds. *BMC Plant Biol.* **2019**, *19*, 120. [CrossRef]
4. Kim, S.-Y.; Kim, E.; Shin, B.K.; Seo, J.-A.; Kim, Y.-S.; Lee, D.Y.; Choi, H.-K. NMR-Based metabolic profiling discriminates the geographical origin of raw sesame seeds. *Food Control* **2020**, *112*, 107–113. [CrossRef]
5. Pathak, N.; Rai, A.; Kumari, R.; Bhat, K. Value addition in sesame: A perspective on bioactive components for enhancing utility and profitability. *Pharmacogn. Rev.* **2014**, *8*, 147–155.
6. Jeon, H.; Kim, I.H.; Lee, C.; Choi, H.D.; Kim, B.H.; Akoh, C.C. Discrimination of origin of sesame oils using fatty acid and lignan profiles in combination with canonical discriminant analysis. *J. Am. Oil Chem. Soc.* **2013**, *90*, 337–347. [CrossRef]
7. Williamson, K.S.; Morris, J.B.; Pye, Q.N.; Kamat, C.D.; Hensley, K. A survey of sesamin and composition of tocopherol variability from seeds of eleven diverse sesame (*Sesamum indicum* L.) genotypes using HPLC-PAD-ECD. *Phytochem. Anal.* **2008**, *19*, 311–322. [CrossRef]
8. Castro-Puyana, M.; Herrero, M. Metabolomics approaches based on mass spectrometry for food safety, quality and traceability. *TrAC Trends Anal. Chem.* **2013**, *52*, 74–87. [CrossRef]

9. Castro-Puyana, M.; Pérez-Míguez, R.; Montero, L.; Herrero, M. Reprint of: Application of mass spectrometry-based metabolomics approaches for food safety, quality and traceability. *TrAC Trends Anal. Chem.* **2017**, *96*, 62–78. [CrossRef]
10. Salvo, A.; Rotondo, A.; Mangano, V.; Grimaldi, M.; Stillitano, I.; D'Ursi, A.M.; Dugo, G.; Rastrelli, L. High-Resolution magic angle spinning nuclear magnetic resonance (HR-MAS-NMR) as quick and direct insight of almonds. *Nat. Prod. Res.* **2017**, *34*, 71–77. [CrossRef]
11. Kim, T.J.; Park, J.G.; Ahn, S.K.; Kim, K.W.; Choi, J.; Kim, H.Y.; Ha, S.-H.; Seo, W.D.; Kim, J.K. Discrimination of adzuki bean (*Vigna angularis*) geographical origin by targeted and non-targeted metabolite profiling with gas chromatography time-of-flight mass spectrometry. *Metabolites* **2020**, *10*, 112. [CrossRef]
12. Park, C.H.; Yeo, H.J.; Park, S.-Y.; Kim, J.K.; Park, S.U. Comparative phytochemical analyses and metabolic profiling of different phenotypes of Chinese cabbage (*Brassica Rapa* ssp. *Pekinensis*). *Foods* **2019**, *8*, 587. [CrossRef]
13. Putri, S.P.; Irifune, T.; Fukusaki, E. GC/MS based metabolite profiling of Indonesian specialty coffee from different species and geographical origin. *Metabolomics* **2019**, *15*, 126. [CrossRef]
14. Mo, C.; Lim, J.; Kwon, S.W.; Lim, D.K.; Kim, M.S.; Kim, G.; Kang, J.; Kwon, K.-D.; Cho, B.-K. Hyperspectral imaging and partial least square discriminant analysis for geographical origin discrimination of white rice. *J. Biosyst. Eng.* **2017**, *42*, 293–300.
15. Luykx, D.M.; Van Ruth, S.M. An overview of analytical methods for determining the geographical origin of food products. *Food Chem.* **2008**, *107*, 897–911. [CrossRef]
16. Vadalà, R.; Mottese, A.F.; Bua, G.D.; Salvo, A.; Mallamace, D.; Corsaro, C.; Vasi, S.; Giofrè, S.V.; Alfa, M.; Cicero, N.; et al. Statistical analysis of mineral concentration for the geographic identification of garlic samples from Sicily (Italy), Tunisia and Spain. *Foods* **2016**, *5*, 20. [CrossRef]
17. Dossa, K.; Wei, X.; Zhang, Y.; Fonceka, D.; Yang, W.; Diouf, D.; Liao, B.; Cissé, N.; Zhang, X. Analysis of genetic diversity and population structure of sesame accessions from Africa and Asia as major centers of its cultivation. *Genes* **2016**, *7*, 14. [CrossRef]
18. Ghimire, B.K.; Yoo, J.H.; Yu, C.Y.; Chung, I.-M. GC–MS analysis of volatile compounds of *Perilla frutescens* Britton var. *Japonica* accessions: Morphological and seasonal variability. *Asian Pac. J. Trop. Med.* **2017**, *10*, 643–651.
19. Korea Statistical Information Service. Available online: http://kosis.kr/statHtml/statHtml.do?orgId=101&tblId=DT_1ET0293&conn_path=I2 (accessed on 26 May 2020).
20. Korea Agricultural Trade Information. Available online: http://www.kati.net/statistics/regionalPerformanceByProduct.do (accessed on 26 May 2020).
21. Horacek, M.; Hansel-Hohl, K.; Burg, K.; Soja, G.; Okello-Anyanga, W.; Fluch, S. Control of origin of sesame oil from various countries by stable isotope analysis and DNA based markers—A pilot study. *PLoS ONE* **2015**, *10*, e0123020. [CrossRef]
22. Choi, Y.H.; Hong, C.K.; Kim, M.; Jung, S.O.; Park, J.; Oh, Y.H.; Kwon, J.-H. Multivariate analysis to discriminate the origin of sesame seeds by multi-element analysis inductively coupled plasma-mass spectrometry. *Food Sci. Biotechnol.* **2017**, *26*, 375–379. [CrossRef]
23. Sa, K.J.; Park, D.H.; Lim, S.E.; Ma, S.J.; Lee, J.K. Geographic variation of morphological characters among cultivated types of perilla crop and their weedy types in Korea and China. *Plant Breed. Biotechnol.* **2019**, *7*, 313–325. [CrossRef]
24. Bang, H.J.; Kim, C.T.; Kim, B.H. Liquid and gas chromatographic analyses of triacylglycerols for Asian sesame oil traceability. *Eur. J. Lipid Sci. Technol.* **2014**, *116*, 1354–1362. [CrossRef]
25. Jin, G.; Kim, J.; Lee, Y.; Kim, J.; Akoh, C.C.; Chun, H.S.; Ahn, S.; Kim, B.H. A nuclear magnetic resonance spectroscopy approach to discriminate the geographic origin of roasted Asian sesame oils. *J. Oleo Sci.* **2017**, *66*, 337–344. [CrossRef]
26. Kim, T.J.; Choi, J.; Kim, K.W.; Ahn, S.K.; Ha, S.H.; Choi, Y.; Park, N.I.; Kim, J.K. Metabolite profiling of peppers of various colors reveals relationships between tocopherol, carotenoid, and phytosterol content. *J. Food Sci.* **2017**, *82*, 2885–2893. [CrossRef]
27. Kim, T.J.; Lee, K.B.; Baek, S.-A.; Choi, J.; Ha, S.-H.; Lim, S.-H.; Park, S.-Y.; Yeo, Y.; Park, S.U.; Kim, J.K. Determination of lipophilic metabolites for species discrimination and quality assessment of nine leafy vegetables. *J. Korean Soc. Appl. Biol. Chem.* **2015**, *58*, 909–918. [CrossRef]

28. Park, P.; Goins, R. In situ preparation of fatty acid methyl esters for analysis of fatty acid composition in foods. *J. Food Sci.* **1994**, *59*, 1262–1266. [CrossRef]
29. Lee, C.M.; Trevino, B.; Chaiyawat, M. A simple and rapid solvent extractionmethod for determining total lipids in fish tissue. *J. AOAC Int.* **1996**, *79*, 487–492. [CrossRef]
30. Eriksson, L.; Byrne, T.; Johansson, E.; Trygg, J.; Vikström, C. *Multi- and Megavariate Data Analysis Basic Principles and Applications*; Umetrics Academy: Umea, Sweden, 2013; Volume 1.
31. Adhikari, P.; Hwang, K.T.; Park, J.N.; Kim, C.K. Policosanol content and composition in perilla seeds. *J. Agric. Food Chem.* **2006**, *54*, 5359–5362. [CrossRef]
32. Jung, D.M.; Lee, M.J.; Yoon, S.H.; Jung, M.Y. A gas chromatography-tandem quadrupole mass spectrometric analysis of policosanols in commercial vegetable oils. *J. Food Sci.* **2011**, *76*, C891–C899. [CrossRef]
33. Skov, T.; Honoré, A.H.; Jensen, H.M.; Næs, T.; Engelsen, S.B. Chemometrics in foodomics: Handling data structures from multiple analytical platforms. *TrAC Trends Anal. Chem.* **2014**, *60*, 71–79. [CrossRef]
34. South, P.F.; Walker, B.J.; Cavanagh, A.P.; Rolland, V.; Badger, M.; Ort, D.R. Bile acid sodium symporter BASS6 can transport glycolate and is involved in photorespiratory metabolism in *Arabidopsis thaliana*. *Plant Cell* **2017**, *29*, 808–823. [CrossRef] [PubMed]

foods MDPI

Article

Food Authentication: Truffle (*Tuber* spp.) Species Differentiation by FT-NIR and Chemometrics

Torben Segelke †, Stefanie Schelm †, Christian Ahlers and Markus Fischer *

Hamburg School of Food Science—Institute of Food Chemistry, University of Hamburg, Grindelallee 117, 20146 Hamburg, Germany; torben.segelke@chemie.uni-hamburg.de (T.S.); stefanie.schelm@chemie.uni-hamburg.de (S.S.); christian.ahlers-2@studium.uni-hamburg.de (C.A.)
* Correspondence: Markus.Fischer@uni-hamburg.de; Tel.: +49-4042-838-43-57
† Authors contributed equally to this work.

Received: 16 June 2020; Accepted: 10 July 2020; Published: 13 July 2020

Abstract: Truffles are certainly the most expensive mushrooms; the price depends primarily on the species and secondly on the origin. Because of the price differences for the truffle species, food fraud is likely to occur, and the visual differentiation is difficult within the group of white and within the group of black truffles. Thus, the aim of this study was to develop a reliable method for the authentication of five commercially relevant truffle species via Fourier transform near-infrared (FT-NIR) spectroscopy as an easy to handle approach combined with chemometrics. NIR-data from 75 freeze-dried fruiting bodies were recorded. Various spectra pre-processing techniques and classification methods were compared and validated using nested cross-validation. For the white truffle species, the most expensive *Tuber magnatum* could be differentiated with an accuracy of 100% from *Tuber borchii*. Regarding the black truffle species, the relatively expensive *Tuber melanosporum* could be distinguished from *Tuber aestivum* and the Chinese truffles with an accuracy of 99%. Since the most expensive Italian *Tuber magnatum* is highly prone to fraud, the origin was investigated and Italian *T. magnatum* truffles could be differentiated from non-Italian *T. magnatum* truffles by 83%. Our results demonstrate the potential of FT-NIR spectroscopy for the authentication of truffle species.

Keywords: truffle; *Tuber* spp.; food authentication; species differentiation; near-infrared spectroscopy; chemometrics

1. Introduction

Today's globalization leads to an increase of known cases of food fraud [1]. At the same time, consumer demand is moving towards food products of higher quality [2]. Many cases of food fraud pose a risk to health if toxic or allergenic substances get into the products through adulteration. However, even in cases of food fraud, which in many cases do not lead to a health hazard, it must be ensured that the consumer is not economically harmed, i.e., that no unjustifiably high prices are charged for inferior goods.

The increasing interest of the consumer in higher quality food [3], and also the willingness to pay more money for it, provides the incentive for criminally motivated actors to stretch high-end products with cheaper ingredients. Since many falsifications cannot be detected immediately by laymen or even by trained personnel in companies, it is becoming increasingly important to have appropriate instrumental detection methods for possible food adulteration at hand [4].

Because of the unique aroma and taste emitted from the fruiting bodies, truffles (*Tuber* spp.) are considered as delicacies. The underground growing ascomycetes represent the most expensive of all edible fungi, whereby the white Piedmont Truffle (*Tuber magnatum*) and the black Périgord

Truffle (*T. melanosporum*) are the most valuable species: prices do range between 3000–5000 €/kg and 700–1200 €/kg, respectively [5–7].

Because of their high price, truffles are often subject to fraud, especially when the species are very similar in their morphological appearance: *T. borchii* (syn. *Tuber albidum* Pico) is a truffle morphologically and biochemically similar to *T. magnatum*, both are classified as white truffles. The latter is the most expensive truffle species of all, so it is obvious that it is the subject of an intended counterfeit [8,9]. However, even unintentional cases of fraud are reported when other truffles, such as *T. borchii* are harvested, although the roots have initially been colonized by *T. magnatum* [10,11].

Amongst black truffles, the species *T. melanosporum* is the most expensive and highly valued for its organoleptic properties [12]. The Asian black truffles (e.g., *Tuber indicum*, *Tuber himalayense*, and *Tuber sinense*) form fruiting bodies morphologically very similar to *T. melanosporum* [13]. In view of the higher price of *T. melanosporum*, there is also a risk of fraud, especially since Asian black truffles are imported into Europe from China [14–16].

Due to the above-mentioned potential fraud cases, analytical authentication techniques are necessary, which must also be time-efficient due to the short-term storage of the industry.

In 2006, Zhao et al. compared five Chinese truffle fruiting bodies using Fourier transform infrared (FT-IR) spectroscopy [17] and successfully differentiated *T. magnatum*, *T. indicum*, and *Tuber excavatum* from each other. More recently, El Karkouri et al. proposed a matrix-assisted laser desorption/ionisation time of flight mass spectrometry (MALDI-TOF-MS) strategy, analysing proteins and applying database search algorithms [5]. In 2020, Krauß et al. analysed different tuber species regarding their geographical origin and species authentication via stable isotope ratio analysis showing that a differentiation with this method is possible [18]. However, these techniques still require costly instrumentation, maintenance and sophisticated handling. Instead, our practical approach is, to our knowledge, the first Fourier transform near-infrared (FT-NIR) spectroscopy study addressing the authentication of truffles with a relatively large number of samples.

FT-NIR spectroscopy is a simple and cost-effective approach, nowadays widely used for the monitoring as well as for the controlling of product quality and safety [19] alike the evaluation of the freshness [20] or of pesticide residues of fruits and vegetables [21]. FT-NIR spectroscopy is widely used for the authentication of foodstuffs [9,22–24] or for controlling the intentionally or unintentionally adulteration of exogenous substances or process by-products [25–27] and was recently used to monitor the post-harvest ripening of white truffles [28].

Data pre-processing of the obtained data is a crucial step in spectroscopic analysis. Therefore, pre-processing techniques, such as scatter correction, smoothing, or detrending steps are used in order to reduce the variability between samples due to scattering caused e.g., by heterogeneous sample size of powdery samples. Furthermore, additive and multiplicative effects in the spectra are removed and a subsequent exploratory analysis, a bi-linear calibration model or a classification model is improved [29]. It is essential to carefully compare and select the data pre-processing techniques to avoid misleading results and overfitting [29–31]. The decision on the classification model is crucial as well, and therefore, similarly to the evaluation of different data pre-treatment steps, we have examined and compared various classification models.

The aim of this study was to develop a reliable, easy-to-handle and low-cost method using the FT-NIR technology coupled to chemometric tools for the differentiation and authentication of five economically relevant truffle species. In this regard, we concentrated on the real truffles of the genus *Tuber* defined in the German Guidelines for mushrooms and mushroom products [32] and used in foodstuffs: the expensive species *T. melanosporum* and *T. magnatum*, as well as the less expensive species *T. aestivum*, *T. borchii*, and *T. indicum*. In this study, 75 truffle samples from three years of harvest and eleven growing countries were analysed. Different common pre-processing techniques were applied to the raw spectra and the results were compared using various classification models.

2. Materials and Methods

2.1. Sample Acquisition

In total, 75 truffle samples of relevant, market available white and black truffle species (harvest years 2017–2020) from 11 different countries were analysed in this study.

More precisely, the sample set consisted of two white species *T. magnatum* (20 samples) and *T. borchii* (5 samples) and three black species *T. melanosporum* (10 samples), *T. aestivum* (synonym *T. uncinatum* [33], 29 samples), and *T. indicum* (11 samples).

Regarding the *T. aestivum* species, molecular biological analyses have shown that *T. aestivum* and *T. uncinatum* are one species. Both terms should therefore be regarded as synonymous. Since *T. aestivum* was described before *T. uncinatum*, the species should be named *T. aestivum* [33]. Based on these molecular biological findings, *T. aestivum* and *T. uncinatum* were subsumed and named *T. aestivum* in this study.

An overview of the collected samples is given in Table S1. Some samples were commercially purchased and, therefore, considered as non-origin-authentic, so the origin is stated as 'unknown' in Table S1. Still, information regarding the truffle species were secured for all samples either by personal participation in harvest or by DNA analysis carried out within the Hamburg School of Food Science [34]. On arrival, all samples were frozen in liquid nitrogen and stored at −80 °C until further treatment.

2.2. Sample Preparation

Per sample, several fruiting bodies, at least 75 g, were cleaned with pure water obtained by a Direct-Q purifying system (Merck Millipore, Burlington, MA, USA) for removing remaining soil. Subsequently, the fruiting bodies were milled using a knife mill (Grindomix GM 300, Retsch, Haan, Germany) with dry ice at a ratio of 1:1 (*w*/*w*) and freeze-dried for 72 h [24]. The truffles were freeze-dried because of two reasons, which are more discussed in Section 3.1: (i) FT-NIR spectra of fresh truffles showed unspecific spectra with large water bands. (ii) It was known from the literature that such a freeze-drying step can enhance the accuracy of the classification models [35]. Freeze-dried material was crushed using a mortar and a pestle to obtain a fine homogeneous powder.

2.3. Spectra Acquisition

For the acquisition of near-infrared spectra, a TANGO FT-NIR spectrometer with an integrating sphere (Bruker Optics, Bremen, Germany) was used. The signals were recorded between 11550–3950 cm^{-1}, collecting 50 scans at a resolution of 4 cm^{-1}. All spectra were acquired at room temperature of 22 ± 2 °C. Samples of 300 mg, weighed in a glass vial (52.0 × 22 mm × 1.2 mm, Nipro Diagnostics Germany GmbH, Ratingen, Germany), were analysed in triplicate, in-between individual spectra recordings the lyophilisate was shaken in the glass vial.

2.4. Spectra Pre-Processing

FT-NIR spectra were pre-processed using MATLAB R2019a (The MathWorks Inc., Natick, MA, USA). After having omitted a specific range of higher wavenumbers (see Table 1 and discussion below), different pre-processing techniques or combinations of them were applied and compared (see Table 1) [36].

Multiplicative scatter correction (MSC) using the average of all spectra as the reference spectra was performed to eliminate scatter effects for all approaches i–vi. First order derivate (approach ii) was calculated to eliminate offset, baseline drifts and additive scattering effects, and second order derivate (approach iii) was calculated to remove multiplicative scattering effects in beyond. Detrending (polynomial order = 1) was applied for approach iv and vii. The effect of smoothing (moving average, span = 5) before MSC was investigated for approach v–vii.

Table 1. Pre-processing steps to the raw spectra in the order 1–2–3. For all approaches, a binning was added as a last step. MSC, multiplicative scatter correction.

Approach No.	Cut	Smoothing	MSC	1st Derivative	2nd Derivative	Detrending
(i)	>9000		1			
(ii)	>9000		1	2		
(iii)	>9000		1		2	
(iv)	>9000		1			2
(v)	>9000	1	2	3		
(vi)	>6000	1	2		3	
(vii)	>9000	1	2			3

After the pre-processing methods stated in Table 1, a binning by averaging 10 adjacent features was carried out with all spectra. Lastly, the triplicate spectra were averaged [24,25,36,37]. For certain issues (e.g., only black or white truffles or origin determination of *T. magnatum* samples), the MSC correction was only applied to the selected spectra.

2.5. Multivariate Data Analysis

For data investigation and visualization, principal component analysis (PCA) and line plots were calculated using MATLAB R2019a after applying spectra pre-treatments and mean centring the data.

For the different pre-processing approaches i–vii (see Table 1) it was each evaluated which classification model achieved the best prediction accuracy using MATLAB R2019a. The classification models examined in this context are stated in Table 2.

Table 2. Overview of the classification models examined in this study.

	Classification Models	Hyperparameters Used	References
a	Linear Discriminant Analysis (LDA)	discrimination type: linear	[38]
b	Linear Support vector machine (lin. SVM)	kernel function: polynomial polynomial order = 1 kernel scale = 1 box constraint level = 1	[23,24,35,37,39]
c	Quadratic Support vector machine (quad. SVM)	kernel function: polynomial polynomial order = 2 kernel scale = 1 box constraint level = 1	
d	Subspace Discriminant (SSD)	method: subspace learners: discriminant number learning cycles = 30	[40]
e	Random Forest (RF)	split criterion: Gini's diversity index max. number of splits = 100:	[41]
f	*k*-nearest neighbour (*k*-NN)	number of neighbours = 1 distance: Euclidean distance weight: equal	[22,42,43]

For optimising the model parameters and for obtaining an unbiased estimate of the model's performance, stratified nested cross-validation was used [44,45]. Therefore, the whole data set was split into four parts whereby the samples were stratified by the species to ensure a representative and balanced training set (three fourths) and test set (one fourth). For the training set, 10-fold cross validation was applied to select the optimal model parameters, referred to as inner cross-validation.

The performance of the calculated model was then evaluated by predicting the test set. This process was repeated for all four folds, so every part of the four-fold outer cross validation was once used as the test set.

Finally, since the results by a single nested cross validation can vary, the entire nested cross-validation and the prediction of the test set were repeated 100 times, of which the mean accuracy and the standard deviation are reported.

3. Results and Discussion

3.1. Spectra Investigation

Figure 1A shows all untreated spectra of the raw data, coloured in accordance to the different truffle species. As anticipated and seen from Figure 1A, the absorbance rises towards lower wavenumbers because of the transition probability which is higher for the first transition than for higher overtones [46].

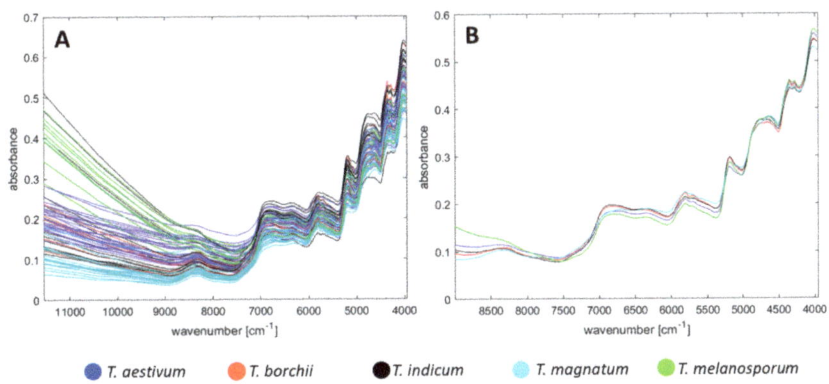

Figure 1. (**A**) Raw Fourier transform near-infrared (FT-NIR) spectra, triplicate measurements from all 75 samples, coloured by truffle species. (**B**) Mean FT-NIR spectra for each truffle species after omitting the >9000 cm^{-1} range, MSC and binning.

However, in the range from 11,550–9000 cm^{-1} some spectra show strong absorbance. Calculating the corresponding wavelength, this region from 11,000–9000 cm^{-1} relates to the region from 1111–909 nm, which is close to the visual region. Here, the 4th overtone of the –OH bond occurs, and the colour of the truffle lyophilisate itself might cause an offset, which could have increased the absorbance [47]. Since the spectra vary in a strong way for this region, chemometric analyses, such as PCA, would excessively focus on this region and would neglect the information that is present in the spectra for smaller wavenumbers, so we excluded the >9000 cm^{-1} region. In fact, the range >9000 cm^{-1} is often excluded in various FT-NIR studies—also because this region is prone to noise when performing data pre-processing methods, such as first or second derivative [37,43].

Regarding the exclusion of some regions in the FT-NIR spectra, special care has to be taken to bands, which can be affected by the water content. Particularly in the region around 5312 cm^{-1} (O–H stretching, first overtone) and around 7142 cm^{-1} (O–H deformation, second overtone), water can affect the absorbance of protein or carbohydrate specific bands [43]. The analysis of fresh truffle samples has shown that a drying step is necessary, as otherwise large water bands and unspecific spectra are obtained which superimpose the information beneath. Thus, the truffle samples were freeze-dried because such a sample preparation can enhance the accuracy of the classification models [35]. Due to the freeze-drying step, the water content in the samples can be seen as negligibly small and in the same range, so it should have no impact on the differentiation with chemometric models in the following steps. In addition, in the region 6500–5300 cm^{-1}, not only water molecules absorb electromagnetic radiation, but also C–H vibrations do, which could be a useful parameter for the differentiation of the truffle species. In order to avoid the loss of useful information, we have not excluded other regions for this non-targeted approach, as several other research groups do in practice [24,37].

For powdered samples, multiplicative scatter effects occur due to differences in the materials' particle size, and have to be corrected for a reasonable data interpretation. To overcome such scattering effects, two approaches are commonly used: MSC and standard normal variate (SNV). According to Dhanoa et al., both pre-processing steps are two alternative approaches, which lead to similar results [48]. In the present study, MSC was chosen to correct for scattering effects. It should be noted that the sequence of the various pre-processing steps is always decisive. In Figure S1, the effect of applying MSC on the raw data, after having omitted the >9000 cm^{-1} region, is shown. On the contrary, applying MSC first and omitting the >9000 cm^{-1} region afterwards will have misleading results, as shown in Figure S1B on the right: the unwanted variance in the >9000 cm^{-1} region is not excluded, but persists in the spectra as an error propagation. By applying pre-processing steps, it is therefore important to examine and to compare the impact of different orders, noted in the same way by Gerretzen et al. [49]. Any further pre-processing steps will be investigated and discussed more deeply in Section 2.4.

3.2. Spectra Interpretation and Assignment of Bands

The FT-NIR spectra reflect the major constituents of the truffles. Naturally low in fat, lyophilised truffle samples are rich in dietary fibre and proteins [50]. These components can be recognised in the spectrum by their characteristic bands; however, it should be noted that an exact assignment of bands for complex samples is difficult due to overlapping effects. For the sake of clarity, the mean spectra have been calculated for each truffle species, and the resulting representation is shown in Figure 1B. At around 6667 cm^{-1} a vast band can be located induced by N−H stretching (first overtone) that can be attributed to proteins and amino acids. Furthermore, N−H combinations are also present around 4687 cm^{-1} and the bands at 4859 cm^{-1} and 4600 cm^{-1} are caused by amide groups [24,38,47].

Regarding the carbohydrates, the double peak at 4338 cm^{-1} and 4257 cm^{-1} can be assigned to $-CH_2$ asymmetric stretching and symmetric stretching, respectively [51]. In addition, C−H stretching (first overtone) and $-CH_2$ vibration lead to peaks at 5760 cm^{-1} and 5742 cm^{-1}, respectively [24,37,47].

In order to put these observations into context, the work of Saltarelli et al. with an analysis of the protein and carbohydrate content of *T. magnatum*, *T. borchii*, *T. aestivum*, and *T. melanosporum* is of great importance. Although their work did not emphasise the species differentiation but storage effects, they have already noticed differences in the major constituents for the truffle species [52]. This can be illustrated well e.g., by the protein fraction: In ascending order, *T. melanosporum*, *T. aestivum*, *T. borchii*, and *T. magnatum* have a soluble protein content of 8.7, 11, 13, and 24%, respectively [52]. Such an order can be found at the wavenumber 6318 cm^{-1}: *T. melanosporum* showing the lowest absorbance for this protein-specific region and *T. magnatum* the highest, so the above-mentioned study and our FT-NIR analysis is therefore consistent. Admittedly, this order is not properly given over the entire protein-specific range, especially *T. magnatum* shows an individual curve, but it should be noted that FT-NIR analysis is not capable of specifically measuring soluble proteins, as Saltarelli et al. (2008) did in their approach. Instead, it returns a general parameter, so the amount of scleroprotein and non-soluble protein fractions could cause the discrepancy. Consequently, it should be possible to distinguish species by—albeit very costly—quantitation of soluble protein and carbohydrate content. FT-NIR analysis, on the other hand, enables the indirect and rapid identification of these major constituents.

3.3. Principal Component Analysis

PCA is widely used for visualising high dimensional data by transforming them into a low dimensional space. As an unsupervised approach, it is useful for the qualitative data exploration, checking for potential outliers and rechecking the research hypothesis before using supervised classification models [53,54].

Figure 2A shows the score-plot for all 75 truffle samples. Tendencies of cluster formations according to the truffle species can be identified: the *T. magnatum* samples are located in the lower-left, whilst the *T. melanosporum* samples are located to the right and the *T. aestivum* samples are in the centre of the plot. *T. borchii* und *T. indicum* samples scatter across the plot. These intermediate results give reason to assume that a classification of truffle species is possible. However, with a differentiation of all five species we do not address real issues in the incoming goods inspection: the truffle's colour can be checked visually; thus, it only makes sense to consider the white and black truffles separately especially because falsification occurs within the white and within the black truffle, and these are not adulterated with each other. Therefore, PCA was calculated only for white and black truffle species and the score-plots are shown in Figure 2B–D, respectively. The trends from the score-plot in Figure 2A are also noticeable here, and FT-NIR analysis appears to be an appropriate method for differentiating truffle species. For the *T. indicum* samples in Figure 2C, some samples are spread over the entire score-plot, but tend to higher PC2 values in the PC4 vs. PC2 plane, already indicating the need for multivariate, non-linear analysis tools hereinafter. Moreover, the fact that there is still cluster formation shows that the important information for the species differentiation is not only contained in the >9000 cm^{-1} region, which was omitted, but is present over the whole spectra.

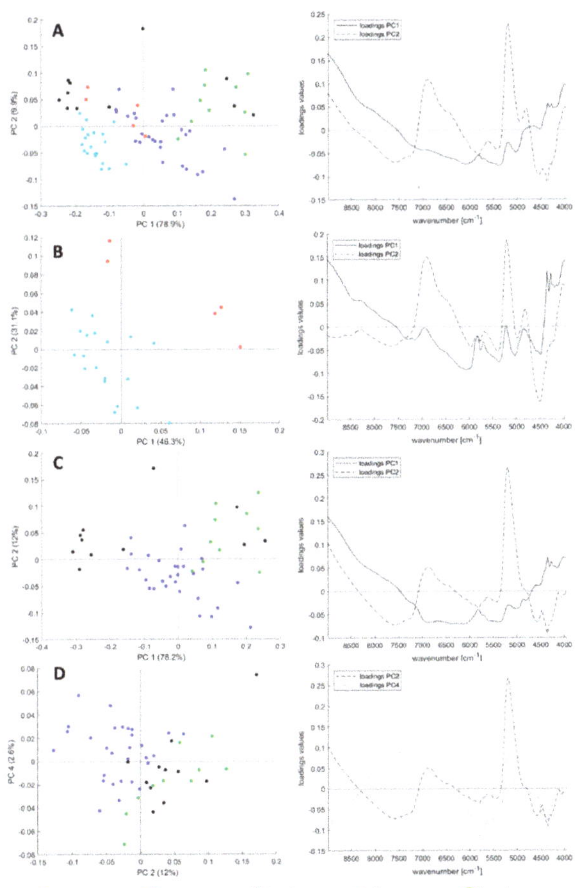

Figure 2. Principal component analysis (PCA) score-plots with their respective loadings plots after pre-processing approach No. i of (**A**) all five truffle species, (**B**) only white truffle species, and only black truffle species in the (**C**) PC2 vs. PC1 plane and (**D**) in the PC4 vs. PC2 plane.

3.4. Evaluation of Pre-Processing and the Suitability for the Species Classification

Whereas applying MSC or SNV correction is necessary without question and is common practice in FT-NIR studies, the need and the impact of any further pre-processing steps should be investigated experimentally [55]. For evaluating the quality of such steps, only visual comparison of 'before-and-after' PCA plots is unlikely to find the most suitable pre-processing strategy and may mislead to an approach, which is not appropriate for a supervised model, so we calculated classification models and compared the prediction accuracy [36,49].

Spectra comparison of different pre-processing approaches examined are shown in Figure 3. The effect of smoothing is not recognisable visually. In addition, it turned out that neighbouring wave numbers show almost identical absorbance values. In order to avoid redundant data and overfitting, a binning was carried out by calculating the mean value of the absorbance for 10 adjacent wavenumbers and combining the measuring points into 248 variables.

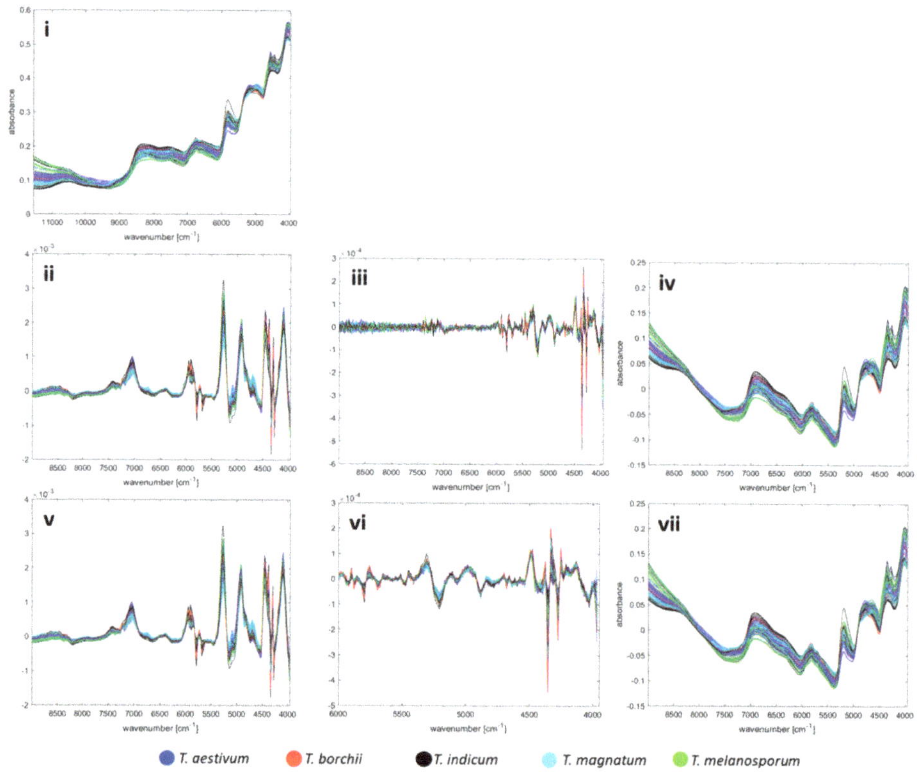

Figure 3. Spectra comparison of different pre-processing approaches, also refer to Table 1. First row: one-step pre-processing: (**i**) MSC. Second row: two-step pre-processing: (**ii**) MSC, 1st derivative. (**iii**) MSC, 2nd derivative. (**iv**) MSC, detrending. Third row: three-step pre-processing: (**v**) smoothing, MSC, 1st derivative. (**vi**) Smoothing, MSC, 2nd derivative. (**vii**) Smoothing, MSC, detrending.

For every pre-processing approaches, all five classification models stated in Table 2 were calculated and validated using stratified nested cross-validation. As the main result parameter for comparing the approaches, we used the mean accuracy instead of the overall accuracy to account for the different size of the groups. The classification accuracies and precision for the test set for the differentiation of white and black truffles are given in Tables 3 and 4, respectively. For the training set used for validation, the classification accuracies and precisions are given in Tables S2 and S3, respectively.

Table 3. Mean accuracy and precision of the prediction of the external test set for different pre-treatments and classification models for the differentiation of the white truffle species (20 *T. magnatum* samples and 5 *T. borchii* samples, all values in %).

		Classification Models					
		(a) LDA	**(b) lin. SVM**	**(c) quad. SVM**	**(d) SSD**	**(e) RF**	**(f) *k*-NN**
pre-processing	(i) MSC	99.2 ± 1.1	91.2 ± 3.9	98.8 ± 1.4	98.7 ± 1.5	98.7 ± 3.2	99.1 ± 1.2
	(ii) MSC, 1st derivative	99.9 ± 0.4	**100 ± 0.0**	100 ± 0.0	99.8 ± 0.7	94.7 ± 3.4	98.7 ± 1.4
	(iii) MSC, 2nd derivative	98.0 ± 1.7	87.1 ± 1.9	94.9 ± 2.1	97.7 ± 2.0	96.3 ± 4.2	93.8 ± 2.0
	(iv) MSC, detrend	98.9 ± 1.3	94.2 ± 3.5	99.5 ± 0.9	99.0 ± 1.4	97.9 ± 2.6	99.1 ± 1.2
	(v) smoothing, MSC, 1st derivative	99.8 ± 0.6	99.5 ± 0.9	99.8 ± 0.6	98.9 ± 1.3	95.0 ± 3.2	98.5 ± 1.5
	(vi) smoothing, MSC, 2nd derivative	98.5 ± 1.5	94.7 ± 2.0	97.7 ± 1.8	98.3 ± 1.6	95.8 ± 2.7	98.5 ± 1.5
	(vii) smoothing, MSC, detrend	98.9 ± 1.5	92.8 ± 3.6	99.4 ± 1.0	98.6 ± 1.8	96.1 ± 3.6	98.5 ± 1.5

Table 4. Mean accuracy and precision of the prediction of the external test set for different pre-treatments and classification models for the differentiation of the black truffle species (29 *T. aestivum* samples, 10 *T. melanosporum* samples, and 11 *T. indicum* samples, all values in %).

		Classification Model					
		(a) LDA	**(b) lin. SVM**	**(c) quad. SVM**	**(d) SSD**	**(e) RF**	**(f) *k*-NN**
pre-processing	(i) MSC	98.7 ± 1.2	87.6 ± 1.2	90.4 ± 2.9	99.0 ± 1.0	90.3 ± 2.2	96.3 ± 2.1
	(ii) MSC, 1st derivative	98.9 ± 1.3	92.8 ± 2.2	94.9 ± 1.6	**99.1 ± 1.2**	91.7 ± 1.9	96.4 ± 1.8
	(iii) MSC, 2nd derivative	95.7 ± 2.2	79.5 ± 2.6	93.2 ± 2.3	93.7 ± 2.3	90.7 ± 2.5	90.0 ± 3.4
	(iv) MSC, detrend	98.7 ± 1.1	88.9 ± 1.4	90.5 ± 2.6	98.7 ± 1.1	91.8 ± 2.0	96.6 ± 2.2
	(v) smoothing, MSC, 1st derivative	98.9 ± 1.1	91.7 ± 2.1	95.4 ± 1.4	99.2 ± 1.1	92.0 ± 2.2	96.1 ± 1.9
	(vi) smoothing, MSC, 2nd derivative	98.8 ± 1.2	95.3 ± 2.0	98.8 ± 1.4	99.2 ± 1.1	92.7 ± 2.3	95.0 ± 2.7
	(vii) smoothing, MSC, detrend	98.8 ± 1.0	88.3 ± 1.4	90.4 ± 2.2	99.0 ± 1.0	91.9 ± 1.9	96.3 ± 2.0

As can be seen in Table 3, all classification models provide good accuracy (>90%). Only the second derivation leads to significantly worse results. A pre-treatment of MSC with first derivation with both a linear and a quadratic SVM lead to an error-free classification of 100% (the most appropriate results are marked bold in the corresponding tables). Accordingly, any falsification of the expensive *T. magnatum* with the cheaper *T. borchii* can be detected. Because of the clear result based on the available and analysed truffle samples, the confusion matrix is not needed here, but can be seen in the supplement in Table S4.

A similar trend can be seen for the black truffles: Here too, high accuracies are generally achieved (>90%), only the second derivative without previous smoothing performs worse and a linear model does not seem to be sufficient for this ternary issue. Although the results overlap when the standard deviation is considered, the best accuracies of 99.1 ± 1.2 % are obtained when using MSC and the first derivative with the SSD model. A previous smoothing does not yield a significant improvement. Since every data pre-treatment is also a manipulation of the data, the model with the fewest steps should be preferred. The corresponding confusion matrix is shown in Table 5. In particular, fraud is common with *T. indicum*, which is counterfeited as the high-priced *T. melanosporum* because the two species are morphologically very similar and collected at the same harvesting times. Therefore, it is pleasing that the specificity for *T. melanosporum* is 97.5%—the error rate of mistakes is only 2.5%.

Table 5. Confusion matrix for classification of the black truffle species with the build subspace discriminant model after MSC and 1st derivative; resulting in 99.1 ± 1.2% mean sensitivity. The predictions of 100 repetitions of the test set were accumulated.

		Predicted Species			
		T. indicum	*T. aestivum*	*T. melanosporum*	sensitivity [%]
actual species	*T. indicum*	1073	1	26	97.5
	T. aestivum	3	2897	0	99.9
	T. melanosporum	1	0	999	99.9
	specificity [%]	99.6	100	97.5	

Table S5 shows the classification results for the test set for the differentiation of all five truffle species, indicating that also for this more complex five-class-issue, classification models can be calculated with high accuracy of 99%, and for the training set used for validation, the classification accuracies and precisions are given in Table S6. The corresponding confusion matrix is shown in Table S7.

DNA analysis is often used to authenticate species and varieties, while FT-NIR analysis is widely established in industrial incoming goods inspection. FT-NIR analysis does not require any specialised training for handling and any special, eventually hazardous chemicals for sample preparation and measurement, therefore the FT-NIR analysis is a "green method" [56]. Additionally, possible contamination due to exponential amplification by PCR quickly leads to false positive results. In order to keep this danger to a minimum, separate laboratories for sample preparation and DNA analysis are necessary, whereas NIR does not have such requirements. Optionally, it would be conceivable to use FT-NIR measurement for sample screening and to countercheck any conspicuous results by DNA analysis.

Regarding the determination of the geographical origin, however, DNA analysis cannot provide reasonable answers since the origin rather affects the phenotype. Here, FT-NIR analysis can be a tool for differentiating the origin [35] and the possibilities for the truffle differentiation by origin are examined in the following chapter.

3.5. Influence of Harvest Year and Geographical Origin

As shown in the PCA plot (Figure 2A), the truffle species has the dominant influence on the NIR spectrum, since the scores cluster according to their species in this unsupervised model. This can be demonstrated on the *T. magnatum* samples, which, although dominant from Italy, originate from Bulgaria, Croatia, and Romania, and are clustering together in the unsupervised PCA. This effect is similar for the *T. aestivum* samples originating mainly from Romania, but also from Bulgaria, France, Iran, Italy, Moldovia, and Slovenia. Thus, the species itself seems to have a much greater influence on the metabolome to be measured by FT-NIR spectroscopy than the origin.

One model for the origins of all truffle samples is not advisable for this reason, since most Italian samples are white truffles and most Romanian samples are *T. aestivum* truffles what is linked to their natural areas of origin. Such a model might, therefore, correlate on a false causality. However, the price depends primarily on the species whilst the origin is a second factor in the purchase decision. Accordingly, for the incoming goods inspection it is important especially for the most expensive *T. magnatum* truffle whether it comes from Italy or not, according to the consumer's expectations. For this Italy vs. non-Italy issue, all pre-processing was compared with classification models, analogous to the previous investigations when targeting the species. The results of the test set are shown in Table 6, and for the training set used for validation, the classification accuracies and precisions are given in Table S8. Best classification results of 88.4 % are reached after MSC and 2nd derivative in combination with a Random Forest (RF) classification model. However, we have decided not to pursue this pre-processing strategy because the spectra line plots in Figure 3iii have shown that a lot of noise occurs in the range of wavenumbers above 6000 cm^{-1} and a smoothing an omitting this range is preferable. This alternative approach leads to a slightly worse accuracy of 82.8 ± 8.1% and the corresponding confusion matrix is shown in Table 7. The accuracy results provided by the LDA classification only differ by a few percentage points, and are even better in some cases. However, we chose the RF model since the PCA plots have arouse the impression that non-linear classification models might be more suitable for this issue.

Table 6. Mean accuracy and precision of the prediction of the external test set for different pre-treatments and classification models for the differentiation of Italian vs. non-Italian *T. magnatum* truffles (all values in %).

		Classification Model					
		(a) LDA	**(b) lin. SVM**	**(c) quad. SVM**	**(d) SSD**	**(e) RF**	**(f) *k*-NN**
pre-processing	(i) MSC	82.4 ± 4.5	51.8 ± 2.5	74.5 ± 5.0	82.6 ± 4.7	72.8 ± 5.4	80.2 ± 4.3
	(ii) MSC, 1st derivative	80.5 ± 4.5	50.8 ± 2.3	80.9 ± 3.4	82.0 ± 4.3	79.2 ± 5.6	82.5 ± 4.5
	(iii) MSC, 2nd derivative	83.6 ± 4.0	58.0 ± 3.6	82.9 ± 2.4	83.2 ± 3.1	88.4 ± 5.0	80.8 ± 4.5
	(iv) MSC, detrend	82.6 ± 4.7	51.7 ± 2.5	74.4 ± 4.9	81.3 ± 4.7	78.9 ± 6.6	81.6 ± 3.8
	(v) smoothing, MSC, 1st derivative	81.1 ± 4.1	51.4 ± 2.1	80.8 ± 3.0	81.1 ± 4.2	79.3 ± 5.4	80.9 ± 4.4
	(vi) smoothing, MSC, 2nd derivative	83.7 ± 4.2	63.8 ± 3.8	82.3 ± 3.6	82.5 ± 4.5	**82.8 ± 8.1**	81.8 ± 4.6
	(vii) smoothing, MSC, detrend	82.6 ± 5.0	51.4 ± 2.4	74.3 ± 4.9	81.8 ± 4.5	80.5 ± 6.6	81.5 ± 4.2

Table 7. Confusion matrix for classification for the differentiation of Italian vs. non-Italian T. magnatum truffles with the build RF model after smoothing. MSC and 2nd derivative; resulting in 82.8 ± 8.1% mean sensitivity. The predictions of 100 repetitions of the test set were accumulated.

		Predicted Origin		
		Italian	non-Italian	sensitivity [%]
actual origin	Italian	1247	153	89.1
	non-Italian	141	459	76.5
	specificity [%]	89.8	75.0	

Additionally, the PCA-plots for the *T. magnatum* samples were calculated and are shown in Figure 4, indicating and confirming that a non-linear classification model, such as RF, is more suited for this issue. Still, there are two aspects to consider: first, the standard deviation is remarkably high and second, the PCA plots show that the variance within the Italian samples is at least as large as the variance of the other origins. An origin model with acceptable accuracy is chemometrically possible, but should be checked with additional samples.

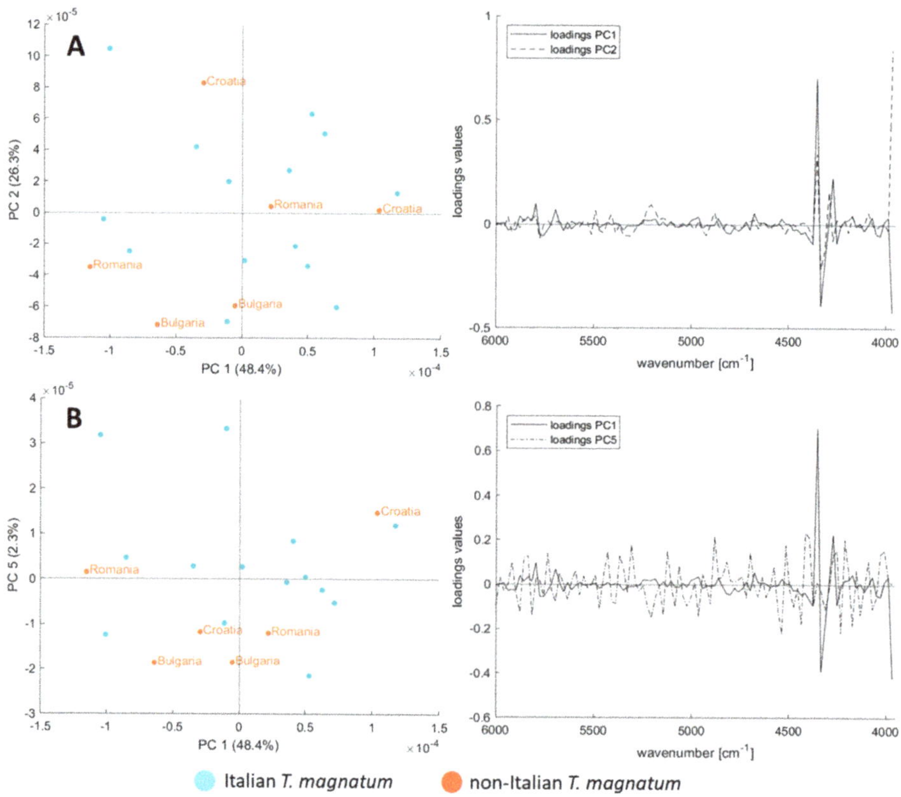

Figure 4. PCA score-plots with their respective loadings plots after pre-processing approach No. vi of the *T. magnatum* samples from Italy and other countries (**A**) PC2 vs. PC1, (**B**) PC5 vs. PC1.

As the results show, FT-NIR can be used for the differentiation of black and white truffles, and Italian and non-Italian truffles of the species *T. magnatum*. Since FT-NIR is a simple and cheap method, it is suitable for industrial applications, for example, for the incoming goods inspection or authenticity checks on truffles. The process of authentication using FT-NIR is shown schematically in Figure 5.

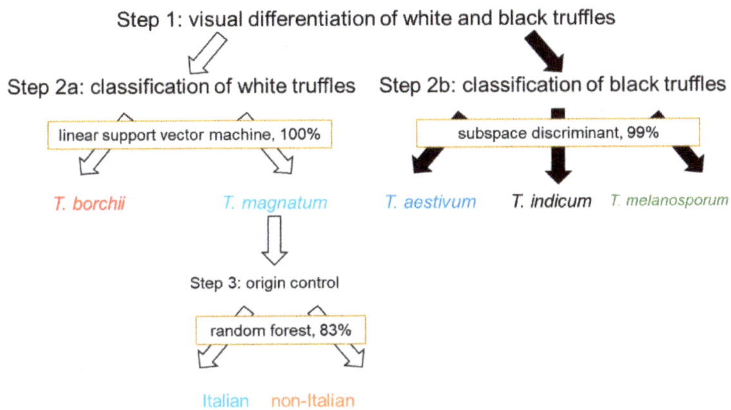

Figure 5. Authentication protocol for the stepwise authentication assessment of truffles with FT-NIR and chemometrics.

4. Conclusions

FT-NIR spectroscopy was combined with chemometrics to distinguish within the white truffles *T. borchii* and *T. magnatum* and the black truffles *T. aestivum*, *T. indicum*, and *T. melanosporum*. Different techniques for pre-processing in combination with various classification models and their effect on the accuracy of the model were compared. Classification accuracies >99% showed that the analysis of truffle samples by FT-NIR spectroscopy is a very suitable tool for species differentiation without sophisticated sample preparation or instruments. When differentiating between Italian and non-Italian *T. magnatum* samples, an accuracy of 83% was achieved. FT-NIR analysis requires no special training for handling and no special, possibly hazardous chemicals for sample preparation and measurement. In addition, most quality assurance laboratories already have FT-NIR instruments. Due to its simple, cost-effective application, FT-NIR analysis is very well suited for industrial screening samples during incoming goods inspection. Considering the number of 75 truffle samples used, we intend to extend the results of our study by analysing further samples, including a research on the potential effects of the harvest year.

Supplementary Materials: The following figures and tables are available online at http://www.mdpi.com/2304-8158/9/7/922/s1, Figure S1: Influence of the order of pre-processing steps. (A) Raw data. (B) MSC and omitting the > 9000 cm^{-1} range. (C) Omitting the >9000 cm^{-1} range first and MSC; Table S1: Overview of the analysed truffle samples with number of samples, harvest year and country; Table S2: Mean accuracy and precision of the training set used for validation for different pre-treatment and classification models for the differentiation of the white truffle species (20 *T. magnatum* samples, 5 *T. borchii* samples, all values in %); Table S3: Mean accuracy and precision of the training set used for validation for different pre-treatment and classification models for the differentiation of the black truffle species (29 *T. aestivum* samples, 10 *T. melanosporum* samples and 11 *T. indicum* samples, all values in %); Table S4: Confusion matrix for classification of the white truffle species with the build linear SVM model after MSC and 1st derivative; resulting in 100% mean sensitivity. The predictions of 100 repetitions of the test set were accumulated; Table S5: Mean accuracy with standard deviation for different pre-treatment and classification models for the prediction of the test set for the differentiation of five truffle species (20 *T. magnatum* samples, 5 *T. borchii* samples, 29 *T. aestivum* samples, 10 *T. melanosporum* samples and 11 *T. indicum* samples, all values in %); Table S6: Mean accuracy and precision of the training set for different pre-treatment and classification models for the differentiation of the five truffle species (20 *T. magnatum* samples, 5 *T. borchii* samples, 29 *T. aestivum* samples, 10 *T. melanosporum* samples and 11 *T. indicum* samples, all values in %); Table S7: Confusion matrix for classification of five truffle species with the build subspace discriminant model after MSC and 1st derivative; resulting in 99.3 ± 0.9% mean sensitivity. The predictions of 100 repetitions of the test set were accumulated; Table S8: Mean accuracy and precision of the training set for different pre-treatment and classification models for the differentiation of Italian vs. non-Italian *T. magnatum* truffles (all values in %), MATLAB function for the creation of stratified parts for the nested cross validation.

Author Contributions: Conceptualization, T.S. and S.S.; methodology, T.S. and S.S.; validation, T.S. and C.A.; formal analysis, T.S. and S.S.; investigation, T.S., S.S. and C.A.; resources, M.F.; data curation, T.S. and S.S., writing—original draft preparation, T.S. and S.S.; writing—review and editing, T.S., S.S., C.A., and M.F.;

visualization, T.S.; supervision, M.F.; project administration, M.F.; funding acquisition, M.F. All authors have read and agreed to the published version of the manuscript.

Funding: This study was performed within the project "Food Profiling—Development of analytical tools for the experimental verification of the origin and identity of food". This Project (Funding reference number: 2816500914) is supported by means of the Federal Ministry of Food and Agriculture (BMEL) by a decision of the German Bundestag (parliament). Project support is provided by the Federal Institute for Agriculture and Food (BLE) within the scope of the program for promoting innovation.

Acknowledgments: The authors gratefully thank the project partners "LA BILANCIA Trüffelhandels GmbH" and "Trüffelkontor e.K." for providing sample material. We would like to thank Maike Arndt and Bernadette Richter for their helpful discussion on the manuscript.

Conflicts of Interest: The authors declare no conflict of interest. The funders had no role in the design of the study; in the collection, analyses, or interpretation of data; in the writing of the manuscript, or in the decision to publish the results.

References

1. Ellefson, W.; Zach, L.; Sullivan, D. *Improving Import Food Safety*; John Wiley & Sons: New York, NY, USA, 2012.
2. Dimara, E.; Skuras, D. Consumer demand for informative labeling of quality food and drink products: A European Union case study. *J. Consum. Mark.* **2005**, *22/2*, 90–100. [CrossRef]
3. Savelli, E.; Bravi, L.; Murmura, F.; Pencarelli, T. Understanding the consumption of traditional-local foods through the experience perspective. *Brit. Food J.* **2019**, *121*, 1261–1280. [CrossRef]
4. Creydt, M.; Fischer, M. Omics approaches for food authentication. *Electrophoresis* **2018**, *39*, 1569–1581. [CrossRef] [PubMed]
5. El Karkouri, K.; Couderc, C.; Decloquement, P.; Abeille, A.; Raoult, D. Rapid MALDI-TOF MS identification of commercial truffles. *Sci. Rep.* **2019**, *9*, 1–9. [CrossRef] [PubMed]
6. Pegler, D. Useful fungi of the world: Morels and truffles. *Mycologist* **2003**, *17*, 174–175. [CrossRef]
7. Squires, N. Truffle Trouble: Prices of Prized Tuber Double in Italy after Driest Autumn for Decades. Available online: https://www.telegraph.co.uk/news/2017/10/26/truffle-trouble-prices-prized-tuber-double-italy-driest-autumn/ (accessed on 19 September 2019).
8. Lazzari, B.; Gianazza, E.; Viotti, A. Molecular characterization of some truffle species. In *Biotechnology of Ectomycorrhizae*; Springer: Berlin/Heidelberg, Germany, 1995; pp. 161–169.
9. Zambonelli, A.; Iotti, M.; Murat, C. *True Truffle (Tuber Spp.) in the World: Soil Ecology, Systematics and Biochemistry*; Springer: Cham, Switzerland, 2016.
10. Gregori, G. Problems and expectations with the cultivation of Tuber magnatum. In Proceedings of the Second International Conference on Edible Mycorrhizal Mushrooms, Christchurch, New Zealand, 3–5 July 2001.
11. Mello, A.; Fontana, A.; Meotto, F.; Comandini, O.; Bonfante, P. Molecular and morphological characterization of Tuber magnatum mycorrhizas in a long-term survey. *Microbiol. Res.* **2001**, *155*, 279–284. [CrossRef]
12. Hall, I.R.; Yun, W.; Amicucci, A. Cultivation of edible ectomycorrhizal mushrooms. *Trends Biotechnol.* **2003**, *21*, 433–438. [CrossRef]
13. Favre, J.; Parguey Leduc, A.; Sejalon Delmas, N.; Dargent, R.; Kulifaj, M. The ascocarp of Tuber indicum (Chinese truffle) recently introduced in France: Preliminary study. *C. R. Acad. Sci.* **1996**, *319*, 517–521.
14. Bonito, G.; Trappe, J.M.; Donovan, S.; Vilgalys, R. The Asian black truffle Tuber indicum can form ectomycorrhizas with North American host plants and complete its life cycle in non-native soils. *Fungal Ecol.* **2011**, *4*, 83–93. [CrossRef]
15. García-Montero, L.G.; Díaz, P.; Di Massimo, G.; García-Abril, A. A review of research on Chinese Tuber species. *Mycol. Prog.* **2010**, *9*, 315–335. [CrossRef]
16. Murat, C.; Zampieri, E.; Vizzini, A.; Bonfante, P. Is the Perigord black truffle threatened by an invasive species? We dreaded it and it has happened! *New Phytol.* **2008**, *178*, 699–702. [CrossRef]
17. Zhao, D.; Liu, G.; Song, D.; Liu, J.-H.; Zhou, Y.; Ou, J.; Sun, S. Fourier transform infrared spectroscopic study of truffles. In *Proceedings of the ICO20: Biomed Opt*; NCBI: Bethesda, MD, USA, 2006; p. 60260H.
18. Krauß, S.; Vetter, W. Geographical and species differentiation of truffles (Tuber spp.) by means of stable isotope ratio analysis of light elements (H, C, N). *J. Agr. Food Chem.* **2020**. [CrossRef]

19. Qu, J.-H.; Liu, D.; Cheng, J.-H.; Sun, D.-W.; Ma, J.; Pu, H.; Zeng, X.-A. Applications of near-infrared spectroscopy in food safety evaluation and control: A review of recent research advances. *Crit. Rev. Food Sci.* **2015**, *55*, 1939–1954. [CrossRef] [PubMed]
20. Pérez-Marín, D.; Sánchez, M.-T.; Paz, P.; González-Dugo, V.; Soriano, M.-A. Postharvest shelf-life discrimination of nectarines produced under different irrigation strategies using NIR-spectroscopy. *LWT-Food Sci. Technol.* **2011**, *44*, 1405–1414. [CrossRef]
21. Salguero-Chaparro, L.; Gaitán-Jurado, A.J.; Ortiz-Somovilla, V.; Peña-Rodríguez, F. Feasibility of using NIR spectroscopy to detect herbicide residues in intact olives. *Food Control.* **2013**, *30*, 504–509. [CrossRef]
22. Dalle Zotte, A.; Ottavian, M.; Concollato, A.; Serva, L.; Martelli, R.; Parisi, G. Authentication of raw and cooked freeze-dried rainbow trout (Oncorhynchus mykiss) by means of near infrared spectroscopy and data fusion. *Food Res. Int.* **2014**, *60*, 180–188. [CrossRef]
23. Luna, A.S.; da Silva, A.P.; Alves, E.A.; Rocha, R.B.; Lima, I.C.; de Gois, J.S. Evaluation of chemometric methodologies for the classification of Coffea canephora cultivars via FT-NIR spectroscopy and direct sample analysis. *Anal. Methods* **2017**, *9*, 4255–4260. [CrossRef]
24. Richter, B.; Rurik, M.; Gurk, S.; Kohlbacher, O.; Fischer, M. Food monitoring: Screening of the geographical origin of white asparagus using FT-NIR and machine learning. *Food Control.* **2019**, *104*, 318–325. [CrossRef]
25. Chen, L.; Xue, X.; Ye, Z.; Zhou, J.; Chen, F.; Zhao, J. Determination of Chinese honey adulterated with high fructose corn syrup by near infrared spectroscopy. *Food Chem.* **2011**, *128*, 1110–1114. [CrossRef]
26. Mishra, S.; Kamboj, U.; Kaur, H.; Kapur, P. Detection of jaggery syrup in honey using near-infrared spectroscopy. *Int. J. Food Sci. Nutr.* **2010**, *61*, 306–315. [CrossRef] [PubMed]
27. Quelal-Vásconez, M.A.; Lerma-García, M.J.; Pérez-Esteve, É.; Arnau-Bonachera, A.; Barat, J.M.; Talens, P. Fast detection of cocoa shell in cocoa powders by near infrared spectroscopy and multivariate analysis. *Food Control.* **2019**, *99*, 68–72. [CrossRef]
28. Mandrile, L.; Mello, A.; Vizzini, A.; Balestrini, R.; Rossi, A.M. Near-infrared spectroscopy as a new method for post-harvest monitoring of white truffles. *Mycol. Prog.* **2020**, *19*, 329–337. [CrossRef]
29. Rinnan, Å.; Van Den Berg, F.; Engelsen, S.B. Review of the most common pre-processing techniques for near-infrared spectra. *TrAC-Trend. Anal. Chem.* **2009**, *28*, 1201–1222. [CrossRef]
30. Candolfi, A.; De Maesschalck, R.; Jouan-Rimbaud, D.; Hailey, P.; Massart, D. The influence of data pre-processing in the pattern recognition of excipients near-infrared spectra. *J. Pharm. Biomed.* **1999**, *21*, 115–132. [CrossRef]
31. Vitale, R.; Bevilacqua, M.; Bucci, R.; Magrì, A.D.; Magrì, A.L.; Marini, F. A rapid and non-invasive method for authenticating the origin of pistachio samples by NIR spectroscopy and chemometrics. *Chemom. Intell Lab.* **2013**, *121*, 90–99. [CrossRef]
32. *German Guidelines For Mushrooms and Mushrooms Products*; Deutsche-Lebensmittelbuch-Komission: Bonn, Germany, 2008.
33. Molinier, V.; van Tuinen, D.; Chevalier, G.; Gollotte, A.; Wipf, D.; Redecker, D. A multigene phylogeny demonstrates that Tuber aestivum and Tuber uncinatum are conspecific. *Org. Divers. Evol.* **2013**, *13*, 503–512. [CrossRef]
34. Schelm, S.; Siemt, M.; Pfeiffer, J.; Lang, C.; Tichy, H.-V.; Fischer, M. Food Authentication: Identification and Quantitation of Different Tuber Species via Capillary Gel Electrophoresis and Real-Time PCR. *Foods* **2020**, *9*, 501. [CrossRef]
35. Arndt, M.; Rurik, M.; Drees, A.; Bigdowski, K.; Kohlbacher, O.; Fischer, M. Comparison of different sample preparation techniques for NIR screening and their influence on the geographical origin determination of almonds (Prunus dulcis MILL.). *Food Control.* **2020**, *115*, 107302. [CrossRef]
36. Achten, E.; Schütz, D.; Fischer, M.; Fauhl-Hassek, C.; Riedl, J.; Horn, B. Classification of Grain Maize (Zea mays L.) from Different Geographical Origins with FTIR Spectroscopy—A Suitable Analytical Tool for Feed Authentication? *Food Anal. Method* **2019**, *12*, 2172–2184. [CrossRef]
37. Chen, Q.; Zhao, J.; Fang, C.; Wang, D. Feasibility study on identification of green, black and Oolong teas using near-infrared reflectance spectroscopy based on support vector machine (SVM). *Spectrochim. Acta. A* **2007**, *66*, 568–574. [CrossRef]
38. Cozzolino, D.; Chree, A.; Scaife, J.; Murray, I. Usefulness of near-infrared reflectance (NIR) spectroscopy and chemometrics to discriminate fishmeal batches made with different fish species. *J. Agric. Food Chem.* **2005**, *53*, 4459–4463. [CrossRef] [PubMed]

39. Cortes, C.; Vapnik, V. Support-vector networks. *Mach. Learn.* **1995**, *20*, 273–297. [CrossRef]
40. Bachmann, R.; Klockmann, S.; Haerdter, J.; Fischer, M.; Hackl, T. 1H NMR spectroscopy for determination of the geographical origin of hazelnuts. *J. Agr. Food Chem.* **2018**, *66*, 11873–11879. [CrossRef]
41. Liaw, A.; Wiener, M. Classification and regression by randomForest. *R. News* **2002**, *2*, 18–22.
42. Cover, T.; Hart, P. Nearest neighbor pattern classification. *IEEE Trans. Inf. Theory* **1967**, *13*, 21–27. [CrossRef]
43. Teye, E.; Huang, X.; Dai, H.; Chen, Q. Rapid differentiation of Ghana cocoa beans by FT-NIR spectroscopy coupled with multivariate classification. *Spectrochim. Acta A* **2013**, *114*, 183–189. [CrossRef]
44. Krstajic, D.; Buturovic, L.J.; Leahy, D.E.; Thomas, S. Cross-validation pitfalls when selecting and assessing regression and classification models. *J. Cheminformatics* **2014**, *6*, 1–15. [CrossRef]
45. Varma, S.; Simon, R. Bias in error estimation when using cross-validation for model selection. *BMC Bioinform.* **2006**, *7*, 91. [CrossRef]
46. Günzler, H.; Böck, H. *IR-Spektroskopie*, 2nd ed.; VCH Weinheim: Weinheim, Germany, 1990.
47. Workman Jr, J.; Weyer, L. *Practical Guide and Spectral Atlas for Interpretive Near-Infrared Spectroscopy*; CRC press: Boca Raton, FL, USA, 2012.
48. Dhanoa, M.; Lister, S.; Sanderson, R.; Barnes, R. The link between multiplicative scatter correction (MSC) and standard normal variate (SNV) transformations of NIR spectra. *J. Near. Infrared. Spec.* **1994**, *2*, 43–47. [CrossRef]
49. Gerretzen, J.; Szymanska, E.; Jansen, J.J.; Bart, J.; van Manen, H.-J.; van den Heuvel, E.R.; Buydens, L.M. Simple and effective way for data preprocessing selection based on design of experiments. *Anal. Chem.* **2015**, *87*, 12096–12103. [CrossRef]
50. Souci, S.W.; Fachmann, W.; Kraut, H. *Mushrooms, Truffles, Tuber Melanosporum Vitt. Food Composition and Nutrition Tables (p. 869)*, 6th ed.; Medpharm GmbH Scientific Publishers: Stuttgart, Germany, 2000.
51. Veselá, A.; Barros, A.S.; Synytsya, A.; Delgadillo, I.; Čopíková, J.; Coimbra, M.A. Infrared spectroscopy and outer product analysis for quantification of fat, nitrogen, and moisture of cocoa powder. *Anal. Chim. Acta.* **2007**, *601*, 77–86. [CrossRef] [PubMed]
52. Saltarelli, R.; Ceccaroli, P.; Cesari, P.; Barbieri, E.; Stocchi, V. Effect of storage on biochemical and microbiological parameters of edible truffle species. *Food Chem.* **2008**, *109*, 8–16. [CrossRef] [PubMed]
53. Kemsley, E.K.; Defernez, M.; Marini, F. Multivariate statistics: Considerations and confidences in food authenticity problems. *Food Control.* **2019**, *104*, 102–112. [CrossRef]
54. Wold, S.; Esbensen, K.; Geladi, P. Principal component analysis. *Chem. Intell. Lab.* **1987**, *2*, 37–52. [CrossRef]
55. Engel, J.; Gerretzen, J.; Szymańska, E.; Jansen, J.J.; Downey, G.; Blanchet, L.; Buydens, L.M. Breaking with trends in pre-processing? *TrAC-Trend Anal. Chem.* **2013**, *50*, 96–106. [CrossRef]
56. Anastas, P.; Eghbali, N. Green chemistry: Principles and practice. *Chem. Soc. Rev.* **2010**, *39*, 301–312. [CrossRef]

foods

MDPI

Article

Procedures for DNA Extraction from Opium Poppy (*Papaver somniferum* L.) and Poppy Seed-Containing Products

Šarlota Kaňuková [1], Michaela Mrkvová [1], Daniel Mihálik [1,2] and Ján Kraic [1,2,*]

1 Department of Biotechnology, Faculty of Natural Sciences, University of SS. Cyril and Methodius, Námestie J. Herdu 2, SK-917 01 Trnava, Slovakia; sarlota.kanukova@gmail.com (Š.K.); michaela.mrkvova@ucm.sk (M.M.); daniel.mihalik@ucm.sk (D.M.)

2 Research Institute of Plant Production, National Agricultural and Food Center, Bratislavská cesta 122, SK-921 68 Piešťany, Slovakia

* Correspondence: jan.kraic@nppc.sk; Tel.: +421-337-947-168; Fax: +421-337-726-306

Received: 4 September 2020; Accepted: 2 October 2020; Published: 9 October 2020

Abstract: Several commonly used extraction procedures and commercial kits were compared for extraction of DNA from opium poppy (*Papaver somniferum* L.) seeds, ground seeds, pollen grains, poppy seed filling from a bakery product, and poppy oil. The newly developed extraction protocol was much simpler, reduced the cost and time required for DNA extraction from the native and ground seeds, and pollen grains. The quality of extracted DNA by newly developed protocol was better or comparable to the most efficient ones. After being extended by a simple purification step on a silica membrane column, the newly developed protocol was also very effective in extracting of poppy DNA from poppy seed filling. DNA extracted from this poppy matrix was amplifiable by PCR analysis. DNA extracted from cold-pressed poppy oil and suitable for amplifications was obtained only by methods developed previously for olive oil. Extracted poppy DNA from all tested matrices was analysed by PCR using primers flanking a microsatellite locus (156 bp) and two different fragments of the reference tubulin gene (553 bp and 96 bp). The long fragment of the reference gene was amplified in DNA extracted from native seeds, ground seeds, and pollen grains. Poppy DNA extracted from the filling of bakery product was confirmed only by amplification of short fragments (96 bp and 156 bp). DNA extracted from cold-pressed poppy oil was determined also only by amplification of these two short fragments.

Keywords: DNA extraction; opium poppy; seed; pollen grains; bakery product; oil; PCR

1. Introduction

Extraction of nucleic acids from various matrices is the first and crucial step in analysis of biological materials generally. Methods of DNA extraction have evolved over time [1], but still contain several basic and necessary steps such as cell disruption, removal of undesirable molecules (lipids, proteins, polyphenols, and others), and purification. In addition to the cell wall disruption, the chemical diversity of metabolites contained in the plant cell is a major complication in the DNA isolation process. There is no available universal protocol for extraction of DNA which would be applicable independently of plant species, plant tissues, and plant matrix [2,3]. Generally, extraction of DNA from young, fast-growing, and healthy tissues is much easier. However, it is often necessary to extract DNA from plant tissues rich in polysaccharides, lipids, secondary metabolites, or even from very complex matrices (processed seeds, oils, foods, feeds). This is also the case of oilseeds where extraction of DNA is considered more demanding than from vegetative plant tissues (e.g., young leaves). Lipids usually prevent the action of solvents during removal of polysaccharides and phenolic compounds. Secondary metabolites can bind and precipitate with DNA and reduce efficiency of isolation procedure. Nevertheless, extraction of DNA from mature seeds may be often preferred over extraction from foliar tissues. Moreover, processing of plant seeds into foods is associated with determination of authenticity and traceability of foods what have recently become very important for various reasons [4–6]. The quantity and quality of DNA extracted from foods and oils tends to decrease to the extent in which the food/oil is processed [7,8]. Processing affects the DNA and may lead to degradation or removal of DNA from sample due to its hydrolysis, oxidation, and deamination [9]. Considering the DNA degradation and the presence of PCR inhibitors, DNA extraction from processed matrices is often a compromise between high yield and high purity [9–11]. The most appropriate extraction method should be chosen case by case. Extracted DNA is used for authentication of foods and feeds and detection of falsifications (e.g., blending of low-quality oil into high-quality oil) [12–14].

Oilseed crop with an interesting position in the world agriculture is the opium poppy (*Papaver somniferum* L.) grown under control only in some countries [15]. In addition to the production of alkaloids extracted from poppy straw, edible seeds are in great demand in cuisine. However, trading with poppy seeds, products (cake fillings, spreads), and oils suffers sometimes from adulteration practices [16]. Sometimes, high quality poppy seeds with a blue colour and a sweet taste are adulterated with technical poppy seed (grey-black colour, no taste). In addition to quality, they differ significantly in price. The consumer may be deceived in both quality and price. Such practices are then transferred to the food industry (poppy bakery products). Falsification is also a serious problem in the production of vegetable oils, especially the more expensive ones, including poppy seed oil. Chemical analyses of oils are used to determine the species origin of oil [17], but DNA analyses are appropriate to determine species origin and also the cultivar origin [18–20].

Poppy seeds with a high content of lipids and secondary metabolites are not a simple object for DNA extraction. This is even more complicated with ground seeds, poppy seed fillings from bakery products, and pressed oil. The number of relevant scientific reports in poppy is very limited and DNA extraction procedures have been published only from defatted seeds [21] and heroin samples [22]. Three commercial kits were tested for DNA extraction from seeds [23]. Very useful would be efficient, simple, and universal protocol for extraction of DNA from poppy seeds, grains, and products containing or made from poppy seeds. Therefore, the aim of this study was to test several methods of DNA extraction and try to design a new, effective procedure from different poppy seed matrices (native and ground seeds, pollen grains, poppy filling of the bakery product, poppy oil) with respect to DNA quality and suitability for amplification analyses.

2. Materials and Methods

2.1. Plant and Food Material

Mature seeds and pollen grains of opium poppy (*Papaver somniferum* L.) were collected from registered cultivar Major, cultivated at the Research and Breeding Station in Malý Šariš (Slovakia). They were stored at 4 °C before DNA extraction. Seeds and pollen grains were homogenized by pestle and mortar before the extraction. Seeds were also ground with the poppy seed mill. The poppy seeds roll (Tastino, Slovakia) and cold-pressed poppy seed oil (Juvamed Ltd., Tastino, Slovakia) were purchased in food store and stored at 4 °C before the DNA extraction.

2.2. DNA Extraction from Seeds

Genomic DNA from seeds was extracted from 0.2–0.5 g of seeds by six methods: Dellaporta et al. [24] with and without CTAB; Bayer BioScience N.V. [25]; Monsanto Company [26]; Murray and Thompson [27]; Sagwan et al. [21] and using four commercial kits: DNeasy® Plant Maxi Kit, QIAamp DNA Stool Mini Kit, PowerSoil DNA Isolation Kit (all from QIAGEN N.V., Hilden, Germany) and Plant DNAzol® Reagent (Thermo Fisher Scientific, Waltham, MA, USA).

Another extraction protocol was newly developed protocol designed on the basis of the Bayer BioScience N.V. procedure [25], but containing several modifications. The content of this protocol is as follows. The sample (200 mg) of seeds was ground to a fine powder with mortar and pestle and extracted with 2.7 mL of extraction buffer (50 mM EDTA, 100 mM Tris-HCl, pH 8.0, 500 mM NaCl), 190 μL of 20% SDS, and 10 μL of 2-mercaptoethanol. The mixture was vortexed and incubated at 65 °C for 30 min. During the incubation, the samples were mixed every 10 min. After incubation, 2.3 mL of mixture phenol:chloroform:isoamyl alcohol (25:24:1) was added, the mixture being shaken for 1 min and centrifuged for 20 min at 5500× *g*. The upper aqueous phase was transferred to a new tube, mixed with 2 mL of isopropanol and precipitated 30 min at −20 °C. Precipitated nucleic acids were transferred to Eppendorf tube and washed with 70% and 96% ethanol. Pellet after drying was dissolved in TE buffer (10 mM Tris-HCl, 1 mM EDTA, pH 8.0) and treated with 10 μL of RNase A (10 mg/mL) for 30 min at 37 °C. After the incubation, 800 μL of mixture chloroform:isoamyl alcohol (24:1) was added, shaken vigorously and centrifuged for 10 min in a microcentrifuge at maximum speed. The upper aqueous phase was transferred to a new Eppendorf tube, 600 μL of isopropanol was added and after vortexing was incubated for 20 min at −20 °C. Precipitate DNA was again washed with 70 and 96% ethanol, dried, dissolved in TE buffer, and stored at −20 °C.

2.3. *DNA Extraction from Ground Seeds*

Six different methods used for isolation DNA from 0.2 g of ground seeds were: Bayer BioScience N.V. [25], Monsanto Company [26], two commercial kits (DNeasy® Plant Maxi Kit, QIAamp DNA Stool Mini Kit), and newly developed protocol (described above).

2.4. *DNA Extraction from Pollen Grains*

Three methods used for isolation of DNA from 0.1 g of pollen grains included DNeasy® Plant Maxi Kit, QIAamp DNA Stool Mini Kit. The third was the newly developed protocol (described above). An efficient mechanical homogenization of pollen grains was particularly important.

2.5. *DNA Extraction from Poppy Seed Filling*

DNA was extracted from 0.5–2.0 g of filling of the bakery product using methods: Bayer BioScience N.V. [25], Monsanto Company [26], QIAamp DNA Stool Mini Kit, and newly developed protocol. Extracted DNA was purified through the silica membrane spin-columns [28].

2.6. *DNA Extraction from Poppy Oil*

DNA was extracted from 0.2–15 mL of oil according to Doveri et al. [29]; Monsanto Company [26]; Bayer BioScience N.V. [25]; Consolandi et al. [30]; Giménez et al. [31]; Raieta et al. [4], newly developed protocol, and commercial kit (QIAamp DNA Stool Mini Kit).

2.7. *Qualitative and Quantitative Analysis of Extracted DNA*

Integrity of the extracted DNA from different poppy matrices was assessed by agarose gel electrophoresis. Parameters of extracted DNA were tested by UV spectrophotometry (NanoDrop ND-1000 spectrophotometer, Thermo Fisher Scientific, Waltham, MA, USA) as well as by electrophoresis in 0.8% agarose gel stained with ethidium bromide.

2.8. *PCR Amplification*

Extracted DNA were amplified by PCR using primers for microsatellite locus psSSR69 [32]. Two pairs of primers for reference gene encoding tubulin beta-7 chain (Table 1) was designed from coding sequence (XM_026557633.1, GenBank®, http://www.ncbi.nlm.nih.gov) [33] using the Primer3 Input software (Whitehead Institute for Biomedical Research, USA).

Table 1. Primer pairs used for amplification of opium poppy DNA. Tm: Melting temperature.

Primer Name	Sequences of Primers	Tm (°C)	PCR Product Size (bp)
psSSR69-F	5′-ATAGATTTATTTTGGCCACCT-3′	54.6	156
psSSR69-R	5′-CACCTATTGATTGAGGATGAA-3	55.2	
tubulin beta-7 chain I-F	5′-CGTGGGTCACAGCAATACAG-3′	59.4	96
tubulin beta-7 chain I-R	5′-ATGCCTAGGATCAGCAGCAC-3′	59.4	
tubulin beta-7 chain II-F	5′-AATCGGTGCAAAGTTCTGG-3′	54.5	553
tubulin beta-7 chain II-R	5′-GTTCCCATCCCAGATCCTG-3′	58.8	

PCR reactions were carried out in 15 μL reaction containing 11.7 μL ddH_2O, 1.5 μL 10× PCR buffer, 0.3 μL of both primer (0.20 μM), 0.3 μL each of dNTP (200 μM), 0.2 μL Taq-polymerase (1U/μL), and 1 μL DNA (25 ng/μL). Parameters of PCR for the psSSR69 locus were: 94 °C for 3 min, 45 cycles of 45 s at 94 °C, 1 min at 54 °C, 1 min at 72 °C, and additional 1 cycle at 72 °C for 10 min. The reference gene for tubulin beta-7 chain was amplified using the program: 94 °C for 5 min, 35 cycles of 45 s at 94 °C, 1 min at 59 °C, 1 min at 72 °C, and additional 1 cycle at 72 °C for 5 min. Amplicons were analysed in 2% agarose gels in TBE buffer and stained with ethidium bromide.

3. Results and Discussion

3.1. DNA from Mature Seeds

Poppy seeds are specific commercial commodity used in the food industry particularly in some regions of the world. However, the food quality and related price of seeds vary considerably for different *P. somniferum* L. cultivars. Unfortunately, it is likely that premium quality seeds (sweet taste, blue colour) of some poppy cultivars are intentionally handled while trading. They are usually exchanged with low-quality seeds or mixed with them, whether intentionally or not. Therefore, different protocols for extraction of total DNA from poppy seeds and poppy containing products were tested. DNA analysis should be used to determine poppy seed cultivar origin. In addition to the six extraction protocols and four commercial kits tested (Table 2), the modified extraction procedure ("newly developed protocol") was proposed in this study. It is based on the results and experiences obtained during testing of ten extraction procedures.

Spectrophotometric analysis as well as gel electrophoresis of DNA from seeds revealed significant differences between used extraction protocols, both in quantity and quality of obtained DNA. The qualitative parameters of DNA were primarily important (Table 2). The protocol of Dellaporta et al. [24] and its modification by incorporation of CTAB showed that mechanical homogenization of seeds directly in the extraction buffer, even without the use of liquid nitrogen, did not lead to deterioration in quality or amount of DNA (Table 2, Figure 1a). It may be concluded that the need to use liquid nitrogen during mechanical homogenization of poppy seeds is not necessary for prevention of degradation of extracted DNA [34–36]. Quality of extracted DNA varied according to the extraction procedure. Procedures according to Sangwan et al. [21], Bayer BioScience N.V. [25], Monsanto Company [26], Murray and Thompson [27], QIAamp DNA Stool Mini Kit, DNeasy® Plant Maxi Kit, as well as the newly developed protocol, provided poppy DNA with $A_{260/280}$ values in range 1.77–2.11. Procedures Dellaporta et al. [24], PowerSoil DNA Isolation Kit, and Plant DNazol® Reagent had the $A_{260/280}$ ratios in range 1.54–1.75 (Table 2).

Table 2. Parameters of DNA extracted from opium poppy native seeds, ground seeds, and pollen grains by different extraction procedures and commercial kits.

P. somniferum L. Matrix	DNA Extraction Method	DNA Yield (µg)	DNA Concentration (ng/µL)	$A_{260/280}$	$A_{260/230}$	Target Locus	
						psSSR69	Tubulin II
Native seeds	Dellaporta et al. [24]	82.76 *	413.78 *	1.66 *	0.87 *	+	+
		90.68	453.39	1.62	0.84	+	+
	Dellaporta et al. [24] with CTAB	53.32 *	266.62 *	1.62 *	0.83 *	+	+
		73.19	365.94	1.55	0.81	+	+
	Bayer BioScience N.V. [25]	115.13	575.63	1.81	1.84	+	+
	Murray, Thompson [27]	289.84	483.07	2.10	2.37	+	+
	Monsanto Company [26]	15.52	62.06	1.77	2.21	+	+
	Sagwan et al. [21]	108.21	1082.14	1.96	2.51	+	+
	Newly developed protocol	248.30	703.85	1.91	2.19	+	+
	DNeasy® Plant Maxi Kit	108.83	145.11	1.96	2.18	+	+
	Plant DNAzol® Reagent	137.40	114.50	1.54	0.28	-	-
	QIAamp DNA Stool Mini Kit	172.68	863.39	2.11	2.32	+	+
	PowerSoil DNA Isolation Kit	1.40	13.97	1.75	0.72	+	-
Ground seeds	Bayer BioScience N.V. [25] (1)	18.43	92.17	1.79	1.12	+	+
	Bayer BioScience N.V. [25] (2)	110.01	550.06	1.90	2.21	+	+
	Monsanto Company [26]	1.35	5.38	1.76	1.86	+	+
	DNeasy® Plant Maxi Kit	6.28	35.88	1.53	0.64	+	+
	QIAamp DNA Stool Mini Kit	37.96	189.79	2.13	2.25	+	+
	Newly modified protocol	175.11	437.77	1.90	2.19	+	+
Pollen grains	Newly developed protocol	210.7	2107.0	1.95	2.17	+	+
	DNeasy® Plant Mini Kit	2.22	22.2	1.92	1.1	+	+
	QIAamp DNA Stool Mini Kit	3.24	16.2	2.15	1.47	+	+

*—Homogenization of seeds with liquid nitrogen; Bayer BioScience N.V. [25] (1)—0.2 g of seeds extracted in volume of extraction buffer for 0.2 g seeds; Bayer BioScience N.V. [25] (2)—DNA extracted from 0.2 g of seeds extracted in volume of extraction buffer for 0.5 g of seeds.

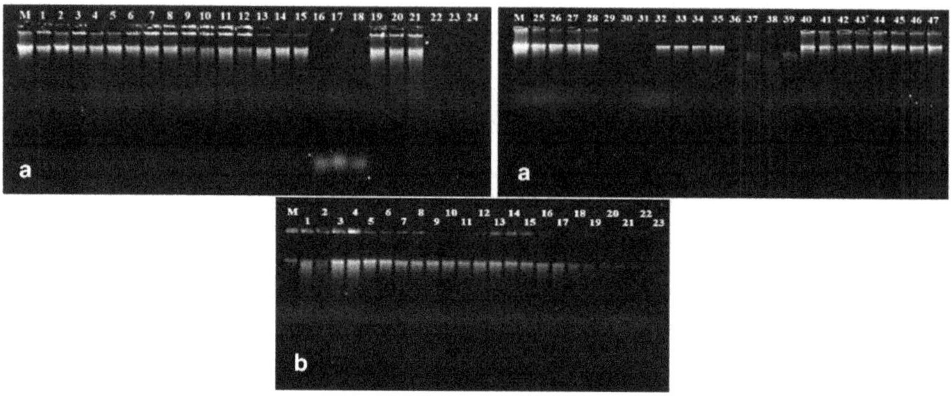

Figure 1. Genomic DNA extracted from opium poppy seeds (**a**) by: Dellaporta et al. [24] (lanes 1–6, where 1–3 homogenization with liquid nitrogen, 4–6 homogenization without liquid nitrogen), Dellaporta et al. [24] with CTAB (lanes 7–12, where 7–9 homogenization with liquid nitrogen, 10–12 homogenization without liquid nitrogen), Bayer BioScience N.V. [25] (lanes 13–15), Murray, Thompson [27] (lanes 16–18), Monsanto Company [26] (lanes 19–21), Sangwan et al. [21] (lanes 22–24), DNeasy® Plant Maxi Kit (lanes 25–28), Plant DNAzol® Reagent (lanes 29–31), QIAamp DNA Stool Mini Kit (lanes 32–35), PowerSoil DNA Isolation Kit (lanes 36–39), newly developed protocol (lanes 40–47). Lane M—λ-phage DNA. (**b**) Ground poppy seeds: DNA extracted by: Bayer BioScience N.V. [25] (lanes 1–7, lines 1–4, 0.2 g of seeds with extraction buffer volume for 0.2 g; lanes 5–7, 0.2 g of seed with extraction buffer volume for 0.5 g of seeds), Monsanto Company (26) (lanes 8–11), newly developed protocol (lanes 12–15), DNeasy® Plant Maxi Kit (lanes 16–19), QIAamp DNA Stool Mini Kit (lanes 20–23). Lane M—λ-phage DNA.

However, the success in amplification of extracted DNA is not guaranteed only by purity, but also by concentration and structural integrity of DNA [37,38]. Although values A_{260} of DNA extracted by protocols Murray and Thomson [27], Sangwan et al. [21], Plant DNAzol® Reagent, and PowerSoil DNA Isolation Kit were high, DNA was not observed in agarose gel (Figure 1a, Table 2). There were probably only limited amounts of poppy DNA and absorbance values have been increased by the presence of RNA and other contaminants. DNA extracted by these protocols also had very low quality. Significant RNA contamination was reported only for the original CTAB method [27] and the QIAamp DNA Stool Mini Kit due to absence of RNase A treatment (Figure 1a).

Amplifications were successful from DNA extracted from mature native seeds by almost all of used protocols and the relevant fragments were generated (Figure 2). The only exception was DNA extracted by the Plant DNAzol® Reagent. DNA extracted by PowerSoil DNA Isolation Kit was probably highly degraded considering that 553 bp fragment of reference tubulin gene was not amplified, but a 156 bp length microsatellite marker was generated (Figure 2).

The only one currently available protocol developed for extraction of DNA from poppy seeds [21] did not provide high quality of DNA within this study (Table 2, Figure 1a). The newly developed protocol has been proven as effective. Compared to the original protocol [25], extraction steps were rearranged, time intervals between steps were changed, and some chemicals/enzymes were eliminated. Both absorbance parameters ($A_{260/280}$ and $A_{260/230}$) as well as electrophoretic profile of DNA predicted very good quality and quantity (Table 2, Figure 1a) that should be suitable for amplification by PCR (Figure 2).

3.2. *DNA from Ground Seeds*

Ground poppy seeds are commonly available in food stores. The sensory values (especially taste and smell) and related varietal origin of high-quality seeds may be easily masked in ground seeds by various additives, mainly by sugar. Analytical testing and confirmation of the poppy seeds varietal origin is necessary in such cases. DNA from ground poppy seeds was extracted by two protocols [25,26], two commercial kits, as well as newly developed protocol. The QIAamp DNA Stool Mini Kit and DNeasy® Plant Maxi Kit produced DNA with the $A_{260/280}$ and $A_{260/230}$ ratios furthest from optimal values (Table 2). Both spectrophotometric ratios of DNA extracted by Monsanto Company [26] protocol indicated high contamination of DNA with proteins, organic solvents, and secondary metabolites, and also very low concentration (Table 2). The yield of DNA was significantly different between tested protocols, but at the same amount of loaded DNA (25 ng/μL) the electrophoretic profiles of all DNA samples were appropriate (Figure 1b). The highest quality and concentration of DNA has been extracted by protocols Bayer Biocience N.V. [25] with changed ratio of sample–extraction buffer (w/v) and the newly developed protocol (Table 2, Figure 1b). Both protocols contained SDS in extraction buffer. It is suggested that SDS-based DNA extractions could be more appropriate for oily plant matrices like ground poppy seeds. The SDS-containing method modified for ground raw soybean seeds had the highest yield of DNA in comparison with the CTAB method and two commercial kits [39]. A lower amount of DNA yielded the CTAB method also from soybean flour [40].

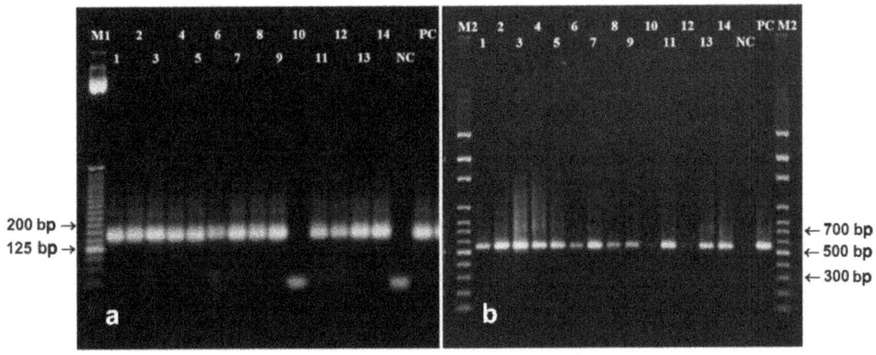

Figure 2. Amplification of 156 bp microsatellite psSSR69 (**a**) and 553 bp fragment of gene for tubulin beta-7 chain (**b**) in DNA extracted from poppy seeds by: Dellaporta et al. [24] with and without liquid N_2 (lanes 1 and 2), Dellaporta et al. [24] with CTAB with and without liquid N_2 (lanes 3 and 4), Bayer BioScience N.V. [25] (5), Murray, Thompson [27] (6), Monsanto Company [26] (7), Sangwan et al. [21] (8), DNeasy® Plant Maxi Kit (9), Plant DNAzol® Reagent (10), QIAamp DNA Stool Mini Kit (11), PowerSoil DNA Isolation Kit (12), newly developed protocol (13–14), NC—negative control, PC—positive control, M1—25 bp ladder (Invitrogen), M2—100 bp DNA ladder (Solis BioDyne).

Amplifications of DNA from ground poppy seeds using primers flanking microsatellite marker psSSR69 and longer fragment of gene for tubulin beta-7 chain resulted in production of both the 156 and 553 bp amplicons in DNA extracted by all used protocols (Figure 3).

3.3. DNA from Poppy Pollen Grains

DNA was extracted by two commercial kits and by newly developed protocol (Table 2). Homogenization by pestle and mortar in liquid nitrogen was efficient for disruption of pollen exine with high structural integrity. Both ratios $A_{260/280}$ and $A_{260/230}$ confirmed that the best quality had DNA extracted by newly developed protocol (Table 2). This simple protocol produced also very high amount of DNA. On the opposite, the QIAamp DNA Stool Mini Kit and DNeasy® Plant Mini Kit extracted the least amount of DNA (Figure 4a). Amplifications of DNA from poppy pollen grains were basically without any complications. All primer pairs were able to amplify relevant amplicons (Figure 4). The genomic DNA is well protected inside the pollen grain therefore, a large fragment of the reference gene (553 bp) was simply amplified (Figure 4b). Amplifications of both shorter, the 156 bp microsatellite marker and 96 bp fragment of reference gene were also easily feasible (Figure 4c,d).

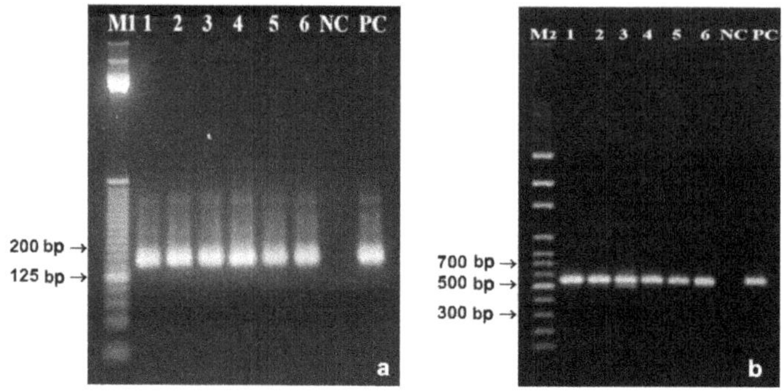

Figure 3. Amplification of 156 bp microsatellite psSSR69 (**a**) and 553 bp fragment of reference tubulin gene (**b**) in DNA extracted from ground seeds extracted by Bayer BioScience N.V. [25] (lanes 1–2, lane 1), Monsanto Company [26] (3), DNeasy® Plant Maxi Kit (4), QIAamp DNA Stool Mini Kit (5), newly developed protocol (6), NC—negative control, PC—positive control, M1—25 bp DNA ladder (Invitrogen), M2—100 bp DNA ladder (Solis BioDyne).

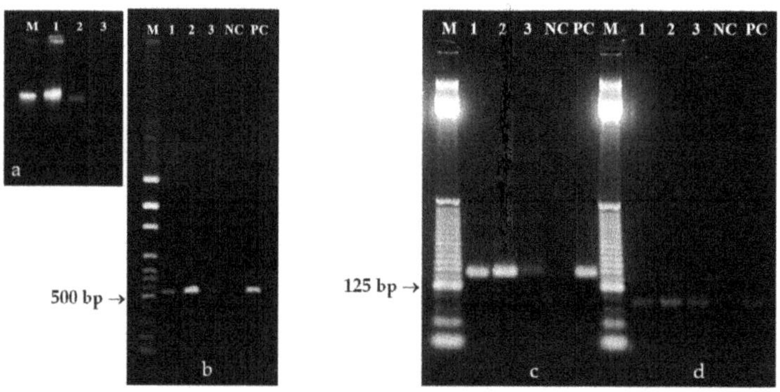

Figure 4. Genomic DNA extracted from opium poppy pollen grains (**a**). Amplification of 553 bp (**b**) and 96 bp (**d**) fragments of reference tubulin gene, and 156 bp (**c**) microsatellite, respectively. (1)—newly developed protocol, (2)—DNeasy® Plant Maxi Kit, (3)—QIAamp DNA Stool Mini Kit, NC—negative control, PC—positive control, M—100 bp DNA ladder (**b**) (Invitrogen) and 25 bp DNA ladder (**c**,**d**) (Solis BioDyne).

Extraction of DNA from pollen grains is needed in different applications including monitoring of pollen grains transfer from transgenic opium poppy plants to the environment [41], detection of pollen species in food (e.g., in honey) for the prevention of allergens [42], forensic palynology [43] and others.

3.4. DNA from Poppy Seed Filling

DNA was extracted by two procedures, one commercial kit, and the newly developed protocol (Table 3). The purification step using the silica membrane spin-columns [28] was added to protocols Monsanto Company [26] and newly developed one. Both ratios $A_{260/280}$ and $A_{260/230}$ confirmed that DNA extracted using almost all extraction protocols had these values out of the optimal range (Table 3). Undamaged high molecular weight DNA extracted from poppy seed filling from the bakery product was not visualizable in agarose gel (data not shown). This reflects fragmentation of poppy DNA to very short fragments due to high degradation during baking. This is common for DNA extracted from a matrix that has undergone processing by high temperature [29] and a combination of grinding, mechanical manipulation, and thermal treatment [44]. However, the objective quality and usability of DNA extracted can only be revealed by its amplification.

Table 3. Parameters of DNA extracted from poppy seed filling and poppy seed oil using different extraction procedures and commercial kits.

P. somniferum L. Matrix	DNA Extraction Method	Weight (g)/Volume (mL) of Sample	DNA Yield (µg)	DNA Concentration (ng/µL)	$A_{260/280}$	$A_{260/230}$	Target Locus	
							psSSR69	Tubulin I
Poppy seed filling	Bayer BioScience N.V. [25]	1	79.91	399.55	1.94	1.92	+	+
	QIAamp DNA Stool Mini Kit	0.5	7.16	35.82	1.96	0.90	+	+
	Monsanto Company [26]	1	241.26	1206.28	2	1.75	-	-
	Monsanto Company [26]	1	0.97 *	32.41 *	1.57 *	0.40 *	+	+
	Newly developed protocol	2	246.55	1232.75	2.04	1.84	-	-
	Newly developed protocol	2	0.81*	26.91 *	1.68 *	0.59 *	+	+
Poppy seed oil	Doveri et al. [29]	1	0.43	7.1	2.53	0.01	+	+
	Consolandi et al. [30] aq	3	3.02	60.3	1.62	0.52	+	+
	Consolandi et al. [30] aq	3	0.94	18.8	1.48	0.54	+	+
	Consolandi et al. [30] oil	3	0.14	2.8	1.44	0.56	+	+
	Consolandi et al. [30] aq	6	2.19	43.8	1.40	0.57	+	+
	Consolandi et al. [30] aq	6	1.20	23.9	1.36	0.45	+	+
	Consolandi et al. [30] oil	6	0.21	4.1	1.22	0.59	+	+
	Giménez et al. [31]	0.5	0.08	3.0	1.75	0.08	+	-
	Giménez et al. [31]	1.5	0.06	2.5	1.28	0.06	+	-
	Raieta et al. [4] oil	3	0.18	6.1	1.26	0.13	+	-
	Raieta et al. [4] aq	3	0.26	8.7	1.61	0.21	+	-
	Raieta et al. [4] oil	1	0.10	3.3	2.07	0.09	-	-
	Raieta et al. [4] aq	1	0.13	4.3	1.26	0.12	+	+
	QIAamp DNA Stool Mini Kit	0.2	1.05	20.9	1.84	0.34	-	+
	QIAamp DNA Stool Mini Kit	1	1.71	34.2	1.78	0.33	-	+
	QIAamp DNA Stool Mini Kit	15	1.83	36.6	1.96	0.09	-	-

Note: *—subsequent purification of extracted DNA through silica membrane spin-columns [28], aq/oil—DNA isolated from the water (aq) or oily (o) phase.

Complex food matrices contain a variety of PCR inhibitors [45]. Other effects of the matrix include degradation, fragmentation, and restricted extractability of DNA, as well as presence of DNA from different organisms [46]. Baking temperature around 200 °C used in processing of bakery goods containing poppy seed filling substantially reduces the size of extracted DNA. Moreover, higher moisture content inside the product, in this case in poppy filling, contributes to greater degradation of DNA [9]. Amplifications of poppy DNA extracted from filling of the baked product were more difficult. As expected, primer pair designed for amplification of 553 bp fragment of reference gene was not able to generate amplicon (data not shown). The Bayer BioScience N.V. method [25] and the QIAamp DNA Stool Mini Kit provided DNA with quality allowing amplification of the 156 bp microsatellite and short (96 bp) fragment of reference gene (Figure 5). Both these methods were effective also without the need of purification in columns. DNA extracted by the Monsanto Company method [26] and newly developed protocol was amplifiable only if the purification step in the silica membrane column [28] was added (Figure 5). Columns were able to bind impurities and inhibitors of polymerase chain reaction from primary DNA extracts.

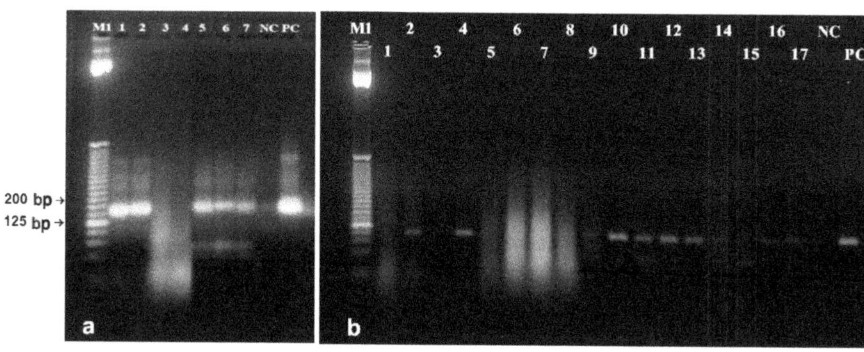

Figure 5. Amplification of 156 bp microsatellite psSSR69 in DNA extracted from the poppy seed filling (**a**) using: Bayer BioScience N.V. [25] (lane 1), QIAamp DNA Stool Mini Kit (2), newly developed protocol (3, 5, 7), Monsanto Company [26] (4, 6) NC—negative control, PC—positive control. Lanes 1–4 represent samples without, lanes 5–7 with purification through silica membrane columns. Amplification of 96 bp fragment of the reference tubulin gene (**b**) using: Bayer BioScience N.V. [25] (lanes 1, 2, 9, 10), QIAamp DNA Stool Mini Kit (lanes 3, 4, 11, 12), newly modified protocol (lanes 5–7, 13–16), Monsanto Company [26] (lane 8, 17) NC-negative control, PC-positive control. Lanes 1–8 represent samples without, lanes 9-17 samples with purification through columns. M1—25 bp DNA ladder (Invitrogen).

3.5. *DNA from Poppy Oil*

Oil from poppy seeds is mainly used for culinary and pharmaceutical purposes, but also for production of cosmetics, paints and varnishes. Cold-pressed oil is quite expensive, so it can sometimes be adulterated by much cheaper vegetable oils (e.g., from rapeseed, sunflower, oil palm). Techniques of analytical chemistry are developing for distinguishing between cheaper oils (e.g., sunflower, oilseed rape) and poppy oil [17]. However, chemical analysis may not be unambiguous [31] due to variation in chemical composition of vegetable oils among growing areas and seasons. Alternative approaches are based on the DNA analysis and require extraction of DNA from oil. Such protocols were developed mainly for olive oil. Four of such methods [4,29–31], the QIAamp DNA Stool Mini Kit as well as Bayer BioScience N.V. [25], Monsanto Company [26], newly developed protocols were tested for different volumes of poppy seed oil. Bayer BioScience N.V. [25], Monsanto Company [26] and newly developed protocol were unable to extract detectable and usable DNA (data not shown). DNA extracted by other protocols had also both absorbance parameters ($A_{260/280}$, $A_{260/230}$) far from the optimal values (Table 3); however, DNA was amplifiable by PCR (Figure 6). DNA in cold-pressed vegetable oil has undergone a process of significant degradation, caused by DNA nucleases released during crushing and malaxation of oily plant material. This will certainly happen when pressing oil from poppy seeds as well. If enzymatic mixtures of proteases are applied during this process, the DNA is prevented to damage and could be extracted with high integrity and concentration, similarly as from vegetative tissues [47]. However, this cannot be ensured in the already pressed oil. Another significant complication in the extraction of DNA is the time since pressing and conditions of the oil storage before the DNA extraction. After a relatively short time interval, a significant decreasing of quality of extracted DNA was observed due to oxidation damage [48]. Following the assumed high degradation, DNA has not even been electrophoretically controlled and only its amplifications revealed the potential utility of the extracted DNA. Statistical analysis did not reveal relationship between concentration, A_{260}/A_{280} ratio, and the ability to undergo amplification by PCR [49].

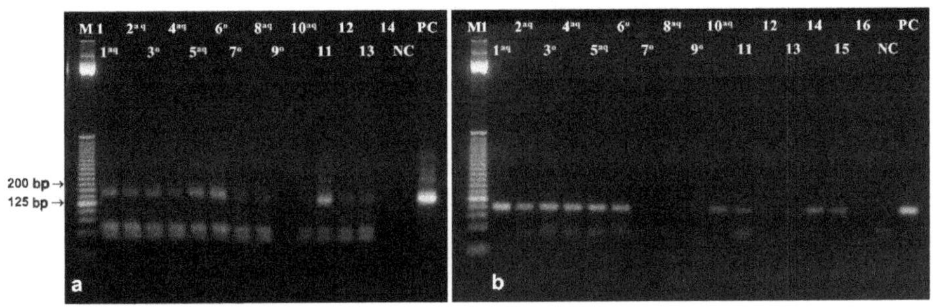

Figure 6. Agarose gel electrophoresis of PCR products obtained by amplification of 156 bp microsatellite psSSR69 (**a**) and 96 bp fragment of reference tubulin gene (**b**). M1—25 bp DNA ladder (Invitrogen). DNA extracted by Consolandi et al. [30] (lines 1–6) from 3 mL (1–3) or 6 mL (4–6) of oil, Raieta et al. [4] (lanes 7–10) from 3 mL (lanes 7,8) or 1 mL (lanes 9–10) oil, Doveri et al. [29] (line 11) extracted from 1 mL of oil, Giménez et al. [31] (lines 12–13) extracted from 0.5 mL (lane 12) or 3 mL (lane 13) of oil, QIAamp DNA Stool Mini Kit (lanes 14–16) extracted from 0.2 mL (lane 14b), 1 mL (lane 15b) or 15 mL (lanes 14a and 16b) of oil, NC—negative control, PC—positive control. $^{aq/o}$—DNA from water (^{aq}) or oily (^{o}) phase.

Four extraction protocols [4,29–31] and the QIAamp DNA Stool Mini Kit provided different results (Figure 6). In addition, DNA extraction was also tested from different starting volumes of poppy seed oil. Extraction protocol developed for authentication of olive oils [30] was efficient either from 3 mL or 6 mL samples of poppy oil. Poppy DNA obtained by this protocol, from both the oily and water phases were amplifiable and provided templates for relevant amplicons. Other used DNA extraction protocols were also developed for olive oil, but based on the CTAB in extraction buffer [4,31]. The resulting poppy DNA behaved unreliably in the PCR reaction. Convincing and reliable amplifications were obtained from DNA extracted by another protocol, modified for olive oil [29] containing guanidine thiocyanate in extraction buffer. The capability of tested QIAamp DNA Stool Mini Kit for DNA extraction from poppy oil has been demonstrated in low oil volumes (0.2–1 mL).

The quality and quantity of DNA extracted from native or processed poppy seeds strongly depended on the character of poppy matrix entering the extraction procedure as well as level of its processing. Amplifications of obtained DNA were also influenced by many factors, especially by the presence of contaminants and inhibitors. Positioning of used primers for PCR analysis considered the expected length of extracted DNA fragments depended on the expected disruption of DNA during processing (baking, pressing) of poppy seed matrix. DNA extracted from different poppy seed matrices by different extraction protocols was amplified using primer pairs flanking the 553-, 156-, and 96 bp fragments, respectively (Table 1). The presence of the longest 553 bp fragment was detected by PCR in poppy DNA extracted from native seeds and ground seeds, but not from processed poppy seed matrices (filling of the bakery product, oil). Both types of poppy seed processing (baking, pressing) reduced the effective concentration of poppy DNA fragments capable of amplification of fragments longer than 100 bp, as was detected in maize cornmeal [50]. DNA from heat-processed and other highly degraded plant matrices should be amplified only in short DNA sequences. This is the strategy also in analysis of DNA from genetically modified organisms in processed foods [9,51]. Analysis of highly degraded DNA by PCR is more advantageous in DNA regions higher in GC content because their stability during heat treatment of the analysed matrices is higher [51].

Specific morphological characteristics, extreme heterogeneity and variation in chemical composition of plant cells cause many problems in DNA extraction. Although numerous protocols for plant DNA extraction have been published, none is found to be universally applicable [52]. Newly developed DNA extraction protocols are usually modifications of already existing protocols. The extraction protocol developed in our study demonstrated a relatively high degree of universality, with respect to poppy matrices. Compared to other DNA extraction protocols, it was quite universal. In comparison with the Bayer BioScience N.V. [25] protocol, from which the most steps were taken, it was approximately one third shorter in time. A significant reduction in time was achieved by adjusting the centrifugation steps. 2-mercaptoethanol ME was added to the first extraction buffer. Some steps during the extraction procedure were eliminated. Along with purification on silica membrane columns, the newly developed extraction protocol was highly efficient and represents a simple and inexpensive alternative to commercial DNA extraction kits. Extraction of DNA from oil required specific extraction protocols that were developed specifically for this type of matrix only.

4. Conclusions

Protocols tested for extraction of DNA from native and ground poppy seeds, pollen grains, poppy seed filling from the bakery product, and poppy oil have been differently effective and suitable depending on individual poppy seed matrices or products. DNA from seeds, ground seeds and pollen grains extracted by almost all extraction procedures had quantity and quality sufficient for PCR analysis of short microsatellite marker (156 bp) and also long fragment of the reference gene (553 bp). The best of these protocols have been tested for DNA extraction from the poppy seed filling from the bakery product. It has been very useful to use silica membrane columns for purification of the extracted DNA. Purified DNA was then amplifiable. Poppy DNA extracted from thermally processed poppy seed filling from the baking product did not amplify long fragment (553 bp) of the reference gene. However, primers designed for amplification of shorter fragment of the reference gene (96 bp) as well as for the microsatellite marker (156 bp) provided the appropriate amplicons. The new extraction protocol developed within this study has proven to be universally applicable to poppy seeds, pollen, and poppy seed containing products. It can be used for various control purposes in poppy breeding programmes, production and distribution of elite poppy seeds for crop production, control of poppy seeds identity as an interesting market commodity, control of products containing poppy seeds during food production. Protocols tested for extraction of poppy DNA from cold-pressed poppy oil were originally developed or modified for olive oil. The most of them [29,30] were effective, and extracted DNA was amplified using primers for the microsatellite marker and the short fragment of the reference gene.

Author Contributions: Conceptualization, J.K.; methodology, J.K., Š.K. and D.M.; investigation, Š.K., M.M. and D.M.; writing—original draft preparation, Š.K.; writing—review and editing, J.K.; visualization, Š.K. and M.M.; supervision, J.K. All authors have read and agreed to the published version of the manuscript.

Funding: This research was funded by the Slovak Research and Development Agency, projects no. APVV-16-0026, APVV-18-0005, APVV-16-0051, and the Operational Programme Research and Development co-financed from the European Regional Development Fund, project No ITMS 26210120039.

Conflicts of Interest: The authors declare no conflict of interest.

References

1. Ali, N.; de Cássia Pontello Rampazzo, R.; Dias Tavares Costa, A.; Krieger, M.A. Current nucleic acid extraction methods and their implications to point-of-care diagnostics. *Biomed Res. Int.* **2017**, *2017*, 9306564. [CrossRef] [PubMed]
2. Di Bernardo, G.; Del Gaudio, S.; Galderisi, U.; Cascino, A.; Cipollaro, M. Comparative evaluation of different DNA extraction procedures from food samples. *Biotechnol. Prog.* **2007**, *23*, 297–301. [CrossRef] [PubMed]
3. Jasbeer, K.; Son, R.; Mohamad Ghazali, F.; Cheah, Y. Real-time PCR evaluation of seven DNA extraction methods for the purpose of GMO analysis. *Int. Food Res. J.* **2009**, *16*, 329–341.
4. Raieta, K.; Muccillo, L.; Colantuoni, V. A novel reliable method of DNA extraction from olive oil suitable for molecular traceability. *Food Chem.* **2015**, *172*, 596–602. [CrossRef]
5. Manning, L.; Soon, J.M. Developing systems to control food adulteration. *Food Policy* **2014**, *49*, 23–32. [CrossRef]
6. Böhme, K.; Calo-Mata, P.; Barros-Velazquez, J.; Ortea, I. Review of recent DNA-based methods for main food authentication topics. *J. Agric. Food Chem.* **2019**, *67*, 3854–3864. [CrossRef] [PubMed]
7. Peano, C.; Samson, M.C.; Palmieri, L.; Gulli, M.; Marmiroli, N. Qualitative and quantitative evaluation of the genomic DNA extracted from GMO and non-GMO foodstuffs with four different extraction methods. *J. Agric. Food Chem.* **2004**, *52*, 6962–6968. [CrossRef]
8. Smith, D.S.; Maxwell, P.W.; De Boer, S.H. Comparison of several methods for the extraction of DNA from potatoes and potato-derived products. *J. Agric. Food Chem.* **2005**, *53*, 9848–9859. [CrossRef]
9. Gryson, N. Effect of food processing on plant DNA degradation and PCR-based GMO analysis: A review. *Anal. Bioanal. Chem.* **2010**, *396*, 2003–2022. [CrossRef]
10. Weighardt, F. GMO quantification in processed food and feed. *Nat. Biotechnol.* **2007**, *25*, 1213–1214. [CrossRef]
11. Demeke, T.; Jenkins, G.R. Influence of DNA extraction methods, PCR inhibitors and quantification methods on real-time PCR assay of biotechnology-derived traits. *Anal. Bioanal. Chem.* **2010**, *396*, 1977–1990. [CrossRef] [PubMed]
12. Wu, Y.; Chen, Y.; Ge, Y.; Wang, J.; Xu, B.; Huang, W.; Yuan, F. Detection of olive oil using the Evagreen real-time PCR method. *Eur. Food. Res. Technol.* **2008**, *227*, 1117–1124. [CrossRef]
13. Ramos-Gómez, S.; Busto, M.D.; Perez-Mateos, M.; Ortega, N. Development of a method to recovery and amplification DNA by real-time PCR from commercial vegetable oils. *Food Chem.* **2014**, *158*, 374–383. [CrossRef] [PubMed]
14. Piarulli, L.; Savoia, M.; Taranto, F.; D'Agostino, N.; Sardaro, R.; D'Agostino, N.; Gadaleta, S.; Fucilli, V.; De Giovanni, C.; Montemurro, C.; et al. A Robust DNA Isolation Protocol from Filtered Commercial Olive Oil for PCR-Based Fingerprinting. *Foods* **2019**, *8*, 462. [CrossRef]
15. Baser, K.H.C.; Arslan, N. Opium poppy (Papaver somniferum). In *Medicinal and Aromatic Plants of the Middle-East. Medicinal and Aromatic Plants of the World*; Yaniv, Z., Dudai, N., Eds.; Springer: Dordrecht, The Netherlands, 2014; Volume 2, pp. 305–332.
16. Bail, S.; Majchrzak, D.; Krist, S.; Elmadfa, I.; Buchbauer, G. Sensory evaluation of poppy oil blended with sunflower seed oil. *Nahrung* **2008**, *32*, 8–15.
17. Krist, S.; Stuebiger, G.; Bail, S.; Unterweger, H. Detection of adulteration of poppy seed oil with sunflower oil based on volatiles and triacylglycerol composition. *J. Agric. Food Chem.* **2006**, *84*, 6385–6389. [CrossRef] [PubMed]
18. Celik, I.; Gultekin, V.; Allmer, J.; Doganlar, S.; Frary, A. Development of genomic simple sequence repeat markers in opium poppy by next-generation sequencing. *Mol. Breed.* **2014**, *34*, 323–334. [CrossRef]
19. Mičianová, V.; Ondreičková, K.; Muchová, D.; Klčová, L.; Hudcovicová, M.; Havrlentová, M.; Mihálik, D.; Kraic, J. Forensic application of EST-derived STR markers in opium poppy. *Biologia* **2017**, *7*, 587–594. [CrossRef]
20. Vašek, J.; Čílová, D.; Melounová, M.; Svoboda, P.; Vejl, P.; Štikarová, R.; Vostrý, L.; Kuchtová, P.; Ovesná, J. New EST-SSR markers for individual genotyping of opium poppy cultivars (*Papaver somniferum* L.). *Plants* **2020**, *9*, 10. [CrossRef]
21. Sangwan, R.S.; Yadav, U.S.; Sangwan, N.S. Isolation of genomic DNA from defatted oil seed residue of opium poppy (Papaver somniferum). *Plant. Mol. Biol. Rep.* **2000**, *18*, 265–270. [CrossRef]

22. Marciano, M.A.; Panicker, S.X.; Liddil, G.D.; Lindgren, D.; Sweder, K.S. Development of a method to extract opium poppy (Papaver somniferum L.) DNA from heroin. *Sci. Rep.* **2018**, *8*, 1–15. [CrossRef] [PubMed]
23. Young, B.; Roman, M.G.; LaRue, B.; Gangitano, D.; Houston, R. Evaluation of 19 short tandem repeat markers for individualization of Papaver somniferum. *Sci. Justice* **2020**, *60*, 253–262. [CrossRef]
24. Dellaporta, S.L.; Wood, J.; Hicks, J.B. A plant DNA mini preparation: Version II. *Plant Mol. Biol. Rep.* **1983**, *1*, 19–21. [CrossRef]
25. Community Reference Laboratory for GM Food and Feed. Seeds Sampling and DNA Extraction of Oilseed Rape. Report on the Validation of an Oilseed Rape DNA Extraction Method from Seeds: Bayer BioScience N.V. Available online: http://gmo-crl.jrc.ec.europa.eu/summaries/T45_DNAExtr_sampl_correctedversion1.pdf (accessed on 24 July 2020).
26. Community Reference Laboratory for GM Food and Feed. Seeds Sampling and DNA Extraction of Oilseed Rape. Report on the Validation of an Oilseed Rape DNA Extraction Method from Seeds: Monsanto Company. Available online: http://gmo-crl.jrc.ec.europa.eu/summaries/CRL-VL-26-04-XP-Corrected-version-1.pdf (accessed on 24 July 2020).
27. Murray, M.G.; Thompson, W.F. Rapid isolation of high molecular weight plant DNA. *Nucleic Acids Res.* **1980**, *8*, 4321–4325. [CrossRef]
28. Borodina, T.A.; Lehrach, H.; Soldatov, A.V. DNA purification on homemade silica spin-columns. *Anal Biochem.* **2003**, *321*, 135–137. [CrossRef]
29. Doveri, S.; O'Sullivan, D.M.; Lee, D. Non-concordance between genetic profiles of olive oil and fruit: A cautionary note to use of DNA markers for provenance testing. *J. Agric. Food Chem.* **2006**, *54*, 9221–9226. [CrossRef]
30. Consolandi, C.; Palmieri, L.; Severgnini, M.; Maestri, E.; Marmiroli, N.; Agrimonti, C.; Baldoni, L.; Donini, P.; De Bellis, G.; Castiglioni, B. A procedure for olive oil traceability and authenticity: DNA extraction, multiplex PCR and LDR-universal array analysis. *Eur. Food Res. Technol.* **2008**, *277*, 1429–1438. [CrossRef]
31. Giménez, M.J.; Pistón, F.; Martín, A.; Atienza, S.G. Application of real-time PCR on the development of molecular markers and to evaluate critical aspects for olive oil authentication. *Food Chem.* **2010**, *118*, 482–487. [CrossRef]
32. Şelale, H.; Çelik, I.; Gültekin, V.; Allmer, J.; Doganlar, S.; Frary, A. Development of EST-SSR markers for diversity and breeding studies in opium poppy. *Plant Breed.* **2013**, *132*, 344–351. [CrossRef]
33. Benson, D.A.; Cavanaugh, M.; Clark, K.; Karsch-Mizrachi, I.; Lipman, D.J.; Ostell, J.; Sayers, E.W. GenBank. *Nucleic Acids Res.* **2013**, *41*, D36–D42. [CrossRef]
34. Sharma, A.D.; Gill, P.K.; Singh, P. DNA isolation from dry and fresh samples of polysaccharide-rich plants. *Plant Mol. Biol. Rep.* **2002**, *20*, 415a–415f. [CrossRef]
35. Ferdous, J.; Hanafi, M.M.; Rafii, M.Y.; Muhammad, K. A quick DNA extraction protocol: Without liquid nitrogen in ambient temperature. *Afr. J. Biotechnol.* **2012**, *11*, 6956–6964.
36. Abdel-Latif, A.; Osman, G. Comparison of three genomic DNA extraction methods to obtain high DNA quality from maize. *Plants Methods* **2017**, *13*, 1–9. [CrossRef]
37. Olden, M.J.; Blasic, J.R.; Bussjaeger, L.; Kao, C.; Shokere, L.A.; Kendall, D.C.; Freese, L.; Jenkins, G.R. Evaluation of Extraction Methodologies for Corn Kernel (Zea mays) DNA for Detection of Trace Amounts of Biotechnology-Derived DNA. *J. Agric. Food. Chem.* **2003**, *51*, 2468–2474. [CrossRef] [PubMed]
38. Cankar, K.; Štebih, D.; Dreo, T.; Žel, J.; Gruden, K. Critical points of DNA quantification by real-time PCR —Effects of DNA extraction method and sample matrix on quantification of genetically modified organisms. *BMC Biotechnol.* **2006**, *6*, 37. [CrossRef]
39. Xia, Y.; Chen, F.; Du, Y.; Liu, C.; Bu, G.; Xin, Y.; Liu, B. A modified SDS-based DNA extraction method from raw soybean. *Biosci. Rep.* **2019**, *39*, BSR20182271. [CrossRef] [PubMed]
40. Mafra, I.; Silva, S.A.; Moreira, E.J.M.O.; da Silva, C.S.F.; Beatriz, M.; Oliveira, P.P. Comparative study of DNA extraction methods for soybean derived food products. *Food Control.* **2008**, *19*, 1183–1190. [CrossRef]
41. Chitti, J.A.; Allen, R.S.; First, A.J.; Larkin, P.J. Genetic transformation in commercial Tasmanian cultivars of opium poppy, *Papaver somniferum*, and movement of transgenic pollen in the field. *Funct. Plant. Biol.* **2003**, *30*, 1045–1058. [CrossRef]
42. Yang, Y.; Wang, H.; Liu, M.; Huang, W.; Wang, Y.; Wu, Y. A multiplex real/time PCR method applied to detect eight pollen species in food for the prevention of allergens. *Eur. Food Res. Technol.* **2019**, *245*, 2195–2207. [CrossRef]

43. Bell, K.L.; Burgess, K.S.; Okamoto, K.C.; Aranda, R.; Brosi, B.J. Review and future prospects for DNA barcoding methods in forensic palynology. *Forensic Sci. Int. Genet.* **2016**, *21*, 110–116. [CrossRef]
44. Nikolić, Z.; Petrović, G.; Panković, D.; Ignjatov, M.; Marinković, D.; Stojanović, M.; Đorđević, V. Threshold level and traceability of Roundup Ready® soybeans in tofu production. *Food Technol. Biotechnol.* **2017**, *44*, 439–444. [CrossRef] [PubMed]
45. Schrader, C.; Schielke, A.; Ellerbroek, L.; Johne, R. PCR inhibitors—Occurrence, properties and removal. *J. Appl. Microbiol.* **2012**, *113*, 1014–1026. [CrossRef] [PubMed]
46. Lipp, M.; Shillito, R.; Giroux, R.; Spiegelhalter, F.; Charlton, S.; Pinero, D.; Song, P. Polymerase chain reaction technology as analytical tool in agricultural biotechnology. *J. AOAC Int.* **2005**, *88*, 136–155. [CrossRef] [PubMed]
47. Muzzalupo, I.; Perri, E. Recovery and characterisation of DNA from virgin olive oil. *Eur. Food Res. Technol.* **2002**, *214*, 528–531. [CrossRef]
48. Pafundo, S.; Busconi, M.; Agrimonti, C.; Fogher, C.; Marmiroli, N. Storage-time effects on olive oil DNA assessed by Amplified Fragments Length Polymorphisms. *Food Chem.* **2010**, *123*, 787–793. [CrossRef]
49. Piskata, Z.; Servusova, E.; Babak, V.; Nesvadbova, B.; Borilova, G. The quality of DNA isolated from processed food and feed via different extraction procedures. *Molecules* **2019**, *24*, 1188. [CrossRef]
50. Murray, S.R.; Butler, R.C.; Hardacre, A.K.; Timmerman-Vaughan, G.M. Use of quantitative real-time PCR to estimate maize endogenous DNA degradation after cooking and extrusion or in food products. *J. Agric. Food Chem.* **2007**, *55*, 2231–2239. [CrossRef]
51. Yoshimura, T.; Kuribara, H.; Matsuoka, T.; Kodama, T.; Iida, M.; Watanabe, T.; Akiyama, H.; Maitani, T.; Furui, S.; Hino, A. Applicability of the quantification of genetically modified organisms to foods processed from maize and soy. *J. Agric. Food Chem.* **2005**, *53*, 2052–2059. [CrossRef]
52. Varma, A.; Padh, H.; Shrivastava, N. Plant genomic DNA isolation: An art or a science. *Biotechnol. J.* **2007**, *2*, 386–392. [CrossRef]

foods

MDPI

Article

Development and Validation of a Real-Time PCR Based Assay to Detect Adulteration with Corn in Commercial Turmeric Powder Products

Su Hong Oh and Cheol Seong Jang *

Plant Genomics Laboratory, Department of Bioresource Sciences, Kangwon National University, Chuncheon 24341, Korea; ohsuhong@kangwon.ac.kr

* Correspondence: csjang@kangwon.ac.kr; Tel.: +82-33-250-6416; Fax: +82-33-259-5558

Received: 9 June 2020; Accepted: 3 July 2020; Published: 5 July 2020

Abstract: Turmeric, or *Curcuma longa*, is commonly consumed in the South East Asian countries as a medical product and as food due to its therapeutic properties. However, with increasing demand for turmeric powder, adulterated turmeric powders mixed with other cheap starch powders, such as from corn or cassava, are being distributed by food suppliers for economic benefit. Here, we developed molecular markers using quantitative real-time PCR to identify adulteration in commercial turmeric powder products. Chloroplast genes, such as *matK*, *atpF*, and *ycf2*, were used to design species-specific primers for *C. longa* and *Zea mays*. Of the six primer pairs designed and tested, the correlation coefficients (R^2) were higher than 0.99 and slopes were −3.136 to −3.498. The efficiency of the primers was between 93.14 and 108.4%. The specificity of the primers was confirmed with ten other species, which could be intentionally added to *C. longa* powders or used as ingredients in complex turmeric foods. In total, 20 blind samples and 10 commercial *C. longa* food products were tested with the designed primer sets to demonstrate the effectiveness of this approach to detect the addition of *Z. mays* products in turmeric powders. Taken together, the real-time PCR assay developed here has the potential to contribute to food safety and the protection of consumer's rights.

Keywords: anti food fraud; *Curcuma longa*; DNA markers; species identification; SYBR-GREEN real-time PCR; *Zea mays*

1. Introduction

Turmeric (*Curcuma longa*) belongs to the ginger family, Zingiberaceae, and is native to Southern Asia and India. Turmeric rhizomes, which have brown skin and a unique flavor, are commonly used as a coloring and flavoring agent in Asian cuisines. Due to its fragrant aroma and slightly bitter taste, turmeric is a common culinary spice in Indian cuisines, especially curry. Additionally, beyond food products, turmeric is commonly consumed as a medical product in South East Asian countries due to its therapeutic properties [1]. The market size of curcumin was valued at USD 58.4 million in 2019 and is expected to experience a CAGR (compound annual growth rate) of 12.7% from 2020 to 2027 [2]. Globally, the demand for turmeric has grown due to its therapeutic functions and low toxicity. Curcumin, (1,7-bis(4-hydroxy-3-methoxyphenyl)-1,6-heptadiene-3,5-dione), also known as diferuloylmethane, is the main natural polyphenol found in rhizomes of *C. longa* (turmeric) and in other *Curcuma* spp. [3]. It has been shown to target multiple signaling molecules while also demonstrating activity at the cellular level, which has helped support its multiple health benefits [4]. It has beneficial effects in inflammatory conditions [3], metabolic syndrome [5], and pain [6], as well as helps in the management of inflammatory and degenerative eye conditions [7]. While there appear to be countless therapeutic benefits of curcumin supplementation, most of them may be due to its antioxidant and anti-inflammatory effects [3].

Reports on the medicinal value of turmeric in treating a variety of ailments have further increased the global demand for turmeric [8]. In the United States, the largest market for turmeric supplements, turmeric was the top-selling herbal supplement, with sales exceeding US $47.6 million in 2016 [8,9]. In addition, turmeric-based dietary supplements, which also include standardized extracts with high concentrations of curcumin, have seen a steady increase in popularity in the United States and elsewhere [10,11]. However, with the increasing demand for turmeric powder, adulterated turmeric powders mixed with other cheap starch powders, such as from corn or cassava, have been distributed by food supplies for economic benefit [12]. According to the United States Grocery Manufacturers Association, food fraud costs $10–15 billion annually in the global food industry and affects approximately 10% of all commercial foods sold [13].

To detect fraudulent ingredients in complicated mixed foods, various technologies, such as sensory-, physicochemical-, chromatographic-, spectroscopic-, and DNA-based assays have been developed. DNA is generally believed to be stable enough to withstand various chemical treatments and high temperatures, and small quantities of DNA can be detected with specific primers using PCR-based methods [14]. DNA-based methods, such as quantitative real-time PCR (real-time PCR), multiplex PCR, and PCR-RFLP have been successfully applied to detect food fraud and adulteration due to their economical and time-saving advantages over other approaches [15,16].

Specifically, real-time PCR (real-time PCR) assay presents with high specificity and sensitivity, capable of detecting very small amounts of target DNA in complex foods. General types of real-time PCR approaches, probe-based real-time PCR (TaqMan assay), and DNA intercalating dye-based real-time PCR (SYBR Green I assay) have been employed for the detection and identification of DNA [17]. Probe-based real-time PCR detects the target sequence with specificity using probes designed to be complementary to a target sequence [18]; however, this approach requires many SNPs or indels to differentiate species, and it is difficult to design probes and optimize real-time PCR conditions [19,20]. Alternatively, SYBR Green I, an intercalating dye that binds to double-stranded DNA in a sequence independent manner, can provide a more flexible, convenient, and inexpensive method over probe-based methods [19].

It is generally believed that the nuclear genome of a cell has a single copy of a particular gene along with a few sequences in low copy numbers; hence, it is difficult to obtain high uniformity in PCR amplification. Especially, DNA extracted from processed commercial foods is of low quality, possibly because of degradation caused by the processes of drying, heating, fermentation, and addition of ingredients. Therefore, markers designed on the extracted nuclear DNA from processed foods exhibit a low ability to discriminate between species because of the low quality of nuclear DNA that has either a single gene or low-copies of genes [21,22]. The chloroplast genome size varies among species, ranging from 107 to 208 kb and consisting of a single circular molecule of DNA that is generally present in hundreds of copies per cell [23]. Chloroplasts are composed of two layers of membranes that enable chloroplasts to persist through decomposition during food processing [24]. The chloroplast genome is generally believed to contain 120–130 genes [22]. Some genes, such as *matK*, *ndhF*, *ycf*2, and *ccsA*, exhibit higher frequencies of single-nucleotide polymorphisms (SNPs) and insertion/deletions (indels) than other chloroplast genes [23]. A variety of chloroplast markers, including *atpF-atpH* spacer, *matK* gene, *rbcL* gene, *rpoB* gene, *rpoC1* gene, *psbK-psbl* spacer, and *trnH-psbA* spacer, have been employed for species identification [25,26].

As described above, cheap corn powder with a similar color to turmeric has been wildly used in adulterated turmeric powders by food suppliers for illegal economic benefit. In this study, we developed SYBR Green-based quantitative real-time PCR assay to identify adulteration in commercial turmeric powder products using turmeric and corn species-specific primer sets. The real-time PCR methodology was optimized for both species-specific primers to correctly identify target species in complex powder products. Subsequently, the designed primers were applied to commercial turmeric products.

2. Materials and Methods

2.1. Plant and Food Sample Preparation

Turmeric (*Curcuma longa*) rhizomes and corn (*Zea mays*) seeds were kindly provided by Gangwondo Agriculture Research and Extension Services (Chuncheon, Korea). Both plants were grown in a stable temperature greenhouse for four weeks with horticulture soil. Samples for DNA isolation were extracted from the leaves of each plant. All *C. longa* commercial products used for the analysis of food complexes were purchased from local markets and stored at room temperature.

2.1.1. Reference Binary Mixtures

To generate a quantitative reference binary mixture model, binary mixtures containing different amounts (2 mg, 0.1%; 20 mg, 1%; 200 mg, 10%; and 2 g, 100%) of turmeric rhizome powders were mixed to prepare a final mixture of 2 g with corn powder, wheat flour, or rice flour purchased from a local market. Additionally, different amounts of corn powder were mixed (2 mg, 0.1%; 20 mg, 1%; 200 mg, 10%; and 2 g, 100%) to prepare final mixtures of 2 g with turmeric rhizome powders. Turmeric rhizomes and corn seeds were dried in a 55 °C dry oven for 48 hours and then ground with a mixing machine.

2.1.2. Blind Samples

Blind powder samples ($n = 20$) were provided by the National Institute of Food and Drug Safety Evaluation of the Ministry of Food and Drug Safety (Cheongju, Korea). The blind samples consisted of different percentages of corn and turmeric rhizome powders. The corn powders were added to turmeric rhizome powders at concentrations of 0–10% *w/w*, to prepare final mixtures of 150 mg.

2.2. DNA Extraction

For the efficiency of the designed primer sets, genomic DNA used for standard curves was extracted from *C. longa* and *Z. mays* leaves using the Dneasy Plant Pro Kit (QIAGEN, Hilden, Germany) according to the manufacturer's protocol. Genomic DNA used to plot standard curves of reference binary mixtures was isolated from the binary mixture samples (2 g each) using a large scale CTAB-based genomic DNA isolation method [27]. Genomic DNA from the commercial turmeric products was extracted using the Dneasy Plant Pro Kit according to the manufacturer's protocol. To obtain high quality genomic DNA, DNA extracted with the large scale CTAB method was purified using the Wizard DNA Clean-Up system (Promega, Madison, USA). DNA quantity and purity were measured using a SPECTROstar Nano reader (BMG Labtech, Ortenberg, Germany). Purity of the DNA extracts was in the range of 1.7–2.

2.3. Sequence Analysis and Primer Design

Sequences of target chloroplast genes such as *matK*, *atpF*, and *ycf2* of two species (*C. longa* for NC_042886.1 and *Z. mays* for NC_001666.2) were downloaded from the National Center for Biotechnology Information (NCBI) and used to design target-specific primers. The nucleotide sequences of the both species were aligned using ClustalW2 (EMBL-EBI, Hinxton, Cambridgeshire, UK) and BioEdit 7.2 (Ibis Biosciences, Carlsbad, CA, USA). Species-specific primer sets were designed based on the variable region between *C. longa* and *Z. mays* using Beacon Designer™ (PRIMER Biosoft, Palo Alto, CA, USA). Species-specific primers were commercially synthesized (Macrogen, Seoul, Korea).

2.4. Quantitative Real-Time PCR

Real-time PCR was performed in a final volume of 20 μL using AccuPower® 2× GreenStar™ real-time PCR Master Mix with SYBR Green (Bioneer, Daejeon, Korea). The real-time PCR reaction mixture consisted of 10 μL 2× GreenStar Master Mix, 0.5 μL 10 pmol each primer, 1 μL of 10 ng/μL genomic DNA, and 0.25 μL ROX Dye. A QuantStudio 3 Real-Time PCR System (Applied Biosystems,

Foster City, CA, USA) was used for real-time PCR amplification. The real-time PCR conditions were as follows: pre-denaturation (10 min at 95 °C), followed by 40 cycles of denaturation for 30 s at 95 °C, annealing for 20 s at 55–60 °C (depending on each targeting primer sequence), and extension for 30 s at 72 °C. All real-time PCRs were performed in technical triplicates for three biological replicates.

2.5. Cloning of PCR Amplicons and Sequencing

Conventional PCR was carried out using TaKaRa Ex TaqTM DNA polymerase (TaKaRa Bio Company, Kusatsu, Shiga, Japan) mixture with 10 ng DNA and 10 pmol each primer using a C1000 Thermal Cycler (BIO-RAD, California, USA). PCR conditions were as follows: pre-denaturation for 5 min at 95 °C, followed by 35 cycles of annealing and denaturation for 30 s at 95 °C, annealing for 20 s at 55–60 °C (depending on primer sequences), and extension 30 s at 72 °C, and final extension for 5 min at 72 °C. PCR products were amplified using target specific primers (CL_matK, CL_atpF, CL_ycf2, ZM_matK, ZM_atpF, and ZM_ycf2) and cloned using the RBC T&A Cloning Vector (Real Biotech Corporation, Taipei, Taiwan). Plasmid DNA was extracted from recombinant plasmids using the DokDo-Prep Plasmid Mini-Kit (ELPISBIOTECH, DaeJeon, South Korea) and sequenced by a commercial service (Macrogen, Seoul, Korea).

2.6. Standard Curve Construction and Data Analysis

The efficiency of the designed primer sets was evaluated using two approaches. First, species-specific PCR products were cloned into the RBC T&A Cloning Vector (Real Biotech Corporation, Taipei, Taiwan), and recombinant clones were then diluted serially (10^7, 10^6, 10^5, 10^4, and 10^3 copies) and used to quantify and confirm the efficiency of equivalent amplification [28,29]. Second, real-time PCR assays were applied to genomic DNA using target and non-target gDNA diluted ten-fold into five series (10 ng to 1 pg).

Each binary mixture with genomic DNAs extracted from the leaves or powder products of each species was diluted to a final concentration of 10 ng/μL. A baseline and a threshold were set for further analysis. The cycle number at the threshold level of log-based fluorescence was defined as the Ct (cycle threshold) number, which was the observed value in the conventional real-time PCR experiments [30]. Correlations between diluted DNAs and cycle threshold (Ct) standard curves were evaluated using a default parameter. The standard curve was calculated as $y = -ax + b$ (a refers to the standard curve slope and b refers to the y-intercept). The efficiency of the reaction (E) was calculated as $E = (10^{-1/a})$, and the percent efficiency was evaluated as $(E - 1) \times 100\%$ [29,30]. For all analyses, three technical replicates of each biological replicate were performed.

To evaluate amplification efficiency and sensitivity, two criteria were used to define an acceptable real-time PCR assay based on previous reports [28,29]: linear dynamic range and amplification efficiency. The linear dynamic range should ideally extend over four $\log_{10}$ concentrations, with the coefficient of determination (R^2) being greater than 0.98, and the amplification efficiency should be in the range of 110-90%, corresponding to a slope between −3.1 and −3.6 [29].

To validate the specificity and sensitivity of the designed target-specific primers, interlaboratory validation was performed in two independent laboratories. Validation was performed in two laboratories using the same PCR conditions and with either an Applied Biosystems 7500 Fast Real-Time PCR Instrument System (Applied Biosystems, Foster City, CA, USA) or a CFX Connect Real-Time PCR Detection System (Bio-Rad, Hercules, CA, USA).

3. Results and Discussion

3.1. Design of Species-Specific Primers

To verify authenticity of *C. longa* commercial food products, we designed species-specific primer pairs for *C. longa* and *Z. mays*. Chloroplast genes, such as *matK*, *atpF*, and *ycf2*, with high frequencies of SNPs and indels between the two species [25] were targeted to design the species-specific primer sets.

For designing species-specific primers, chloroplast genes of both species, as well as those of other starch crops (*Oryza sativa* and *Triticum aestivum*), were aligned using a software program with ClustalW2 (EMBL-EBI, Hinxton, Cambridgeshire, UK) and BioEdit 7.2 (Ibis Biosciences, Carlsbad, CA, USA; Supplementary Figures S1 and S2). We identified a variety of SNPs within three chloroplast genes among four species (Supplementary Figure S1). Food processing, such as heating, drying, and mixing, is known to damage and degrade DNA [31]. If the length of PCR amplicons is long, real-time PCR would be decreased in various food products. Therefore, based on species-specific SNPs, target-specific primers were designed to amplify short products ranging from 80 to 194 bp (Table 1).

Table 1. Primer sets designed for species-specific targeting.

Target Species	Target Gene	Primer	Length (bp)	Sequence (5′→3′)	Size (bp)	Tm (°C)
All plants	18s rRNA region	18s rRNA_F 18s rRNA_R	25 25	TCTGCCCTATCAACTTTCGATGGTA AATTTGCGCGCCTGCTGCCTTCCTT	137	58
Curcuma longa	*matK*	*CL_matK_F* *CL_matK_R*	19 18	CAATCCTATATGGTTGAGA GTCAGAAGACTCTATGGA	171	55
	atpF	*CL_atpF_F* *CL_atpF_R*	20 22	GCATTATTGGTTGATAGAGA GTTTATTTCAAGAATAGGATGG	194	58
	ycf2	*CL_ycf2_F* *CL_ycf2_R*	20 20	GAAGAAGAGGAAGAGGACAT CATATTCTAGGAGCCCAAAC	80	60
Zea mays	*matK*	*ZM_matK_F* *ZM_matK_R*	19 16	TTGATATCGAACATAATGC ACATCTTCTGGAACCT	135	55
	atpF	*ZM_atpF_F* *ZM_atpF_R*	19 18	TGGAAGCAGATGAGTATCG TGTTGTCGGACCTGATTC	160	60
	ycf2	*ZM_ycf2_F* *ZM_ycf2_R*	20 18	AAGAGGATGAGTTGTCAGAG GCAAGAAGTCCGAATCAG	99	59

3.2. Amplification Efficiency of the Designed Primer Sets

Amplification efficiency of the six primer sets (CL_matK, CL_atpF, CL_ycf2, ZM_matK, ZM_atpF, and ZM_ycf2) was evaluated by constructing standard curves using 10-fold serial dilutions (10^7 to 10^3) of each recombinant plasmid DNA, and regression analyses were performed (Figure 1, Supplementary Figure S3). The correlation coefficients (R^2) of the six primer pairs were higher than 0.99, and slopes ranged from −3.14 to −3.50. The efficiency of the primers was between 93.14 and 108.40% (Supplementary Table S1). All values fit the ENGL (European Network of GMO Laboratories) guidelines, with the coefficient of determination (R^2) being greater 0.98 and the amplification efficiency ranging from 110 to 90%, which corresponds to a slope between −3.1 and −3.6 [29]. Subsequently, we evaluated the efficiency of the primers using the 10-fold serially diluted genomic DNAs (from 10 ng to 1 pg) extracted from plant samples (Figure 2, Supplementary Figure S4). Similarly to the results of recombinant plasmid DNAs, standard curves in the gDNA samples also ranged from −3.42 to −3.54, exhibited $R^2 > 0.99$, and efficiency values of 91.78–95.92%, which also conformed to the ENGL guidelines (Supplementary Table S1) [29].

In addition, to evaluate the adaptability of the primes across machines, amplification efficiency was evaluated by two independent laboratories. As a result, the primer sets were found to meet the ENGL criteria ($R^2 > 0.98$ and efficiency ranges of 91.78–108.40; Supplementary Table S2). Based on the evaluation of amplification efficiency of the designed primer sets through three approaches, with recombinant plasmids, genomic DNA, and interlaboratory evaluation, the designed primer sets could be suitable to detect the target species.

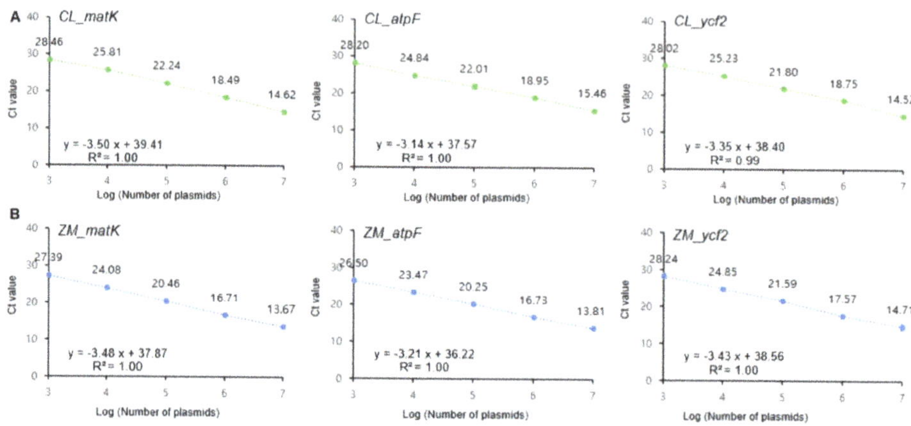

Figure 1. Standard curve of cycle threshold (Ct) values were obtained on the basis of efficiency and correlation of coefficient (R^2) in serial dilution series recombinant plasmids (*C. longa* and *Z. mays*) using species-specific primer sets. The *x*-axis represents log number of plasmids and the *y*-axis represents means of Ct value ± SD. (**A**) *C. longa* targeting primer sets (CL_matK, CL_atpF, and CL_ycf2). Green dots represent serial dilution series of recombinant plasmids (107–103) containing *C. longa* specific target gene (matK, atpF and ycf2) sequences; (**B**) *Z. mays* targeting primer sets (ZM_matK, ZM_atpF, and ZM_ycf2). Blue dots represent serial dilution series of recombinant plasmids (107–103) containing *Z. mays* specific target gene (matK, atpF and ycf2) sequences. The real-time PCRs were carried out in triplicate (n = 3).

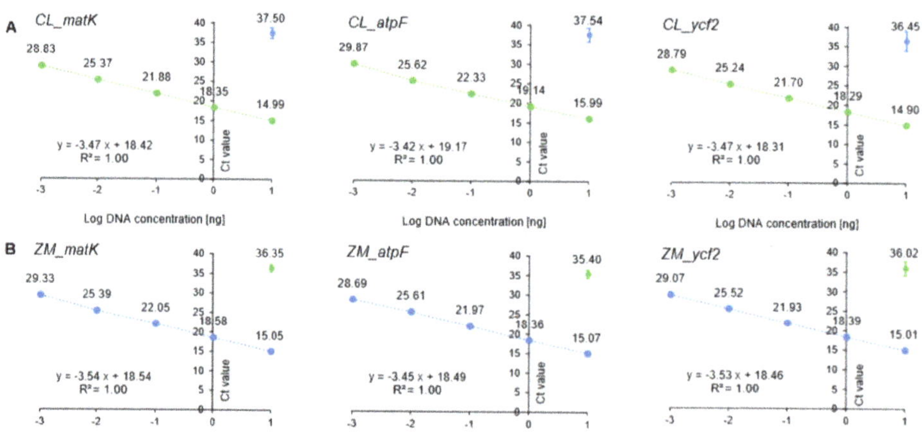

Figure 2. Standard curve of cycle threshold (Ct) values were obtained on the basis of efficiency and correlation of coefficient (R^2) in serial dilution series genomic DNA (*C. longa* and *Z. mays*) using species-specific primer sets. The *x*-axis represents log DNA concentration (ng) and the *y*-axis represents means of Ct value ± SD. (**A**) *C. longa* targeting primer sets (CL_matK, CL_atpF, and CL_ycf2). Green dots represent serial dilution series of genomic DNA in *C. longa* leaves (10ng–1pg) and blue dots represent genomic DNA of *Z. mays* (10ng); (**B**) *Z. mays* targeting primer sets (ZM_matK, ZM_atpF, and ZM_ycf2). Blue dots represent serial dilution series of genomic DNA in *Z. mays* leaves (10ng–1pg) and green dot represents genomic DNA of *C. longa* (10ng). The real-time PCRs were carried out in triplicate (n = 3).

3.3. Sensitivity and Specificity of the Assay

Globally, most *C. longa*-containing foods are prepared with rhizomes and dry powders. Therefore, we tested the sensitivity and specificity of the designed *C. longa* primer sets with binary mixtures

(0.1–100% (*w*/*w*)) of *C. longa* dry rhizome powders containing each of three starch crops, including corn, rice, and wheat (Figure 3A–C). All three *C. longa* primer sets with slopes ranging from −3.35 to −3.550 exhibited $R^2 > 0.99$ and efficiency values of 91.29–98.84% when used on mixed powders of *C. longa* and each starch crop, supporting the high sensitivity of the primer sets for verifying the presence of *C. longa* in mixtures sets (Supplementary Table S3). Subsequently, sensitivity of the three *Z. mays* primer sets was tested with binary mixtures of *Z. mays* and *C. longa* (0.1–100% (*w*/*w*)). Similarly, the three *Z. mays* primer sets with slope ranging from −3.12 to −3.44 exhibited $R^2 > 0.99$ and efficiency values of 95.30–109.18% when used on mixed powders of *C. longa* and *Z. mays*, supporting the high sensitivity of the primer sets for verifying the presence of *Z. mays* in mixtures. Next, we determined the cut-off of Ct values based on the binary mixture standard collinearity equation of each primer set (Supplementary Table S3) to identify intended additions of cheap starch ingredients, such as *Z. mays*, in the *C. longa* powders. Ct values of 0.1% target species were determined as cut-off values for each primer set because additions of less than 0.1% of non-target species were not considered to be intended for illegal economic profit. The cut-off Ct values (0.1% target species in binary mixtures) were established to verify the presence of the target species from the calibration curves (Figure 3). The cut-off Ct values ranged from 26.82 to 29.59 cycles for each primer set targeting *C. longa* and 27.58 to 29.68 cycles for those targeting *Z. mays* (Supplementary Table S3).

Subsequently, we conducted a specificity test using the species-specific primer sets. A total of 10 species of cereals and vegetables were examined to assess cross-reactivity (Table 2). The cheap starch crops such as barley, wheat, oats, rice, sweat potato, and cassava, which are likely to be intentionally mixed as ingredients in complex turmeric foods for illegal economic profits, were included for the specificity test. In addition, one vegetable crop such as cabbage and one oilseed crop such as peanuts were used for the specificity test as out groups. 18S plant rRNA primer sets were used as a positive control [32], which exhibited lower Ct values than the cut-off. As shown in Table 2, Cl_matK, CL_atpF, and CL_ycf2 exhibited *C. longa*-specific amplification but did not amplify the DNA of other species. Similarly, ZM_matK, ZM_atpF, and ZM_ycf2 exhibited *Z. mays* specific amplification did not amplify the DNA of other species. The specificity test demonstrates that the primer sets could be useful for detecting the target species in unknown-ingredient powders and in complex food products.

Table 2. Results of the specificity test with other plants.

NO	Species	Plant Systems	*Curcuma longa*			*Zea mays*		
		18s rRNA	*CL_matK* 28 [a] (Cycles)	*Cl_atpF* 28 (Cycles)	*CL_ycf2* 29 (Cycles)	*ZM_matK* 28 (Cycles)	*ZM_atpF* 29 (Cycles)	*ZM_ycf2* 28 (Cycles)
1	*Curcuma longa* (Turmeric)	+	+ [b]	+	+	-	-	-
2	*Hodeum vulgare* (Barley)	+	- [c]	-	-	-	-	-
3	*Avena sativa* (Oats)	+	-	-	-	-	-	-
4	*Triticum aestivum* (Wheat)	+	-	-	-	-	-	-
5	*Zea mays* (Corn)	+	-	-	-	+	+	+
6	*Oryza sativa* (Rice)	+	-	-	-	-	-	-
7	*Brassica oleracea* var. *capitate* (Cabbage)	+	-	-	-	-	-	-
8	*Ipomoea batatas* (Sweat potato)	+	-	-	-	-	-	-
9	*Arachis hypogaea* (Peanuts)	+	-	-	-	-	-	-
10	*Manihot esculenta* (Cassava)	+	-	-	-	-	-	-

[a] Cycles, conventional PCR cycles based on cut-off (Ct) values of each specific primer sets (Ct values of 0.1% binary mixture); [b] +, detected at less than Ct values of primers; [c] -, not detected before the primers' Ct values.

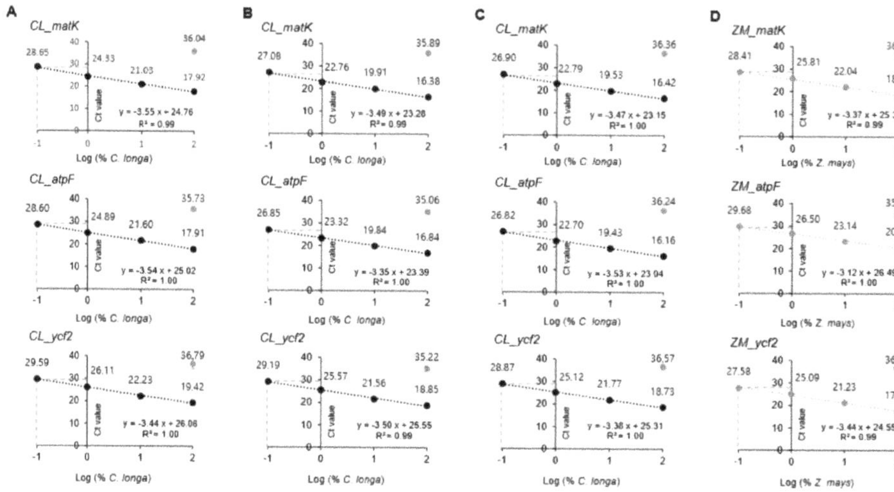

Figure 3. Standard curve of cycle threshold (Ct) values obtained on the basis of efficiency and correlation of coefficient (R^2) by reference binary mixtures. The x-axis represents log percentage of the target species (%) and the y-axis represents means of Ct value ± SD. plotted against the logarithm of the target species concentration (100, 10, 1, and 0.1%). (**A–C**); each *C. longa* rhizome powders were mixed with three different plant powders (*Z. mays*, *O. sativa*, and *T. aestivum*) by ten-fold dilutions (0.1, 1, 10 and 100%, final mass of 2g) and the each mixture gDNA(10 ng/uL) was amplified using the *C. longa* targeting primer sets (CL_matK, CL_atpF, and CL_ycf2). The green dotted line means the 0.1% binary mixture Cts amplified using the *C. longa* targeting primer sets, CL_matK, CL_atpF and CL_ycf2) (**A**) binary mixture of *C. longa* and *Z. mays*; (**B**) binary mixture of *C. longa* and *O. sativa*; (**C**) binary mixture of *C. longa* and *T. aestivum*. (**D**) *Z. mays* powders were mixed with *C. longa* rhizome powders by ten-fold dilutions (0.1, 1, 10 and 100%, final mass of 2g) and each mixture gDNA(10 ng/uL) was amplified using the *Z. mays* targeting primer sets (ZM_matK, ZM_atpF, and ZM_ycf2). The blue dotted line means the 0.1% binary mixture Cts amplified using the *Z. mays* targeting primer sets, ZM_matK, ZM_atpF and ZM_ycf2). The real-time PCRs were carried out in triplicate (n = 3).

3.4. Application of the Developed Real-Time PCR Assay to Blind Samples

A blind test was conducted to estimate the reliability of the developed real-time PCR assays. Twenty unknown powder samples of *C. longa* and *Z. mays* were mixed randomly by an independent research group. The 18S rRNA plant primer sets were used as positive amplification controls [32], which exhibited low Cts (13.27–16.21; Table 3).

Next, we determined whether *Z. mays* powder was present in the samples based on the cut-off Ct values of the designed primer sets (0.1% *Z. mays* in binary mixtures). As a result, we identified four samples (sample 3, 9, 12, and 19) with Ct values exceeding the cut-off Ct values, indicating that the samples did not contain *Z. mays* powder mixed in *C. longa* powder. The other 16 samples were exhibited lower Cts than the cut-off Ct values, indicating that those samples contained *Z. mays* powder. In addition, the ratio of *Z. mays* powder mixed into the 16 samples was predicted using the developed binary mixture assay of the three primer sets (ZM_matK, ZM_atpF, and ZM_ycf2). The predicted percentage of *Z. mays* in each blind sample was extrapolated by inserting the Cts into the standard collinearity equation of each primer set (ZM_matK, ZM_atpF, and ZM_ycf2). As a result, the predicted percentages of *Z. mays* present in each sample were consistent with those of the mixed samples (Table 3). Therefore, the real-time PCR methodologies developed in this study demonstrated high accuracy for detecting the addition of *Z. mays* in *C. longa* rhizome powders.

Table 3. Results of the blind mixture test for evaluating the reliability of the developed primer sets.

No.	Ingredient		PAC[a]	E[b] (%)	*Z. mays* Specific Primer Ct ± SD			A/D
	C. longa (%)	*Z. mays* (%)			ZM_matK	ZM_atpF	ZM_ycf2	
1	99	1	13.57 ± 0.04	0.5–1.5	25.02 ± 0.06	25.61 ± 0.20	26.44 ± 0.21	A[d]
2	98	2	14.73 ± 0.09	1–5	25.09 ± 0.02	24.78 ± 0.12	22.8 ± 0.18	A
3	100	0	14.03 ± 0.01	ND[c]	31.51 ± 0.15	31.32 ± 0.26	35.15 ± 0.05	A
4	98	2	14.31 ± 0.01	1–5	24.40 ± 0.01	24.75 ± 0.15	23.92 ± 0.09	A
5	95	5	14.05 ± 0.01	1–5	23.72 ± 0.09	24.34 ± 0.09	21.94 ± 0.22	A
6	97	3	14.11 ± 0.06	1–5	23.45 ± 0.05	24.90 ± 0.07	22.97 ± 0.25	A
7	99.5	0.5	14.21 ± 0.02	0.5–1.5	25.09 ± 0.05	26.95 ± 0.07	26.34 ± 0.15	A
8	98.5	1.5	14.12 ± 0.05	0.5–1.5	25.59 ± 0.03	26.32 ± 0.10	24.69 ± 0.19	A
9	100	0	14.33 ± 0.08	ND	31.67 ± 0.20	31.01 ± 0.80	34.89 ± 0.10	A
10	98.5	1.5	14.23 ± 0.03	0.3–2	25.13 ± 0.09	25.69 ± 0.10	24.21 ± 0.02	A
11	99.5	0.5	14.27 ± 0.01	0.1–1	26.06 ± 0.09	27.08 ± 0.10	26.19 ± 0.09	A
12	100	0	13.61 ± 0.12	ND	31.33 ± 0.28	30.98 ± 0.11	34.46 ± 0.26	A
13	97	3	14.31 ± 0.10	1–5	24.97 ± 0.09	24.37 ± 0.07	23.86 ± 0.13	A
14	98	2	13.22 ± 0.07	1–5	24.19 ± 0.09	26.14 ± 0.12	23.22 ± 0.09	A
15	96	4	13.33 ± 0.06	1–5	23.92 ± 0.15	24.26 ± 0.03	22.72 ± 0.27	A
16	99	1	14.26 ± 0.05	0.1–1	25.75 ± 0.10	26.75 ± 0.14	25.11 ± 0.10	A
17	93	7	13.95 ± 0.07	5–10	22.11 ± 0.11	23.99 ± 0.13	21.72 ± 0.11	A
18	90	10	13.74 ± 0.02	10–15	22.11 ± 0.10	21.99 ± 0.18	21.23 ± 0.13	A
19	100	0	16.21 ± 0.07	ND	30.82 ± 0.20	31.09 ± 0.01	34.06 ± 0.56	A
20	97	3	13.27 ± 0.1	1–5	23.96 ± 0.05	24.28 ± 0.10	23.22 ± 0.12	A

[a] Positive amplification control (18s rRNA); [b], expected ratio of the *Z. mays*; [C], not detected; [d] accordance.

3.5. Application of the Developed Assay in Commercial Products

To verify adulteration with corn powder of *C. longa* food products, we performed the developed real-time PCR assays on 10 *C. longa* commercial food products (Supplementary Table S4, Table 4). First, the quality of genomic DNA isolated from the food products was evaluated using a spectrometer. As depicted in Table 4, the 18S rRNA primer sets exhibited low Cts (14.01–19.82), indicating that the gDNA from all the commercial products was sufficient to provide amplifiable gDNA. We found that all *C. longa* commercial food products (samples 1–10) were amplified with lower Ct values (from 14.1 to 21.971 cycles) using the *C. longa* species-specific primers (CL_matK, CL_atpF, and CL_ycf2) than the cut-off Ct values (Ct values of 0.1% *C. longa*-specific primer set in binary mixtures) for each primer set (CT values of CL_matK, CL_atpF, and CL_ycf2 were 28.65, 28.60, and 29.59 cycles, respectively; Figure 3, Supplementary Table S3). Additionally, all samples were amplified with higher Ct values (from 30.23 cycles to not detected before 40 cycles) with *Z. mays* targeting primers (ZM_matK, ZM_atpF, and ZM_ycf2) than the cut-off Ct values (Ct values of 0.1% *Z. mays*-specific primer sets in binary mixtures) for each primer set (28.41, 29.68, and 27.58 cycles, respectively; Supplementary Table S2). As a result, the commercial products purchased from local markets did not contain *Z. mays*, suggesting that the developed real-time PCR assays could be successfully applied to detect the presence of *Z. mays* in commercial complex *C. longa* products.

Table 4. Result of the real-time PCR assay using 10 commercial products.

Real Commercial Products Tested that Labeled as 100% *Curcuma longa*							
Sample Number	Plant System (18s rRNA)	*CL_matK*	*CL_atpF*	*CL_ycf2*	*ZM_matK*	*ZM_atpF*	*ZM_ycf2*
1	14.49 ±0.15	14.83 ±0.13	14.10 ±0.17	14.24 ±0.14	ND[a]	34.00 ±0.07	ND
2	15.23 ±0.06	15.27 ±0.18	15.70 ±0.09	14.50 ±0.08	38.26 ±1.1	33.59 ±2.01	ND
3	19.82 ±0.08	22.62 ±0.27	22.65 ±0.04	20.16 ±0.06	36.28 ±0.61	32.32 ±0.54	34.79 ±0.47
4	15.25 ±0.06	15.95 ±0.07	16.33 ±0.03	15.25 ±0.14	33.60 ±0.20	30.05 ±0.07	34.35 ±0.21
5	14.01 ±0.05	15.29 ±0.07	15.46 ±0.01	15.29 ±0.07	31.53 ±0.54	31.41 ±0.56	ND
6	19.19 ±0.03	17.68 ±0.12	21.97 ±0.04	15.28 ±0.04	32.34 ±0.76	31.14 ±0.14	ND
7	16.44 ±0.18	16.10 ±0.03	16.39 ±0.03	14.90 ±0.09	34.89 ±0.35	34.57 ±0.65	ND
8	18.87 ±0.08	15.35 ±0.02	20.24 ±0.12	14.10 ±0.02	32.82 ±0.33	34.55 ±0.95	ND
9	15.31 ±0.09	14.63 ±0.08	15.15 ±0.06	13.26 ±0.02	31.05 ±4.55	34.00 ±0.08	ND
10	16.10 ±0.01	15.84 ±0.03	16.57 ±0.02	14.95 ±0.06	32.02 ±0.24	30.23 ±0.06	33.77 ±0.92

[a] ND indicates not detected at less than 40 cycles.

4. Conclusions

A real-time PCR assay is a highly sensitive, rapid, and specific method to detect target-species in processed food complexes. We designed three chloroplast gene targeted primer sets for both *C. longa* and *Z. mays*. To assess the quantities of the target-species present, standard curves were constructed using recombinant plasmid DNA and binary DNA mixtures. The specificities of the designed primers were confirmed with ten other species. Blind sample analysis and the application to commercial *C. longa* food products supported the effectiveness of the real-time PCR assays to detect *Z. mays* products added for illegal economic profits. Therefore, the developed real-time PCR assay could contribute to food safety and the protection of consumer's rights.

Supplementary Materials: The following are available online at http://www.mdpi.com/2304-8158/9/7/882/s1, Figure S1: Alignment of the target chloroplast gene (*matK*, *atpF* and *ycf2*) nucleotide sequences of *C. longa*, *Z. mays* and starch crops (*O. sativa* and *T. aestivum*) mainly eating as powders amplified by *C. longa* specific primer sets (CL_matK CL_atpF and CL_ycf 2), Figure S2: Alignment of the target chloroplast gene(*matK*, *atpF* and *ycf2*)nucleotide sequences of *Z. mays*, *C. longa* and starch crops(*O. sativa*, and *T. aestivum*) mainly eating as powders, amplified by *Z. mays* s pecific primer sets (ZM_matK, ZM_atpF and ZM_ycf2), Figure S3: Real time PCR with SYBR Green and DNA melting curve analyses.(A) Serial dilution series recombinant plasmids (10^7–10^3) containing *C. longa* specific gene (*matK*, *atpF* and *ycf 2*) sequence were amplified using *C. longa* specific primer sets. (**B**) Serial dilution series recombinant plasmids (10^7–10^3) containing *Z. mays* specific gene (*matK*, *atpF* and *ycf 2*) sequence were amplified using *Z. mays* specific primer sets. The real time PCRs were performed on a QuantStudio 3 Real Time PCR System (Applied Biosystems, Foster City, CA, USA) and carried out in triplicate (n = 3), Figure S4: Real time PCR with SYBR Green and DNA melting curve analyses green lanes mean the *C. longa* blue lanes mean *Z. mays* and pink lanes mean NTC. (**A**) Serial dilution series of *C. longa* genomic DNA (10 ng–1 pg) was amplified using *C. longa* specific primer sets. (**B**) Serial dilution series of *Z. mays* (10 ng–1 pg) was amplified using *Z. mays* specific primer sets. The real time PCRs were performed on a QuantStudio 3 Real Time PCR System (Applied Biosystems, Foster City, CA, USA) and carried out in triplicate (n = 3), Table S1: Evaluation of slope, R^2, and efficiency using the developed primer sets, Table S2: Result of the real-time PCR assay in an interlaboratory experiment, Table S3: Evaluation of the slope, R^2, and efficiency using binary mixtures containing three different intentionally added powders, Table S4: Information on the commercial food products.

Author Contributions: C.S.J. conceived of the overall study; S.H.O. carried out the experiment; S.H.O and C.S.J. wrote the manuscript. All authors have read and agreed to the published version of the manuscript.

Funding: This research received no external funding.

Acknowledgments: This research was supported by a grant (17162MFDS065) from the Ministry of Food and Drug Safety.

Conflicts of Interest: The authors declare that they have no known competing financial interests or personal relationships that could have appeared to influence the work reported in this paper.

References

1. Priyadarsini, K.I. The chemistry of curcumin: From extraction to therapeutic agent. *Molecules* **2014**, *19*, 20091–20112. [CrossRef] [PubMed]
2. Curcumin Market Size, Share & Trends Analysis Report by Application (Pharmaceutical, Food, Cosmetics), by Region (North America, Europe, Asia Pacific, Central & South America, Middle East & Africa), and Segment Forecasts, 2020–2027. Available online: https://www.grandviewresearch.com/industry-analysis/turmeric-extract-curcumin-market (accessed on 26 April 2020).
3. Aggarwal, B.B.; Kumar, A.; Bharti, A.C. Anticancer potential of curcumin: Preclinical and clinical studies. *Anticancer Res.* **2003**, *23*, 363–398.
4. Gupta, S.C.; Patchva, S.; Aggarwal, B.B. Therapeutic Roles of Curcumin: Lessons Learned from Clinical Trials. *AAPS J.* **2013**, *15*, 195–218. [CrossRef]
5. Panahi, Y.; Hosseini, M.S.; Khalili, N.; Naimi, E.; Simental-Mendia, L.E.; Majeed, M.; Sahebkar, A. Effects of curcumin on serum cytokine concentrations in subjects with metabolic syndrome: A post-hoc analysis of a randomized controlled trial. *Biomed Pharmacother.* **2016**, *82*, 578–582. [CrossRef] [PubMed]
6. Kuptniratsaikul, V.; Dajpratham, P.; Taechaarpornkul, W.; Buntragulpoontawee, M.; Lukkanapichonchut, P.; Chootip, C.; Saengsuwan, J.; Tantayakom, K.; Laongpech, S. Efficacy and safety of Curcuma domestica extracts compared with ibuprofen in patients with knee osteoarthritis: A multicenter study. *Clin. Interv. Aging* **2014**, *9*, 451–458. [CrossRef] [PubMed]
7. Mazzolani, F.; Togni, S. Oral administration of a curcumin-phospholipid delivery system for the treatment of central serous chorioretinopathy: A 12-month follow-up study. *Clin. Ophthalmol.* **2013**, *7*, 939–945.
8. Smith, T.; Kawa, K.; Eckl, V.; Morton, C.; Stredney, R. Herbal supplement sales in US increase 7.7% in 2016. *HerbalGram* **2017**, *115*, 56–65.
9. Pina, P. The Rise of Functional Foods. Think with Google Website. April 2016. Available online: www.thinkwithgoogle.com/consumer-insights/2016-food-trends-google/ (accessed on 19 July 2017).
10. Food Trends 2016. Available online: https://think.storage.googleapis.com/docs/FoodTrends-2016.pdf (accessed on 19 July 2017).
11. Turmeric Jumps from the Spice Rack to the Coffee Cup at Peet's. Forbes. Available online: www.forbes.com/sites/michelinemaynard/ (accessed on 3 July 2018).
12. Balakrishnan, K.V. Postharvest technology and processing of turmeric. In *Turmeric: The Genus Curcuma*; Ravindran, P.V., Nirmal Babu, K., Sivaraman, K., Eds.; CRC Press: Boca Raton, FL, USA, 2007; Volume 45, pp. 193–256.
13. Johnson, R. Food Fraud and "Economically Motivated Adulteration" of Food and Food Ingredients, Report, January 10, 2014; Washington DC. Available online: https://digital.library.unt.edu/ark:/67531/metadc276904/ (accessed on 27 May 2020).
14. Hong, E.Y.; Lee, S.Y.; Jeong, J.Y.; Park, J.M.; Kim, B.H.; Kwon, K.S.; Chun, H.S. Modern analytical methods for the detection of food fraud and adulteration by food category. *J. Sci. Food Agric.* **2017**, *97*, 3877–3896. [CrossRef]
15. Kane, D.E.; Hellberg, R.S. Identification of species in ground meat products sold on the US commercial market using DNA-based methods. *Food Control* **2016**, *59*, 158–163. [CrossRef]
16. An, J.; Moon, J.C.; Jang, C.S. Markers for distinguishing Orostachys species by SYBR Green-based real-time PCR and verification of their application in commercial O. japonica food products. *Appl. Biol. Chem.* **2018**, *61*, 499–508. [CrossRef]
17. Arya, M.; Shergill, I.S.; Williamson, M.; Gommersall, L.; Arya, N.; Patel, H.R. Basic principles of real-time quantitative PCR. *Expert Rev. Mol. Diagn.* **2005**, *5*, 209–219. [CrossRef] [PubMed]
18. López-Calleja, I.; González, I.; Fajardo, V.; Martín, I.; Hernández, P.; García, T.; Martin, R. Real-time TaqMan PCR for quantitative detection of cows' milk in ewes' milk mixtures. *Int. Dairy J.* **2007**, *17*, 729–736. [CrossRef]
19. Ergün, Ş.; Ahmet, K. Practical Molecular Detection Method of Beef and Pork in Meat and Meat Products by Intercalating Dye Based Duplex Real-Time Polimerase Chain Reaction. *Int. J. Food Prop.* **2016**, *19*, 31–40. [CrossRef]
20. Muhammad, S.; Yasmeen, J. Development and validation of fast duplex real-time PCR assays based on SYBER Green florescence for detection of bovine and poultry origins in feedstuffs. *Food Chem.* **2015**, *173*, 660–664. [CrossRef]

21. Chase, M.W.; Cowan, R.S.; Hollingsworth, P.M.; van den Berg, C.; Madriñán, S.; Petersen, G.; Seberg, O.; Jorgsensen, T.; Cameron, K.M.; Carine, M.A.; et al. A proposal for a standardized protocol to barcode all land plants. *Taxon* **2007**, *56*, 295–299. [CrossRef]
22. Vijayan, K.; Tsou, C.H. DNA Barcoding in Plants: Taxonomy in a New Perspective. *Curr. Sci.* **2010**, *99*, 1530–1541.
23. Daniell, H.; Lin, C.S.; Yu, M.; Chang, M.J. Chloroplast genomes: Diversity, evolution, and applications in genetic engineering. *Genome Biol.* **2016**, *17*, 134. [CrossRef]
24. Garino, C.; De Paolis, A.; Coisson, J.D.; Bianchi, D.M.; Decastelli, L.; Arlorio, M. Sensitive and specific detection of pine nut (Pinus spp.) by real-time PCR in complex food products. *Food Chem.* **2016**, *194*, 980–985. [CrossRef]
25. Hollingsworth, P.M.; Forrest, L.L.; Spouge, J.L.; Hajibabaei, M.; Ratnasingham, S.; van der Bank, M.; Chase, M.W.; Cowan, R.S.; Erickson, D.L.; CBOL Plant Working Group; et al. A DNA barcode for land plants. *Proc. Natl. Acad. Sci. USA* **2009**, *106*, 12794–12797. [CrossRef]
26. Caroline, P.L.; Anne, C.E. Development and Evaluation of a Real-Time PCR Multiplex Assay for the Detection of Allergenic Peanut Using Chloroplast DNA Markers. *J. Agric. Food Chem.* **2018**, *66*, 8623–8629. [CrossRef]
27. Sasikumar, B.; Remya, R.; John Zachariah, T. PCR Based Detection of Adulteration in the Market Samples of Turmeric Powder. *Food Biotechnol.* **2004**, *18*, 299–306. [CrossRef]
28. Bustin, S.A.; Benes, V.; Garson, J.A.; Huggett, J.; Kubista, M.; Mueller, R.; Nolan, T.; Pfaffl, M.W.; Shipley, G.L.; Vandesompele, J.; et al. The MIQE guidelines: Minimum information for publication of quantitative real-time PCR experiments. *Clin. Chem.* **2009**, *55*, 611–622. [CrossRef]
29. ENGL (European Network of GMO Laboratories). Definition of Minimum Performance Requirements for Analytical Methods of GMO Testing. Available online: http://gmo-crl.irc.ec.ecrl.jrc.ec.europa.eu/doc/MPR%20Report%20Application%2020_10_2015.pdf (accessed on 20 October 2015).
30. Yuan, J.S.; Reed, A.; Chen, F.; Stewart, C.N. Statistical analysis of real-time PCR data. *BMC Bioinform.* **2006**, *7*, 85. [CrossRef] [PubMed]
31. Lo, Y.; Shaw, P. DNA-based techniques for authentication of processed food and food supplements. *Food Chem.* **2018**, *240*, 767–774. [CrossRef] [PubMed]
32. Allmann, M.; Candrian, U.; Höfelein, C.; Lüthy, J. Polymerase chain reaction (PCR): A possible alternative to immunochemical methods assuring safety and quality of food Detection of wheat contamination in non-wheat food products. *Z. Lebensm. Unters. Forch.* **1993**, *196*, 248–251. [CrossRef] [PubMed]

foods

Article

Authentication of *Ginkgo biloba* Herbal Products by a Novel Quantitative Real-Time PCR Approach

Liliana Grazina [1], Joana S. Amaral [2], Joana Costa [1] and Isabel Mafra [1,*]

1 REQUIMTE-LAQV, Faculdade de Farmácia, Universidade do Porto, Rua de Jorge Viterbo Ferreira, 228, 4050-313 Porto, Portugal; li_grazina@hotmail.com (L.G.); jbcosta@ff.up.pt (J.C.)

2 Centro de Investigação de Montanha (CIMO), Instituto Politécnico de Bragança, Campus de Sta. Apolónia, 5301-857 Bragança, Portugal; jamaral@ipb.pt

* Correspondence: isabel.mafra@ff.up.pt; Tel.: +351-220-428-500

Received: 7 August 2020; Accepted: 29 August 2020; Published: 4 September 2020

Abstract: *Ginkgo biloba* is a widely used medicinal plant. Due to its potential therapeutic effects, it is an ingredient in several herbal products, such as plant infusions and plant food supplements (PFS). Currently, ginkgo is one of the most popular botanicals used in PFS. Due to their popularity and high cost, ginkgo-containing products are prone to be fraudulently substituted by other plant species. Therefore, this work aimed at developing a method for *G. biloba* detection and quantification. A new internal transcribe spacer (ITS) marker was identified, allowing the development of a ginkgo-specific real-time polymerase chain reaction (PCR) assay targeting the ITS region, with high specificity and sensitivity, down to 0.02 pg of DNA. Additionally, a normalized real-time PCR approach using the delta cycle quantification (ΔCq) method was proposed for the effective quantification of ginkgo in plant mixtures. The method exhibited high performance parameters, namely PCR efficiency, coefficient of correlation and covered dynamic range (50–0.01%), achieving limits of detection and quantification of 0.01% (*w/w*) of ginkgo in tea plant (*Camellia sinensis*). The quantitative approach was successfully validated with blind mixtures and further applied to commercial ginkgo-containing herbal infusions. The estimated ginkgo contents of plant mixture samples suggest adulterations due to reduction or almost elimination of ginkgo. In this work, useful and robust tools were proposed to detect/quantify ginkgo in herbal products, which suggests the need for a more effective and stricter control of such products.

Keywords: adulteration; authenticity; *Ginkgo biloba*; plant infusions; real-time polymerase chain reaction

---

1. Introduction

Ginkgo (*Ginkgo biloba* L.) is a millenary Chinese tree that belongs to the Ginkgoaceae family whose leaves are widely used for medicinal purposes [1]. Owing to its composition in pharmacologically active compounds, such as flavonol glycosides and terpene trilactones (bilobalides and ginkgolides) [2,3], ginkgo is used for its capacity to improve cognitive impairment in the elderly and quality of life in mild dementia. It is also known for its therapeutic action in peripheral circulatory illnesses, improving blood circulation and preventing clot formation [1,2,4–6]. Currently, different herbal products that have ginkgo as an ingredient are readily available in the global market, including in plant food supplements (PFS) and herbal infusions. According to recent surveys, ginkgo was the most popular botanical in PFS and is used in six European Union countries [7], while in the United States it ranked among the top 10 dietary supplements in the category of herbal/botanicals [8]. Moreover, the global market of *G. biloba* extracts, mainly intended for pharmaceutical and food supplement industries, was estimated to be US $1590.5 million in 2018 and projected to reach US $2379.2 million by 2028 [9].

The high demand of ginkgo in the global market and the increased value of ginkgo products make them potential targets for economically motivated adulteration. Frauds can be performed by the total or partial replacement of ginkgo with other plant species or by adding pure flavonols/flavonol glycosides or extracts (rich in flavonol glycosides) from other plant species, such as *Styphnolobium* japonicum (syn: *Sophora japonica*) and *Fagopyrum esculentum* Moench, belonging to the Fabaceae and Polygonaceae families, respectively [6].

Both pharmaceuticals and traditional herbal medicinal products (THMP) (either final products or the extracts used for their production) must comply with the Pharmacopeia standards, established for ginkgo leaves or extracts, to ensure the product's quality [3,10]. However, in the case of other ginkgo-containing products, such as herbal infusions and PFS that are legally considered as foods, they do not have to comply with those standards. Moreover, in these type of products, previous studies have reported adulterations associated with the partial or complete replacement of ginkgo with other plants [1,3]. Thus, it is crucial to provide analytical tools that allow the identification and quantification of *G. biloba* in herbal products classified as foods, making possible the verification of compliance with label statements.

Several analytical methodologies have been proposed for authenticity assessment of ginkgo-containing herbal products based on liquid chromatography coupled to mass spectrometry (LC-MS), high performance thin layer chromatography (HPTLC), HPTLC coupled with nuclear magnetic resonance and spectroscopy [1,3,11–14]. Those methodologies rely on the identification of bioactive compounds and/or chemical profile, which can be affected by several external factors, such as the plant part/tissue, plant age, environmental conditions, geographical location, and storage conditions, among others. Furthermore, chemical approaches can be less adequate when the formulation includes several plant species. On the contrary, DNA-based methodologies have been shown to be suitable tools for the identification/discrimination of species due to their high specificity and sensitivity, with different works reporting successful applications in the authentication of herbal products, namely food supplements or herbal infusions [15–17]. In this regard, different approaches including species-specific polymerase chain reaction (PCR), multiplex PCR, real-time PCR, high resolution melting (HRM) analysis, sequence characterization of amplified regions (SCAR), DNA barcoding, and next generation sequencing (NGS), among others, have been proposed to authenticate medicinal plants in herbal products [18]. Among them, real-time PCR offers the advantage of providing quantitative information, being a very sensitive, specific, and fast tool.

So far, only a few works regard the identification of *G. biloba* in herbal products and PFS using DNA-based approaches. Little [19] proposed the use of DNA barcoding targeting a short region of *matK* gene to identify gingko in PFS. Despite using a DNA mini-barcode (166 bp), 3 out of 40 samples were not successfully amplified. Besides, it should be noticed that this approach is not adequate for samples containing mixtures of ingredients/medicinal plants. Liu et al. [20] developed a rapid identification method to detect both gingko and a possible adulterant (*Sophora japonica*) in herbal products using a recombinase polymerase amplification (RPA) approach, which relied on the use of species-specific primers and a probe with high specificity, though with limited cross-reactivity testing. More recently, Dhivya et al. [21] developed a real-time PCR assay using a species-specific hydrolysis probe to identify *G. biloba* in natural health products. The method allowed the specific and sensitive detection of *G. biloba*, but without any quantitative analysis that should rely on the development of an adequate calibration model. Besides, the authors did not demonstrate its applicability in the analysis of processed/complex products. Therefore, the present work aimed at filling this gap by providing a specific, sensitive, high-throughput and cost-effective real-time PCR method that, besides establishing the unequivocal identification of *G. biloba* in herbal products, enables its quantification in plant mixtures. For this purpose, a normalized quantitative method was proposed, which was further validated and applied to assess the authenticity of ginkgo-containing commercial herbal infusions and to verify their labelling compliance.

2. Materials and Methods

2.1. Plant Species and Commercial Samples

Leaves from *G. biloba* were kindly provided by the Botanical Garden of University of Porto, Botanical Garden of Bern, Serralves Garden and Botanical Garden of Madeira (Table S1, Supplementary Material). Leaves or seeds of 73 plant species corresponding to medicinal plants, fruits and spices were used for cross-reactivity testing (Table S1, Supplementary Material). A total of 20 herbal infusions were bought at local stores, including specialized herbalists, and from the internet (Table 1).

For method development, model mixtures with known amounts of dried leaves of *G. biloba* in *Camellia sinensis* were prepared to contain 50%, 10%, 5%, 1%, 0.5%, 0.1%, 0.05% and 0.01% (*w*/*w*). Firstly, a reference mixture with 50% of *G. biloba* was prepared by adding 10 g of ground *G. biloba* leaves to 10 g of ground plant material of *C. sinensis*. All the subsequent mixtures were prepared by sequential additions of *C. sinensis* plant material up to the level of 0.01% (*w*/*w*). For method validation, blind mixtures were independently prepared as described for the reference mixtures, with the proportion of 20%, 8%, 2%, and 0.2% (*w*/*w*) of *G. biloba* in *C. sinensis* plant material and were further analyzed as unknown samples.

Seeds were ground with a mortar, while the leaves and herbal infusions were ground in a laboratory mill Grindomix GM200 (Retsch, Haan, Germany).

2.2. DNA Extraction

The NucleoSpin Plant II kit (Macherey-Nagel, Düren, Germany) was chosen to perform the DNA extraction from 50 mg of each sample, according to the manufacturer's instructions with slight modifications, as described by Costa et al. [22]

2.3. DNA Quality and Purity

The yield and purity of DNA extracts were assessed by UV spectrophotometry using a Take3 micro-volume plate accessory, on a Synergy HT multi-mode microplate reader (BioTek Instruments, Inc., Winooski, VT, USA). The nucleic acid protocol was set for double-strand DNA in the Gen5 data analysis software version 2.01 (BioTek Instruments, Inc., Winooski, VT, USA), which was applied to absorbance data measured at 260 and 280 nm.

The quality of DNA extracts was further assessed by electrophoresis with 1% of agarose gel as previously described [22].

2.4. Target Gene Selection, Oligonucleotide Primers and Probes

A set of primers (Gkb2-F/Gkb2-R) and a specific probe (Gkb2-P) labelled with fluorescein (FAM) as a fluorescent reporter and black hole quencher 1 BHQ-1 as quencher, were designed to target the Internal Transcribed Space (ITS) region of *G. biloba* (GenBank: Y16892.1) (Table 2). In silico analysis of sequences and primers was performed using the BLAST and Primer-BLAST tools to verify fragment and primer specificity, respectively. OligoCalc software was used to check primer properties and ensure the absence of primer hairpins and self-hybridization.

Table 1. Results of the application of qualitative PCR and normalized quantitative real-time PCR systems targeting the ITS1 region (primers Gkb2-F/Gkb2-R) for the detection and quantification of *G. biloba* in samples of herbal infusions and a universal gene (primers EG-F/EG-R).

	Samples	Label	Qualitative PCR [a]		Quantitative Real-Time PCR		
			EG-F/EG-R	Gkb2-F/Gkb2-R	Cq ± SD [b] EG-F/EG-R	Cq± SD [b] Gkb2-F/Gkb2-R	Estimated Ginkgo (%, *w*/*w*, Mean ± SD) [c]
1	Ginkgo infusion (leaves)	ginkgo	+	+	NA [d]	NA	NA
2	Ginkgo infusion (leaves)	ginkgo	+	+	NA	NA	NA
3	Ginkgo infusion (leaves)	ginkgo	+	+	NA	NA	NA
4	Herbal infusion (sachets)	15% ginkgo	+	+	17.03 ± 0.05	24.53 ± 0.32	2.98 ± 0.60
5	Ginkgo infusion (sachets)	100% ginkgo	+	+	NA	NA	NA
6	Ginkgo infusion (sachets)	100% ginkgo	+	+	NA	NA	NA
7	Ginkgo infusion (leaves)	ginkgo	+	+	NA	NA	NA
8	Ginkgo bio-infusion (leaves) (Agr. non EU)	100% ginkgo	+	+	NA	NA	NA
9	Ginkgo infusion (leaves) (Agr. non EU)	100% ginkgo	+	+	NA	NA	NA
10	Tisana MC (Plant mixture)	15% ginkgo	+	+/−	17.85 ± 0.49	34.00 ± 1.11	0.01 ± 0.01
11	Tisana TB (Plant mixture)	15% ginkgo	+	+	15.50 ± 0.18	23.50 ± 0.05	1.98 ± 0.06
12	Ginkgo infusion (sachets)	30% ginkgo	+	+	18.46 ± 0.13	24.12 ± 0.12	9.95 ± 0.79
13	Ginkgo infusion (sachets)	100% ginkgo	+	+	NA	NA	NA
14	Herbal infusion (sachets)	ginkgo (% NL [e])	+	+	17.19 ± 0.07	24.80 ± 0.13	2.68 ± 0.24
15	Herbal infusion (sachets)	ginkgo (% NL)	+	+	16.71 ± 0.02	24.75 ± 0.02	2.00 ± 0.26
16	Herbal infusion (sachets) (Agr. non EU)	15% ginkgo	+	+	16.33 ± 0.11	26.64 ± 0.35	0.47 ± 0.03
17	Ginkgo infusion (leaves)	ginkgo	+	+	NA	NA	NA
18	Ginkgo infusion (leaves)	ginkgo	+	+	NA	NA	NA
19	Ginkgo infusion (sachets)	100% ginkgo	+	+	NA	NA	NA
20	Ginkgo infusion (leaves)	ginkgo	+	+	NA	NA	NA

[a] (+) Positive amplification; (−) Negative amplification; +/− Doubtful amplification. [b] Mean cycle of quantification (Cq) values ± standard deviation (SD) ($n = 4$). [c] Mean percentage (%) values ± SD ($n = 4$). [d] NA—not applicable. [e] NL—not labelled.

Table 2. Data of primers used, targeting the ITS1 region of Ginkgo biloba and a conserved eukaryotic region.

Species	Target	Primer	Sequence (5′→3′)	Length	Reference
G. biloba	ITS1	Gkb2-F Gkb2-R Gkb2-P	GCGGTAAGCCCATCTCTCGA CCGAAGCGAACCCGAACAAC FAM-ATGCCAAGGTCGCCGGACCGTC-BHQ1	175 bp	This work
Eukaryote	Nuclear 18S rRNA	EG-F EG-R EG-P	TCGATGGTAGGATAGTGGCCTACT TGCTGCCTTCCTTGGATGTGGTA FAM-ACGGGTGACGGAGAATTAGGGT TCGATTC-BHQ-1	109 bp	[23] [24]

To ensure the presence of amplifiable DNA, a universal eukaryotic primer pair (EG-F/EG-R), targeting a conserved 18S rRNA nuclear region, was used [23]. The same primer pair together with a probe (EG-P) was used as an endogenous control gene for developing the normalized real-time PCR system [24]. The primers and probes were synthesized by Eurofins MWG Operon (Ebersberg, Germany).

2.5. Qualitative PCR

PCR amplification was performed using a total reaction volume of 25 μL which contained 20 ng of DNA, buffer (67 mM Tris-HCl, pH 8.8, 16 mM $(NH_4)2SO_4$, 0.01% Tween 20), 3 mM of $MgCl_2$, 1.0 U of SuperHot Taq DNA Polymerase (Genaxxon Bioscience GmbH, Ulm, Germany), 280 nM of each primer and 200 μM of dNTP (Grisp, Porto, Portugal) (Table 2). The reactions were carried out in a MJ Mini™ Gradient Thermal Cycler (Bio-Rad Laboratories, Hercules, CA, USA), using the following optimized programs: initial denaturation at 95 °C for 5 min; 35 or 40 cycles (for EG-F/EG-R or Gkb2-F/Gkb2-R primers, respectively) of amplification at 95 °C for 30 s, 63 °C or 62 °C (for EG-F/EG-R or Gkb2-F/Gkb2-R primers, respectively) for 30 s and extension at 72 °C for 30 s; and a final extension at 72 °C for 5 min.

PCR products were further analyzed by electrophoresis in a 1.5% agarose gel stained with 1× Gel Red (Biotium, Hayward, CA, USA) and running in 1× SGTB buffer (GRISP, Porto, Portugal) for 20–25 min at 200 V. Each extract was amplified in at least two independent assays.

2.6. Real-Time PCR

The reactions were performed using 20 μL of total reaction volume, containing 2 μL of DNA (20 ng), 1× SsoFast Probes Supermix (Bio-Rad Laboratories, Hercules, CA, USA), 300 nM or 400 nM of each primer set (EG-F/EG-R or Gkb2-F/Gkb2-R, respectively) and 200 nM of each probe (EG-P or Gkb2-P, for eukaryotic and *G. biloba* genes, respectively). A fluorometric thermal cycler CFX96 Real-time PCR Detection System (Bio-Rad Laboratories, Hercules, CA, USA) was used to amplify, simultaneously and in parallel reactions, each target sequence, under the following conditions: 95 °C for 5 min, 45 cycles at 95 °C for 15 s and 65 °C for 45 s, and the fluorescence signal was collected at the end of each cycle. The data evaluation, from each real-time PCR assay, was made using the software Bio-Rad CFX Manager 3.1 (Bio-Rad Laboratories, Hercules, CA, USA). Real-time PCR assays were performed, at least, in two independent runs using $n = 3$ or $n = 4$ replicates in each one.

For the construction of a calibration curve and for the determination of the absolute limits of detection (LOD) and quantification (LOQ), 10-fold serially diluted ginkgo DNA extracts (20 ng–0.002 pg) were amplified by real-time PCR. Additionally, a normalized calibration model was constructed based on the parallel amplification of the ITS1 region of *G. biloba* (target sequence) and the 18S rRNA gene (reference for eukaryotes) using the model mixtures (0.01–50%) of *G. biloba* in *C. sinensis*. The acceptance criteria established for real-time PCR assays were the PCR efficiency between 90–110%, the slope within −3.6 and −3.1 and the correlation coefficient (R^2) above 0.98 [25,26]. The lowest amplified level for 95% of the replicates was considered as the LOD and the LOQ was set as the lowest amplified level within the linear dynamic range of the calibration curve, which should cover a minimum of 4 orders of magnitude and should extend to ideally 5 or 6 $\log_{10}$ concentrations [25,26].

3. Results and Discussion

3.1. DNA Quality and Selection of Target Region

In general, DNA extracts from the leaves, seeds and commercial samples showed adequate yields and purities, being in the range of 17.6–270.8 ng/μL and 1.4–2.1, respectively. Before the *G. biloba* specific amplification of target region, all extracts were tested by PCR targeting a universal eukaryotic region (EG-F/EG-R) to check the capacity of DNA amplification and avoid false negatives [23]. All DNA extracts used for reactivity testing were amplified (Table S1, Supplementary Material).

So far, different regions have been assessed, either as a single locus or in combination, for their adequacy as barcode markers in plant species, which include *matk*, *rbcL*, ITS and ITS2, among others [27]. In this work, the non-coding ITS region of nuclear ribosomal DNA was selected due to its high power of species discrimination over plastid regions, allowing the differentiation of closely related species [27–29]. This region has been previously proposed for the development of PCR assays using species-specific primers aiming at identifying medicinal plant species, with high specificity and sensitivity [30,31]. The specificity of the newly designed primers (Gkb2-F/Gkb2-R) was initially in silico verified and subsequently assayed experimentally against different DNA extracts from several plant species ($n = 73$). As expected, the primers proved to be specific since only the DNA extracts from *G. biloba* were amplified (Table S1, Supplementary Material). Afterwards, the optimized species-specific PCR assay, using a 10-fold serially diluted *G. biloba* DNA extract (20 ng), showed a sensitivity down to 0.002 ng (Figure S1, Supplementary Material) and was further applied in the analysis of the commercial samples (Table 1). The achieved sensitivity was much higher than that obtained by the RPA-lateral flow strip device reported by Liu et al., which was approximately 1 ng of purified DNA. Moreover, only a few plant species were used for cross-reactivity testing (*Crataegus pinnatifida*, *Epimedium brevicornu*, *Selaginella tamariscina* and *Arisaema heterophyllum*) by those authors. In the same work, a species-specific PCR assay targeting *G. biloba* DNA was also developed, but again with very limited specificity testing (only against *S. japonica*).

3.2. Quantitative Real-Time PCR

3.2.1. Method Development

Following the demonstrated suitability of the proposed primers for *G. biloba* specific detection, a real-time PCR method was developed using a newly designed hydrolysis probe (Gkb2-P), increasing the sensitivity and specificity of the assay. Figure 1 presents the real-time PCR amplification curves and respective calibration curve using a 10-fold serially diluted ginkgo DNA extract. The average parameters of PCR efficiency (101.4%), slope (−3.284) and R^2 (0.988) were all within the acceptance criteria (Figure 1B), suggesting a high performance of the assay [25,26]. The dynamic range covered six orders of magnitude of the target analyte (20 ng to 0.02 pg of ginkgo DNA) and the absolute LOD of the real-time PCR assay was established as 0.02 pg of *G. biloba* DNA, corresponding to 0.285 genomic DNA copies (using the mean value of the Plant DNA C-value database [32]) and considering the amplification of all replicates ($n = 6$ from two independent assays). Since the LOD value was within the linear dynamic range of the calibration curve, the LOQ value was set at the same value (0.02 pg) [25,26].

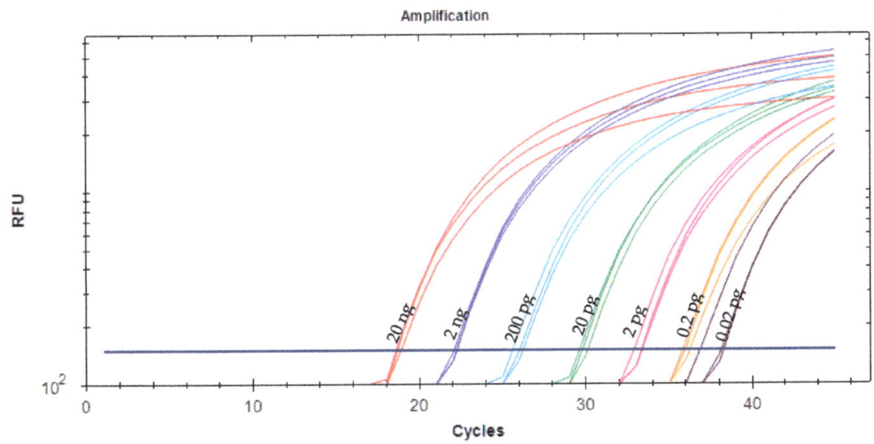

(**A**)

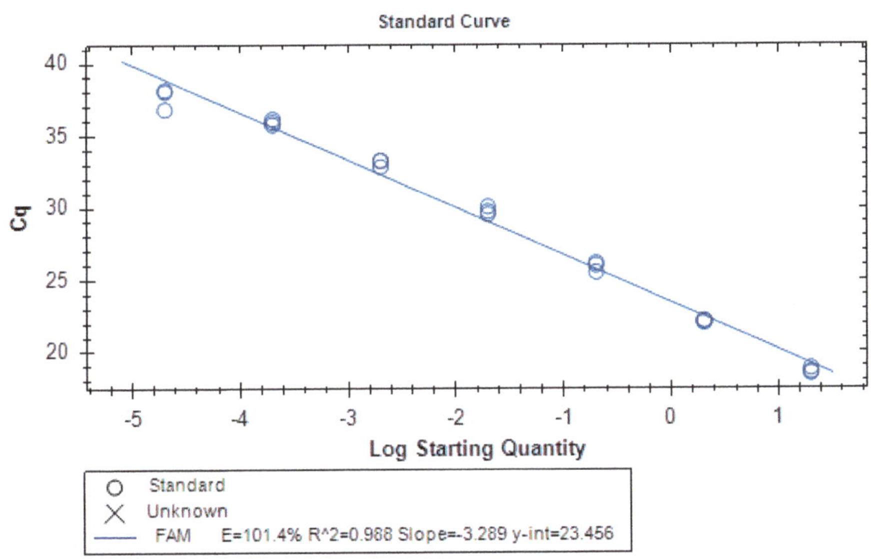

(**B**)

Figure 1. Amplification curves (**A**) and respective calibration curve (**B**) of a real-time PCR assay with a hydrolysis probe targeting ITS1 region of *G. biloba*. The amplified extracts correspond to 10-fold serially diluted ginkgo DNA from 20 ng to 0.002 pg (n = 3 replicates). Cq (cycle of quantification, also known as Ct, cycle threshold).

For establishing a quantitative model of ginkgo in herbal material, a normalized real-time PCR assay using the ΔCt method was developed. This approach accounts with amplification variations due to inconsistent DNA recovery and quality/degradation among extracts as a result of processing [24,33–35]. It relies on the construction of a normalized calibration curve using the cycle of quantitation (Cq) values from the target region (ITS1) and a reference endogenous gene (nuclear 18S rRNA) by applying the expression ΔCq = Cq (ginkgo)−Cq (universal gene). The normalized calibration curve was obtained by plotting the calculated ΔCq values versus the logarithm of the gingko concentration, using the binary mixtures with known quantities of *G. biloba* in *C. sinensis* (50.0%, 10.0%, 5.0%, 1.0%, 0.5%, 0.1%, 0.05%, and 0.01%, *w*/*w*) (Figure 2). The choice of *C. sinensis*, also commonly known as the "tea plant", to prepare the reference mixtures was based on the high frequency of its use in mixed herbal infusions. The developed normalized real-time PCR approach exhibited high performance, as inferred from the obtained parameters of PCR efficiency (96.2%), R^2 (0.982) and slope (−3.417) (mean values from 6 independent assays), covering 7 magnitude orders, which were all within the acceptable criteria. The approach enabled an LOD and LOQ down to 0.01% (*w*/*w*) (n = 12 from 3 independent assays), corresponding to 0.1 g of *G. biloba* per 1 kg of *C. sinensis*.

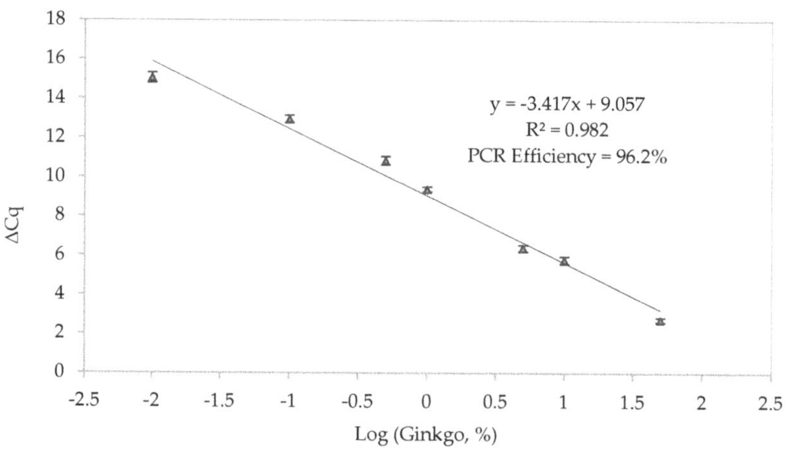

Figure 2. Normalized calibration curves obtained by real-time PCR, targeting the ITS1 region of ginkgo, using the binary mixtures of *G. biloba* in *C. sinensis* (50%, 10%, 5%, 1%, 0.5%, 0.1%, 0.05% and 0.01% (*w*/*w*)). The normalized ΔCq method was performed by the parallel amplification of a eukaryotic sequence (18S rRNA) as reference (mean values of six independent assays with n = 3 replicates).

Compared with the recent report of Dhivya et al. [21], describing a species-specific real-time PCR with a hydrolysis probe targeting the *matk* gene, the present approach achieved similar performance parameters in terms of PCR efficiency and R^2 using serially diluted leaf DNA of ginkgo. However, the proposed real-time PCR method provides a much wider dynamic range (seven orders of magnitude) and a higher sensitivity (0.02 pg of ginkgo DNA) than that obtained by Dhivya et al. [21] (five orders of magnitude and 10 pg of ginkgo DNA). Regarding specificity, the proposed primers and probe targeting the ITS region do not provide any cross-reactivity with any of the known potential adulterants (*Sophora japonica* and *Fagopyrum esculentum* Moench) (Figure S2), while the method of Dhivya et al. [21] was reactive with *S. japonica* at late amplification cycles, which compromised its sensitivity, and the potential reactivity with *F. esculentum* Moench was not verified by the referred authors. Therefore, the proposed method demonstrated full specificity and high sensitivity for gingko detection, with the important achievement of providing, for the first time, a normalized quantitative real-time PCR approach to enable a determination of the proportion of ginkgo in herbal products.

3.2.2. Method Validation

To proceed with the validation of the method, the precision and accuracy should also be evaluated [25,26]. Therefore, blind mixtures containing 20.0%, 8.0%, 2.0%, and 0.2% (*w*/*w*) of *G. biloba* in *C. sinensis* were used. The results regarding the estimated values (%) of ginkgo and the comparative analysis with the real values are presented in Table 3. The obtained values exhibited adequate coefficients of variation (CV), which were between 5.6–17.9% and, therefore, lower than the maximum acceptable (25%), demonstrating the high precision of the method over the considered dynamic range.

Regarding the accuracy, three out of the four blind mixtures presented bias values in the range of 5.6–17.9%, being within the recommended range (±25%) [26]. Although the mixture with 0.2% (*w/w*) presented a slightly higher error (−27.4%), this is the lowest tested level, not likely to occur due to adulteration, but rather from contamination. Besides, according to Kang [36], bias within 25–30% have been considered as acceptable in real-time PCR methods for food analysis.

Table 3. Results of the validation assays using the normalized quantitative PCR system applied to blind mixtures of *G. biloba* in *C. sinensis*.

Samples	Ginkgo (%, *w/w*)		CV [b] (%)	Error [c] (%)
	Real Value (%)	Estimated Value [a] (%)		
CG–A	20.0	23.58 ± 2.22	9.41	17.9
CG–B	8.0	6.86 ± 0.66	9.55	−14.3
CG–C	2.0	2.11 ± 0.15	7.03	5.6
CG–D	0.2	0.15 ± 0.02	16.88	−27.4

[a] Mean values ± standard deviation (SD) (n = 4) of three independent assays. [b] Coefficient of variation (CV). [c] Error = ((mean estimated value—real value)/real value) × 100.

3.2.3. Analysis of Commercial Herbal Infusions

For assessing the applicability of the method, the normalized real-time PCR system was used to analyze and further verify the authenticity and labelling compliance of several commercial herbal products (herbal infusions). The analyzed herbal infusions were all labelled as containing ginkgo, wholly or partially (Table 1). All the samples produced amplifiable DNA extracts, which were positive for the ginkgo-specific PCR assay. The samples of mixed herbal species were further assayed by quantitative real-time PCR to assess their ginkgo content. The quantitative results demonstrated that, out of five samples of herbal mixtures with labelled ginkgo contents, four samples (#4, #10, #11 and #16) declared 15% of ginkgo, but the obtained contents were within 0.01–2.98%. In particular, sample #10 had only trace amounts (0.01%) of gingko, suggesting its complete substitution with other plant(s). Sample #12 declared 30% of ginkgo, but the obtained content was 9.95%. Consequently, the results of samples #4, #11, #12, and #16 suggest the partial substitution of ginkgo with other plant(s). The other two mixed herbal samples (#14, #15) did not provide any quantitative information regarding gingko, having low estimated amounts (<3%), suggesting again its reduced use. Therefore, the results of mixed herbal products strongly suggest the practice of adulterations, probably due to the high market price of *G. biloba* and its increasing demand, with the industries using less quantity than they declared to raise their profits.

4. Conclusions

In the herein presented work, a new molecular marker of the ITS region was identified for the species-specific detection of *G. biloba* by both qualitative PCR and real-time PCR with a TaqMan probe, providing high specificity and sensitivity, down to 0.02 pg of DNA (0.285 genomic DNA copies). For the effective quantification of ginkgo in herbal products, a novel normalized real-time PCR system based on the ΔCq method was successfully developed using reference herbal mixtures. The method exhibited high performance parameters, namely PCR efficiency, coefficient of correlation and covered dynamic range (50–0.01%), achieving a LOD and LOQ of 0.01% (*w/w*) of ginkgo in tea plant. The quantitative approach was further validated with blind mixtures, demonstrating accuracy, repeatability, and trueness within the range of 20–2%. The applicability of the PCR approaches was demonstrated using a set of commercial ginkgo-containing herbal infusions (n = 20), confirming the presence of ginkgo in all the products. However, the obtained quantitative results regarding the estimated ginkgo content of seven herbal mixture samples suggest adulterations due to reduction or almost elimination of ginkgo. The proposed system was demonstrated to be a powerful and robust tool for control laboratories and regulatory authorities to ensure labelling compliance of ginkgo-containing herbal products. Since it was demonstrated that the developed method has a high specificity and sensitivity, it can potentially be useful for further detecting *G. biloba* in other processed herbal products or foods.

Supplementary Materials: The following are available online at http://www.mdpi.com/2304-8158/9/9/1233/s1, Table S1: Results of cross-reactivity testing of ITS1 primers are presented. Figure S1: Sensitivity by qualitative PCR. Figure S2: Analytical specificity by real-time PCR.

Author Contributions: Conceptualization, I.M. and J.S.A.; methodology, L.G. and J.C.; validation, L.G. and J.C.; formal analysis, L.G. and J.C.; investigation, I.M. and J.S.A.; writing—original draft preparation, L.G.; writing—review and editing, I.M. and J.S.A.; supervision, I.M. and J.S.A.; project administration, I.M.; funding acquisition, I.M. All authors have read and agreed to the published version of the manuscript.

Funding: This work was supported by FCT (Fundação para a Ciência e Tecnologia) under the Partnership Agreements UIDB 50006/2020 and UIDB 00690/2020. L. Grazina is grateful to FCT grant (SFRH/BD/132462/2017) financed by POPH-QREN (subsidised by FSE and MCTES).

Acknowledgments: The authors are grateful for the supply of leaves from the Botanical Garden of University of Porto (Porto, Portugal), Botanical Garden of Bern (Bern, Switzerland), Serralves Garden (Porto, Portugal), Botanical Garden of Madeira and Botanical garden of UTAD (Vila Real, Portugal), as well as to the voucher seeds from the USDA Grin by the University of Arizona Herbarium (Tucson, AZ, USA) and the RBG (Kew, Ardingly, West Sussex, UK).

Conflicts of Interest: The authors declare no conflict of interest.

References

1. Booker, A.; Frommenwiler, D.; Reich, E.; Horsfield, S.; Heinrich, M. Adulteration and poor quality of Ginkgo biloba supplements. *J. Herbal Med.* **2016**, *6*, 79–87. [CrossRef]
2. Singh, B.; Kaur, P.; Gopichand; Singh, R.D.; Ahuja, P.S. Biology and chemistry of *Ginkgo biloba*. *Fitoterapia* **2008**, *79*, 401–418. [CrossRef] [PubMed]
3. Wohlmuth, H.; Savage, K.; Dowell, A.; Mouatt, P. Adulteration of *Ginkgo biloba* products and a simple method to improve its detection. *Phytomedicine* **2014**, *21*, 912–918. [CrossRef]
4. WHO. New WHO Guidelines to Promote Proper Use of Alternative Medicines. 2004. Available online: https://www.who.int/mediacentre/news/releases/2004/pr44/en/ (accessed on 2 December 2019).
5. EMA. Ginkgo Folium—Committee on Herbal Medicinal Products (HMPC/321097/2012). 2015. Available online: https://www.ema.europa.eu/en/medicines/herbal/ginkgo-folium (accessed on 2 December 2019).
6. Gafner, S. Botanical Adulterants Prevention Program—Adulteration of Ginkgo Biloba Leaf Extract. 2018. Available online: http://cms.herbalgram.org/BAP/BAB/GinkgoBulletin.html (accessed on 3 December 2019).
7. Garcia-Alvarez, A.; Egan, B.; de Klein, S.; Dima, L.; Maggi, F.M.; Isoniemi, M.; Ribas-Barba, L.; Raats, M.M.; Meissner, E.M.; Badea, M.; et al. Usage of plant food supplements across six European countries: Findings from the PlantLIBRA consumer survey. *PLoS ONE* **2014**, *9*, e92265. [CrossRef] [PubMed]
8. CRN. Consumer Survey on Dietary Supplements. Available online: https://www.crnusa.org/sites/default/files/images/CRN-2017-ConsumerSurvey-4-page-highlights.pdf (accessed on 18 December 2019).
9. Market.us. *Ginkgo biloba* Extract Market by Form (Tablets, Capsules, Liquid Extracts, Others), by Application (Food and Beverages, Cosmetics, Pharmaceuticals), and by Region—Global Forecast to 2029. 2019. Available online: https://www.marketwatch.com/press-release/global-ginkgo-biloba-extract-market-expected-to-observe-major-growth-with-cagr-of-41-2019-08-29?mod=mw_quote_news (accessed on 18 December 2019).
10. Pharmacopeia, E. *European Directorate for the Quality of Medicines and HealthCare*, 8th ed.; Oxford Public International Law: Strasbourg, France, 2013.
11. Ko, Y.-C.; Lee, R.-J.; Feng, H.-T.; Lee, M.-R. Development of a liquid chromatography tandem mass spectrometry method for identification of flavonoids in *Ginkgo biloba*. *J. Chin. Chem. Soc.* **2013**, *60*, 1333–1338. [CrossRef]
12. Avula, B.; Sagi, S.; Gafner, S.; Upton, R.; Wang, Y.-H.; Wang, M.; Khan, I.A. Identification of *Ginkgo biloba* supplements adulteration using high performance thin layer chromatography and ultra high performance liquid chromatography-diode array detector-quadrupole time of flight-mass spectrometry. *Anal. Bioanal. Chem.* **2015**, *407*, 7733–7746. [CrossRef]

13. Frommenwiler, D.A.; Booker, A.; Vila, R.; Heinrich, M.; Reich, E.; Cañigueral, S. Comprehensive HPTLC fingerprinting as a tool for a simplified analysis of purity of ginkgo products. *J. Ethnopharmacol.* **2019**, *243*, 112084. [CrossRef]
14. Walkowiak, A.; Ledziński, Ł.; Zapadka, M.; Kupcewicz, B. Detection of adulterants in dietary supplements with Ginkgo biloba extract by attenuated total reflectance Fourier transform infrared spectroscopy and multivariate methods PLS-DA and PCA. *Spectrochim. Acta Part A Mol. Biomol. Spect.* **2019**, *208*, 222–228. [CrossRef]
15. Franz, C.; Chizzola, R.; Novak, J.; Sponza, S. Botanical species being used for manufacturing plant food supplements (PFS) and related products in the EU member states and selected third countries. *Food Funct.* **2011**, *2*, 720–730. [CrossRef]
16. Madesis, P.; Ganopoulos, I.; Sakaridis, I.; Argiriou, A.; Tsaftaris, A. Advances of DNA-based methods for tracing the botanical origin of food products. *Food Res. Int.* **2014**, *60*, 163–172. [CrossRef]
17. Grazina, L.; Amaral, J.S.; Mafra, I. Botanical origin authentication of dietary supplements by DNA-based approaches. *Compr. Rev. Food Sci. Food Saf.* **2020**, *19*, 1080–1109. [CrossRef]
18. Ganie, S.H.; Upadhyay, P.; Das, S.; Prasad Sharma, M. Authentication of medicinal plants by DNA markers. *Plant Gene* **2015**, *4*, 83–99. [CrossRef] [PubMed]
19. Little, D.P. Authentication of *Ginkgo biloba* herbal dietary supplements using DNA barcoding. *Genome* **2014**, *57*, 513–516. [CrossRef]
20. Liu, Y.; Wang, X.-Y.; Wei, X.-M.; Gao, Z.-T.; Han, J.-P. Rapid authentication of *Ginkgo biloba* herbal products using the recombinase polymerase amplification assay. *Sci. Rep.* **2018**, *8*, 8002. [CrossRef] [PubMed]
21. Dhivya, S.; Ragupathy, S.; Kesanakurti, P.; Jeevitha, S.; Noce, I.D.; Newmaster, S.G. Validated identity test method for *Ginkgo biloba* NHPs using dna-based species-specific hydrolysis pcr probe. *J. AOAC Int.* **2019**, *102*, 1779–1786. [CrossRef] [PubMed]
22. Costa, J.; Campos, B.; Amaral, J.S.; Nunes, M.E.; Oliveira, M.B.P.P.; Mafra, I. HRM analysis targeting ITS1 and matK loci as potential DNA mini-barcodes for the authentication of *Hypericum perforatum* and *Hypericum androsaemum* in herbal infusions. *Food Control* **2016**, *61*, 105–114. [CrossRef]
23. Villa, C.; Costa, J.; Oliveira, M.B.P.P.; Mafra, I. Novel quantitative real-time PCR approach to determine safflower (*Carthamus tinctorius*) adulteration in saffron (*Crocus sativus*). *Food Chem.* **2017**, *229*, 680–687. [CrossRef]
24. Villa, C.; Costa, J.; Gondar, C.; Oliveira, M.B.P.P.; Mafra, I. Effect of food matrix and thermal processing on the performance of a normalised quantitative real-time PCR approach for lupine (*Lupinus albus*) detection as a potential allergenic food. *Food Chem.* **2018**, *262*, 251–259. [CrossRef]
25. Bustin, S.A.; Benes, V.; Garson, J.A.; Hellemans, J.; Huggett, J.; Kubista, M.; Mueller, R.; Nolan, T.; Pfaffl, M.W.; Shipley, G.L.; et al. The MIQE Guidelines: Minimum Information for Publication of Quantitative Real-Time PCR Experiments. *Clin. Chem.* **2009**, *55*, 611–622. [CrossRef]
26. ENGL. Definition of Minimum Performance Requirements for Analytical Methods of GMO Testing European Network of GMO Laboratories. European Network of GMO Laboratories, Join Research Centre, EURL, 2015. Available online: https://gmo-crl.jrc.ec.europa.eu/doc/MPR%20Report%20Application%2020_10_2015.pdf (accessed on 3 December 2019).
27. Hollingsworth, P.M.; Graham, S.W.; Little, D.P. Choosing and using a plant DNA barcode. *PLoS ONE* **2011**, *6*, e19254. [CrossRef]
28. Li, X.; Yang, Y.; Henry, R.J.; Maurizio, R.; Yitao, W.; Shilin, C. Plant DNA barcoding: From gene to genome. *Biol. Rev.* **2015**, *90*, 157–166. [CrossRef] [PubMed]
29. Cheng, T.; Xu, C.; Lei, L.; Li, C.; Zhang, Y.; Zhou, S. Barcoding the kingdom Plantae: New PCR primers for ITS regions of plants with improved universality and specificity. *Mol. Ecol. Resour.* **2016**, *16*, 138–149. [CrossRef] [PubMed]
30. Wu, L.; Sun, W.; Wang, B.; Zhao, H.; Li, Y.; Cai, S.; Xiang, L.; Zhu, Y.; Yao, H.; Song, J.; et al. An integrated system for identifying the hidden assassins in traditional medicines containing aristolochic acids. *Sci. Rep.* **2015**, *5*, 11318. [CrossRef] [PubMed]
31. Doganay-Knapp, K.; Orland, A.; König, G.M.; Knöss, W. The potential of three different PCR-related approaches for the authentication of mixtures of herbal substances and finished herbal medicinal products. *Phytomedicine* **2018**, *43*, 60–67. [CrossRef] [PubMed]

32. RBG (Royal Botanic Garden). Kew, Plant DNA C-Values Database, Surrey, Canada. Available online: http://data.kew.org/cvalues/ (accessed on 12 February 2020).
33. López-Andreo, M.; Lugo, L.; Garrido-Pertierra, A.; Prieto, M.I.; Puyet, A. Identification and quantitation of species in complex DNA mixtures by real-time polymerase chain reaction. *Anal. Biochem.* **2005**, *339*, 73–82. [CrossRef]
34. Soares, S.; Amaral, J.S.; Oliveira, M.B.P.P.; Mafra, I. Quantitative detection of soybean in meat products by a TaqMan real-time PCR assay. *Meat Sci.* **2014**, *98*, 41–46. [CrossRef]
35. Costa, J.; Amaral, J.S.; Grazina, L.; Oliveira, M.B.P.P.; Mafra, I. Matrix-normalised real-time PCR approach to quantify soybean as a potential food allergen as affected by thermal processing. *Food Chem.* **2017**, *221*, 1843–1850. [CrossRef]
36. Kang, T.S. Basic principles for developing real-time PCR methods used in food analysis: A review. *Trends Food Sci. Technol.* **2019**, *91*, 574–585. [CrossRef]

foods

MDPI

Article

A Chip Digital PCR Assay for Quantification of Common Wheat Contamination in Pasta Production Chain

Caterina Morcia [1,†], Raffaella Bergami [2,†], Sonia Scaramagli [2], Roberta Ghizzoni [1], Paola Carnevali [3] and Valeria Terzi [1,*]

1 Council for Agricultural Research and Economics, Research Centre for Genomics and Bioinformatics, Via San Protaso 302, I-29017 Fiorenzuola d'Arda PC, Italy; caterina.morcia@crea.gov.it (C.M.); roberta.ghizzoni@crea.gov.it (R.G.)

2 Coop Italia, Via del Lavoro, 6/8, I-40033 Casalecchio di Reno BO, Italy; raffaella.bergami@coopitalia.coop.it (R.B.); sonia.scaramagli@coopitalia.coop.it (S.S.)

3 Barilla S.p.A., Via Mantova 166, I-43122 Parma PR, Italy; paola.carnevali@barilla.com

* Correspondence: valeria.terzi@crea.gov.it; Tel.: +39-0523-983758

† These two authors contributed equally to the work.

Received: 27 May 2020; Accepted: 7 July 2020; Published: 10 July 2020

Abstract: Pasta, the Italian product par excellence, is made of pure durum wheat. The use of *Triticum durum* derived semolina is in fact mandatory for Italian pasta, in which *Triticum aestivum* species is considered a contamination that must not exceed the 3% maximum level. Over the last 50 years, various electrophoretic, chemical, and immuno-chemical methods have been proposed aimed to track the possible presence of common wheat in semolina and pasta. More recently, a new generation of methods, based on DNA (DeoxyriboNucleic Acid) analysis, has been developed to this aim. Species traceability can be now enforced by a new technology, namely digital Polymerase Chain Reaction (dPCR) which quantify the number of target sequence present in a sample, using limiting dilutions, PCR, and Poisson statistics. In our work we have developed a duplex chip digital PCR (cdPCR) assay able to quantify common wheat presence along pasta production chain, from raw materials to final products. The assay was verified on reference samples at known level of common wheat contamination and applied to commercial pastas sampled in the Italian market.

Keywords: pasta; *Triticum aestivum*; *Triticum durum*; genetic traceability; digital PCR; semolina; species

1. Introduction

Pasta production is a strategic chain in the Italian agri-food sector, covering around the 6% of total industrial output [1]. Italy is at the same time the world's leading pasta producer, with an annual production around 3.2 million tons and, in the same time, is the largest consumer of pasta (26 kg per capita). A pillar of Italian pasta production chain is the grain identity: The use of *Triticum durum* derived semolina is in fact mandatory for Italian pasta, in which *Triticum aestivum* species is considered a contamination that must not exceed the 3% maximum level, as indicated by Law n.580 of 1967 [2] and by subsequent Decreto del Presidente della Repubblica (D.P.R.) 187, 9 February 2001 [3] and D.P.R. 41, 5 March 2013 [4]. Traditional Italian pasta, according to such regulations, is therefore the result of the extrusion, rolling and drying of dough made exclusively from durum wheat and water. The choice of *Triticum durum* is based on its peculiarities, among others the hardiness of the caryopsis, the intense yellow color due to carotenoids, the gluten composition. Thanks to such specific properties, starch is not lost during cooking, avoiding sticking and ensuring a unique and authentic taste to pasta.

Beyond fraudulent behavior, dictated by the lower price of common wheat compared to durum, the purity of the semolina can also be compromised during the various processing stages of the supply chain, which range from harvesting in the field to storing the grains. Analytical methods have been proposed aiming at the detection and quantification of the possible presence of common wheat in semolina and pasta. In this perspective, over the last 50 years, various electrophoretic, chemical and immuno-chemical methods have been proposed aimed at detecting the purity of the semolina [5–9]. Such methods are based on the identification and quantification of specific protein, which, however, can be degraded by the high temperatures nowadays used to dry pasta. To overcome this gap and taking advantage of the remarkable thermic stability of DNA (DeoxyriboNucleic Acid), a new generation of methods, based on DNA analysis, has been developed during the last two decades. PCR (Polymerase Chain Reaction) based assays to identify common wheat by distinguishing it from durum one has been developed by Bryan et al. [10], by Arlorio et al. [11] and by Sonnante et al. [12], using respectively *Dgas44* gene sequence, puroindoline B and SSR (Simple Sequence Repeats) related sequences. Untargeted DNA fingerprinting through tubulin-based polymorphism (TBP) have been optimized by Casazza et al. [13] and by Silletti et al. [14] for the authentication of cereal species, including wheat and farro. qPCR assays for the quantification of *Triticum aestivum* species have been proposed by Alary et al. [15], Terzi et al. [16], Matsuoka et al. [17], and by Imai et al. [18]. These two last assays have been in-house verified and compared by Paterno' et al. [19], with the aim to select a taxon-specific assay useful for unauthorized GM (Genetically Modified) wheat detection in wheat samples. An inter-laboratory validation in collaboration with public and private laboratories has been even reported by Morcia et al. [20] to determine the performance parameters of a qPCR assay based on the primers designed on puroindoline-b gene by Alary et al. [15] and on low molecular weight glutenin encoding sequence by Terzi et al. [16].

Species traceability can be now enforced by a new technology, namely digital PCR (dPCR) which quantify the number of target sequence present in a sample, using limiting dilutions, PCR and Poisson statistics [21]. The PCR mix is compartmentalized across a large number of partitions or droplets containing zero, one or more copies of the target sequence. After endpoint PCR amplification, a partition can be positive ("1", the presence of PCR product) or negative ("0", the absence of PCR product). The absolute number of target nucleic acid molecules contained in the original sample before partitioning can be calculated directly from the ratio of the number of positive to total partitions, obtained using Poisson statistics. It is an absolute quantification strategy because there is not the need to have a standard curve as reference for quantification. In the past several years, dPCR has achieved progress in in agri-food sector, especially for GMO (Genetically Modified Organism) testing [21,22] and for pathogen diagnostics and, at more limited extent, to the detection of animal- and plant-derived ingredients in food adulteration control [23].

The aim of this work has been to develop a chip digital PCR (cdPCR) assay able to quantify common wheat presence along pasta production chain, from raw materials to final products. The assay was verified on reference samples at known level of common wheat contamination and applied to commercial pastas sampled in the Italian market.

2. Materials and Methods

2.1. Mono-Species Flour Samples Preparation and DNA Extraction

Certified *Triticum durum* (Claudio variety) and *Triticum aestivum* (Eureka variety) seeds were obtained from CREA DC (Tavazzano, Italy). Such first-reproduction seeds are controlled and certified both at species and variety levels. In major details, at species purity level, the maximum admitted contamination is of 7 seeds belonging to different cereal species/500 g of certified seeds, according to the Italian D.P.R. n. 1065, 8 October 1973. The seeds were milled using a Cyclotec (FOSS Italia S.r.l., Padova, Italy) at 0.2 mm grid diameter, avoiding any contamination between samples. Samples of

100% durum wheat semolina and 100% common wheat flour were separately stored at controlled temperature and humidity conditions until further use.

DNA were extracted from three biological replicates of milled *Triticum aestivum* and *Triticum durum* seeds using the DNeasy mericon Food Kit (Qiagen, Milan, Italy), that is based on an improved cetyltrimethylammonium bromide (CTAB) extraction of total cellular nucleic acids. The flour samples (2 g) were extracted according to manifacturer's instructions. The evaluation of quality and quantity of extracted DNA was done using Qubit™ fluorometer in combination with the Qubit™ dsDNA BR Assay kit (Invitrogen by Thermo Fisher Scientific, Monza, Italy).

2.2. Mixed Species DNA Samples Preparation

Triticum aestivum and *Triticum durum* DNA, extracted from the mono-species flours described in point 2.1, were mixed to obtain the following samples:

- *T.durum* DNA 99.7% + *T.aestivum* DNA 0.3%;
- *T.durum* DNA 98.5% + *T.aestivum* DNA 1.5%;
- *T.durum* DNA 97% + *T.aestivum* DNA 3%;
- *T.durum* DNA 95.5% + *T.aestivum* DNA 4.5%;
- *T.durum* DNA 70% + *T.aestivum* DNA 30%

2.3. Mixed Species Flour Samples Preparation and DNA Extraction

Common wheat flour was used to contaminate durum wheat semolina with the aim to produce durum wheat samples containing 0.3, 1.5, 3, 4.5, and 30% of common wheat. After weighing the common and durum wheat flour, samples containing different percentages of the two species were homogenized for 10 min. DNA were extracted from flours (2 g) with the DNeasy mericon Food Kit (Qiagen, Milan, Italy), as previously described. The evaluation of quality and quantity of extracted DNA was done using Qubit™ fluorometer in combination with the Qubit™ dsDNA BR Assay kit (Invitrogen by Thermo Fisher Scientific, Monza, Italy).

2.4. Reference and Commercial Pasta Samples and DNA Extraction

Four reference pasta samples were prepared by mixing tap water and wheat flours containing the following common wheat percentages: 1.5%, 3%, 4.5%, 10%. The samples were dried in oven at 80 °C for 1 hour, followed by 3 hours at decreasing temperature. Such desiccation thermal profile is those commonly used for commercial pasta preparation. DNA were extracted from two biological replicates of reference pasta using the DNeasy mericon Food Kit (Qiagen, Milan, Italy), Twenty commercial pasta samples of different brands were purchased from the market. The pasta samples were milled with M20 Universal Mill (IKA). Samples (2 g) were extracted in single replicate with the DNeasy mericon Food Kit (Qiagen, Milan, Italy), as previously described. The DNA obtained was measured using Qubit™ fluorometer in combination with the Qubit™ dsDNA BR Assay kit (Invitrogen by Thermo Fisher Scientific, Monza, Italy).

2.5. Primers and Probes

Primers and probes (Table 1) were designed using Primer Express 3.0.1 Software (Life Technologies Corporation). Each primer was checked for absence of self-complementarity and primer dimer formation with other primer pairs using the online tool Multiple Primer Analyzer (Thermo Fisher Scientific, Monza, Italy). Primer specificity was checked by blasting in EnsemblPlants (https://plants.ensembl.org/index.html) against the *Triticum aestivum* database.

Table 1. Primers and probes.

Name	Primer Sequence (5′-3′)	Gene	Target
GranoCO2- Forward	TGCTAACCGTGTGGCATCAC	*Triticum TaHd1*	*Triticum genus*
GranoCO2 Reverse	GGTACATAGTGCTGCTGCATCTG		
GranoCO2 probe	VIC- CATGAGCGTGTGCGTG -MGB		
TritA_APX Forward	AGGAGCGGCCGAAGCT	*Pinb-D1*	*Triticum aestivum*
TritA_APX Reverse	TGTGAAACATCGCTCCATCAC		
TritA_APX probe	FAM-AGCTCTTGCAAGGAT -MGB		

2.6. Real-Time PCR

The reaction mixture was prepared in a final volume of 25 μL consisting of 12.5 μL of SYBR Green PCR, 2× GoTaq qPCR Master Mix (Promega Italia, Milan, Italy), 0.25 μl of 100× Reference Dye (Promega Italia, Milan, Italy), 0.5 μL of each primer at 10 μM (final concentration 200 nmol), 4 μL of DNA template serial dilution (10, 5, 2.5, 0.5, 0.25 and 0.025 ng/μL) and water to 25 μL. Three technical real-time PCR replicates were done for each sample and control. The PCR mixture was activated at 95 °C for 10 min. Forty amplification cycles were carried out at 95 °C for 15 s followed by 60 °C for 1 min. A melting curve analysis was included in each run.

2.7. Chip Digital PCR

Chip digital PCR was performed using QuantStudioTM 3D Digital PCR System (Applied Biosystems by Life Technologies, Monza. Italy). The reaction mixture was prepared in a final volume of 16 μLconsisting of 8 μL QuantStudioTM 3D Digital PCR 2X Master Mix, 0.72 μL of each primer at 20 μM (final concentration 900 nmol), 0.32 μL of FAM and VIC-MGB probes at 10 μM (final concentration 200 nmol), 2 μl of DNA (40 ng/μL) and nuclease free-water. Also, a negative control with nuclease free-water as template was added. A total volume of 15 μL reaction mixture was loaded onto the QuantStudioTM 3D Digital PCR chips using QuantStudioTM 3D Digital chip loader, according to manufacturer protocol. Amplifications were performed in ProFlexTM 2Xflat PCR System Thermocycler (Applied Biosystems by Life Technologies, Monza, Italy) under the following conditions: 96 °C for 10 min, 45 cycle of 55 °C annealing for 2 min and 98 °C denaturation for 30 s, followed by 60 °C for 2 min and 10 °C. End-point fluorescence data were collected in QuantStudioTM 3D Digital PCR Instrument and files generated were analyzed using cloud-based platform QuantStudioTM 3D AnalysisSuite dPCR software, version 3.1.6. Each sample was analyzed in triplicate.

2.8. *Triticum aestivum* Percentage Calculation

For the common wheat percentage calculation, we start from the absolute copies/μL yielded by the QuantStudioTM 3D Analysis Suite dPCR software. In our assay the *T. aestivum* target sequence is marked with FAM, whereas the taxon target sequence is marked in VIC. Equation 1 was used to calculate the percentage of common wheat copies in the sample, in which FAM stands for the number of FAM copies/μL and VIC for the number of VIC copies/μL:

$$\frac{FAM}{\frac{VIC-3*FAM}{2}+FAM}*100 \quad (1)$$

3. Results

3.1. Reference Samples

Several factors are important for accurate quantification of multiplexed assays, including target linkage, probe specificity and differential PCR efficiencies.

The absence of linking between the two targets has been evaluated through literature and bioinformatic analysis. Nemoto et al. [24] demonstrated, through Southern blot analysis, that the *Triticum TaHd1* gene is present in single copy on each A, B and D genomes of wheat and maps on long arm of chromosome 6. *Pinb-D1*gene maps in D sub-genome and is located on chromosome 5 at the Hardness (Ha) locus. The two targets are therefore not linked.

Primers/probes specificity have been preliminarily evaluated in qPCR, finding that TritA_APX assay gives a signal only in hexaploid wheat, whereas GranoCO2 assay gives a signal both in hexaploid and tetraploid wheats (including farro dicoccum and Kamut).

Amplification efficiency and reproducibility for each primer set were examined through a standard curve qPCR assay, using bread and durum wheat DNA dilutions (Figure 1). Efficiency of reactions were calculated from the slope using the formula E = 10−1/slope. The slope values obtained were of −3.44 for GranoCO2 primers, and of −3.17 was obtained for TritAPX primers. Amplification efficiencies were of 99.6 and 104%, respectively.

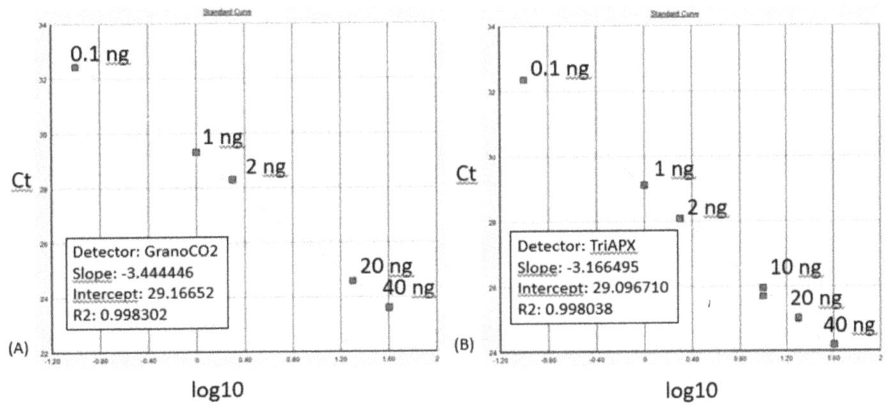

Figure 1. qPCR standard curves obtained after amplification of the DNA dilutions reported in the graph with GranoCO2 primers (**A**) and with TriAPX primers (**B**).

The duplex method was then optimized in cdPCR system for specificity on the reference samples described in Materials and Methods. The concentrations of primers and probes were optimized at 900 nmol and 200 nmol respectively and the annealing temperature was fixed at 55 °C. The resolution of the clusters (Figure 2) was obtained in absence of restriction digestion of the samples, therefore this time-consuming procedure was omitted from the protocol.

The mean common wheat percentages experimentally determined in "mixed flour" and "mixed DNA" samples in comparison with actual percentages are reported in Table 2. The SD values reported in the same table express the precision of the method, i.e., the closeness of agreement between replicate measurements. At 3% level, the SD values are <35% for all the samples and therefore the precision is acceptable, according to Codex Alimentarius Commission/Guidelines 74–2010 [25]. In Table 2 are even reported some values informative about the precision and the accuracy of the method, such as the coefficient of variation (CV), the absolute error and the relative error.

The trueness of the method is usually defined as the degree of agreement of the expected value with the true value or accepted reference value. In GMO testing the trueness must be within 25% of the accepted reference value [25]. The trueness of our method fits the purpose: The estimated concentrations over the dynamic range tested were within the ± 25% acceptable bias as recommended by GMO analytical guidelines [26]. In particular, at 3% level the experimentally determined percentages are very close to the true one. In the evaluated dynamic range, the LOD (Limit of Detection) of the method has been found at 0.3% common wheat contamination, whereas the LOQ (Limit of Quantification) at 1.5% level.

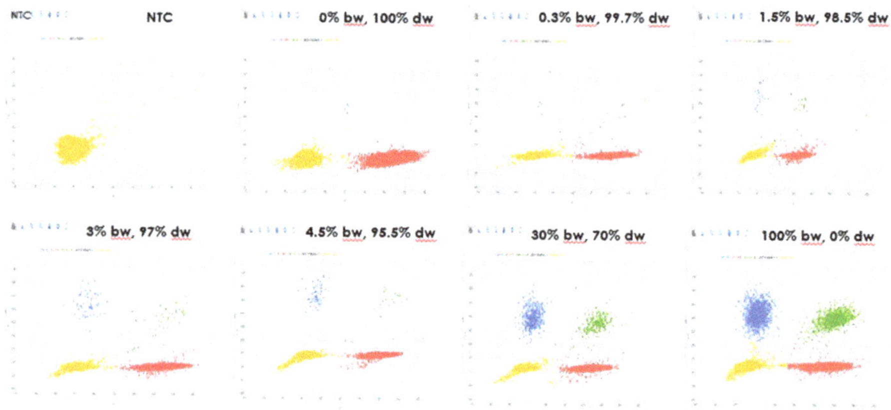

Figure 2. Two-dimensional scatter graphs generated by chip digital PCR (cdPCR) analysis of eight different samples. NTC (No Template Control) is a blank sample without DNA; The other samples are made of durum wheat (dw) DNA or common wheat (bw) DNA or a mix of the two, as indicated in the figure; In this graph a partition can fall into one of four possible clusters: negative partition that contain no amplified targets (yellow), single positive partition for *Triticum* genus (red), single positive partition for common wheat (blue) and positive partitions that contain a positive signal for both targets (green, double-positive partitions).

Table 2. Actual common wheat percentages in comparison with those experimentally determined in two different classes of samples. "Mixed DNA" samples were obtained by mixing DNA extracted from pure common and durum wheat species. "Mixed flour" samples were obtained by extracting DNA from of common and durum wheat flours mixed at different percentages). CV: Coefficient of variation.

Actual Common Wheat %	Mixed DNA					Mixed Flour				
	Mean Common Wheat %	Std Dev	CV	Absolute Error	Relative Error	Mean Common Wheat %	Std Dev	CV	Absolute Error	Relative Error
0	0.12	0.05	0.39	0.12		0.09	0.06	0.65	0.09	
100	105.00	7.00	0.07	5.00	0.05	94.40	6.85	0.07	5.60	0.06
0.3	0.43	0.05	0.12	0.13	0.43	0.37	0.12	0.34	0.07	0.23
1.5	1.37	0.07	0.05	0.13	0.09	1.43	0.28	0.19	0.07	0.05
3	3.06	0.05	0.01	0.06	0.02	2.86	0.32	0.11	0.14	0.05
4.5	4.50	0.04	0.01	0.00	0.00	3.93	0.51	0.13	0.57	0.13
30	25.90	0.46	0.02	4.10	0.14	24.90	1.68	0.07	5.10	0.17

The Pearson's r between the expected and calculated common wheat percentages were determined in mixed DNA samples and in mixed flour samples. The correlation values found are respectively of 0.9985 and of 0.9993. Extracting DNA from mixed flours and their subsequent amplification is much more realistic model of real foods, rather than mixing DNA from different species/samples. However, the preparation of mixtures of flours can be potentially affected by weighting errors and by heterogeneity problems, due, for example, to variation in granulometry, in mixing and blending. On the other hand, DNA mixtures can be affected by errors in DNA quantification and mixing. Therefore, with the intent to minimize the inaccuracy of the reference materials we decided to prepare two series of blends using the two different options. After analyses, the two classes of reference materials gave the same results. No statistically significant differences were found among mean common wheat % values determined from mixed DNA samples and from mixed flours. It is therefore possible to conclude that the two classes of reference materials prepared worked in agreement.

Since 3% common wheat threshold is in percentage of mass ratio (% *m/m*) and since the analytical output is in number of common wheat and taxon target copies, a conversion factor is needed. This conversion factor, CF, mainly depends on the zygosity, but even on differences linked to DNA extraction and varieties. CF for GMO detection is available for each CRM (Certified Reference Material) [21].

For our homozygous samples, for which certified reference materials are not available, a conversion from % (copy/copy) to % (m/m) can be hypothesized. This same approach has been used in the study of Dong et al. [23] aimed to quantify kidney bean in lotus seed paste.

In 3% common wheat reference samples, a mean percent recovery of 100.44 has been obtained, that fully fits with the acceptable range for major components in low complexity matrices (95–105%).

3.2. *Reference and Commercial Pasta Samples*

The applicability of duplex dPCR assay to pasta was evaluated in two different groups of samples: 4 reference pasta samples prepared in our laboratory and contaminated with 1.5%, 3%, 4.5%, and 10% common wheat and on 20 pasta samples of different brands commercialized in Italy.

The results are reported in Figure 3, from which it can be observed that the duplex dPCR assay performs well on reference pasta, with a correlation value of 0.99 among actual and measured percentages and a mean relative error of 0.07.

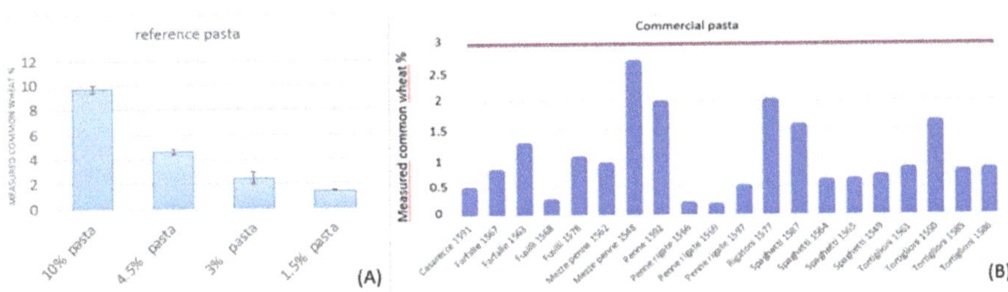

Figure 3. Common wheat percentages determined in 4 reference pasta (**A**) and in 20 commercial pasta samples (**B**) with duplex digital PCR (dPCR) assay. In (A) the percentages values before the word "pasta" indicate the common wheat contaminations. In (B) the red horizontal line indicates the maximum level of common wheat contamination allowed by law.)

As previously introduced, a body of Italian laws and regulations rule the product named "pasta" [2–4]. The denomination "pasta" strictly defines a product obtained after drawing, rolling and subsequent drying of a dough exclusively made from durum wheat (flour or semolina or whole semolina) and water. In the final product the humidity must not exceed 12.50%. The production of pasta with common wheat flour is forbidden, but a maximum level of 3% common wheat flour is tolerated as result of accidental contamination during the production chain. The inclusion of ingredients different from durum wheat and water is reserved to "special pasta". The special pastas must be offered for sale in Italy with the name durum wheat semolina pasta supplemented by the mention of the ingredient used and, in the case of several ingredients, of that or the characterizing ones. Anyway, even in special pasta, common wheat is a contaminants. The special pasta represents a minor sector of Italian pasta production and consumption. Therefore, as representative of the market, pasta of different brands has been considered in this study. The analyzed samples were all labelled as "pasta" and all reported, as ingredients, durum wheat and water. According to Italian laws, a maximum 3% common wheat presence is expected. All the commercial samples have been found below the 3% common wheat contamination threshold. The analytical data confirm that all the samples comply with the Italian laws.

4. Discussion

We have developed a duplex chip digital PCR analytical protocol to identify and quantify common wheat contamination in pasta production chain. The reason for developing such new assay is related to dPCR particularities. In comparison with conventional end-point PCR and qPCR, this technique has been reported to have many advantages (reviewed by Demeke et al. [27]), the major the absolute quantification of a target without reference to a standard/calibration curve. This fact reduces the errors deriving from the comparison of different matrices, i.e., the calibrant and the test sample. Moreover, because of the high-level sample partitioning, dPCR is less sensitive to PCR inhibitors and the results obtained are potentially very precise and accurate [27,28]. Thanks to the high resilience to inhibitors, the efficiency and the reproducibility on different platforms, dPCR is candidate as higher-order reference measurement methods and as the method for value assignment of reference materials [28]. On the other hand, a limitation of such approach is that it is more expensive than qPCR, but the use of multiplex approaches moves the scales in favor of dPCR [27]. From a technology transfer point of view, both the pasta industry and the large consumer cooperative, between the other involved in this work, expressed interest in developing and applying a dPCR strategy for control pasta chain. The key control points are in the passage of the grains from stackers to the mills, of semolina batches from the mills to the pasta factory and in the final product, the pasta. The pasta chain stakeholders interested in such analytical tool are therefore the farmer associations, the stackers, the mills, the pasta industry, the consumer associations and the public and private control bodies. All the stakeholders have the interest to share a method for common wheat contamination control in grains, semolina, and pasta. Several assays has been developed and validated for such purpose, but are all dependent on a calibration curve and suffer from the loss of certified reference materials for the construction of such curves. DigitalPCR, that works without the need of calibrants, can fill this gap. It can in fact be proposed as method for the validation of reference materials to be used for qPCR standard curves and as higher order reference measurement method. This hypothesis to apply dPCR technology to prepare reference materials has been advanced by other authors, e.g., Mehle et al. [29] in plant pathogen detection, by Dong et al. [30] in environmental microbiology and by Pavšič et al. [31] in microbial diagnostics. The potential for synergy of qPCR and dPCR has been underlined by Debski et al. [32]. in the field of medical diagnostics. In conclusion, the opportunity to complement and strengthen the cheaper qPCR analyses justify the higher cost of dPCR assays.

Our cdPCR assay is based on duplex non-competing reactions: two amplicons are generated from two primer sets and the signal generated from a probe specific for each amplicon enable to distinguish the two targets within a single reaction. Such concurrent amplifications reduce technical errors, reagent and time needed. One of the target is a D-genome specific genic sequence and the other a *Triticum* specific genic sequence present in A, B and D genomes. This taxon-specific assay was designed on *TaHd1* gene sequence. Such gene, involved in the photoperiodic flowering pathway, has been demonstrated to be present in single copy in each of the A, B and D *Triticum* genomes [20]. The bread wheat specific assay was designed on *Pinb-D1*, a single-copy gene encoding for puroindoline b protein [15,33]. This gene belongs to the Ha locus, occurring only on chromosome 5D in common wheat [26]. Accordingly, we have developed the formula reported in Materials and Methods for the common wheat % calculation. In the formula we have considered:

- The different level of ploidy between common wheat (hexaploid, with the three A, B and D genomes) and durum wheat (tetraploid, with A and B genomes);
- the fact that *TaHd1* gene is present in single copy/A, B and D haploid genomes;
- the fact that *Pinb-D1* gene is present in single copy/D haploid genome; and
- the comparable amplification efficiency of the two targets

The *Pinb-D1* gene sequence has been used to target common wheat in cqPCR assays previously developed, whereas the *TaHd1* gene sequence has never been used in pasta authenticity assessment.

As verified on reference samples, the proposed protocol highly performs to track 3% common wheat contamination, that is the critical value fixed by law as limit between accidental contamination and fraud. Its applicability has been evaluated on reference and commercial pasta samples. In conclusion, a cdPCR duplex assay has been developed to control pasta production chain from an economically motivated adulteration, that is the use of cheaper ingredient (i.e., common wheat) instead of durum wheat for pasta manufacturing. It is possible to quantify the mass of common wheat directly in flours and in highly processed food, such as pasta. The inter-laboratory validation of the method can be proposed as further step.

Author Contributions: Conceptualization, S.S., P.C. and V.T.; Data curation, R.B.; Funding acquisition, V.T.; Methodology, C.M. and R.B.; Validation, R.G.; Writing—original draft, C.M. and V.T.; Writing—review and editing, S.S. All authors have read and agreed to the published version of the manuscript.

Funding: This research was partially funded by Horizon 2020 INVITE project, grant number 817970 and by METROFOOD project, Horizon 2020 grant number 739568.

Conflicts of Interest: The authors declare no conflict of interest.

References

1. Cacchiarelli, L.; Sorrentino, A. Pricing strategies in the Italian retail sector: The case of pasta. *Soc. Sci.* **2019**, *8*, 113. [CrossRef]
2. LEGGE 4 luglio 1967, n. 580 Disciplina per la lavorazione e commercio dei cereali, degli sfarinati, del pane e delle paste alimentari. (GU Serie Generale n.189 del 29-07-1967). Available online: https://www.gazzettaufficiale.it/gazzetta/serie_generale/caricaDettaglio?dataPubblicazioneGazzetta=1967-07-29&numeroGazzetta=189 (accessed on 7 July 2020).
3. DECRETO DEL PRESIDENTE DELLA REPUBBLICA 9 febbraio 2001, n. 187. Regolamento per la revisione della normativa sulla produzione e commercializzazione di sfarinati e paste alimentari, a norma dell'articolo 50 della legge 22 febbraio 1994, n. 146. (GU Serie Generale n.117 del 22-05-2001). Available online: https://www.gazzettaufficiale.it/atto/serie_generale/caricaDettaglioAtto/originario?atto.dataPubblicazioneGazzetta=2001-05-22&atto.codiceRedazionale=001A5636&elenco30giorni=false (accessed on 7 July 2020).
4. DECRETO DEL PRESIDENTE DELLA REPUBBLICA 5 marzo 2013, n. 41. Regolamento recante modifiche al decreto del Presidente della Repubblica 9 febbraio 2001, n. 187, concernente la revisione della normativa sulla produzione e commercializzazione di sfarinati e paste alimentari. (13G00082) (GU Serie Generale n.95 del 23-04-2013). Available online: https://www.politicheagricole.it/flex/cm/pages/ServeBLOB.php/L/IT/IDPagina/6230 (accessed on 7 July 2020).
5. Resmini, P.; De Bernardi, G. Il metodo elettroforetico rapido per il riconoscimento ed il dosaggio del grano tenero nel grano duro, negli sfarinati e nelle paste alimentari. *Tecnica Molitoria* **1976**, *10*, 97–101.
6. Cantagalli, P.; Piazzi, S.E.; Sordi-Galli, S. Il controllo della genuinità delle semole di frumento duro e delle paste alimentari mediante immunologia. *Tecnica Molitoria* **1969**, *4*, 79–85.
7. Boney, M. Rapid diagnostics technologies for food industrie. *Food Sci. Technol. Today* **1998**, *12*, 50–52.
8. Kobrenel, K.; Gaga, O.; Autran, J.C. Possibilite de detection de la presence de bles tendres dans les pates alimentaires ayant subi des traitements thermiques a haute temperature. *Annales des Falsifications et de l'Expertise Chimique et Toxicologique* **1985**, *78*, 109–117.
9. Caboni, M.F.; Iafelice, G.; Pelillo, M.; Marconi, E. Analysis of fatty steryl esters in tetraploid and hexaploid wheats: Identification and comparison between chromatographic methods. *J. Agric. Food Chem.* **2005**, *53*, 7465–7472. [CrossRef]
10. Bryan, G.J.; Dixon, A.; Gale, M.D.; Wiseman, G. A PCR-based method for the detection of hexaploid bread wheat adulteration of durum wheat and pasta. *J. Cereal Sci.* **1998**, *39*, 175–180. [CrossRef]
11. Arlorio, M.; Coisson, J.D.; Cereti, E.; Travaglia, F.; Papasso, M.; Martelli, A. Polymerase chain reaction (PCR) of puroindoline b and ribosomal/puroindoline b multiplex PCR for the detection of common wheat (*Triticum aestivum*) in Italian pasta. *Eur. Food Res. Technol.* **2003**, *216*, 253–258. [CrossRef]

12. Sonnante, G.; Montemurro, C.; Morgese, A.; Sabetta, W.; Blanco, A.; Pasqualone, A. DNA microsatellite region for a reliable quantification of soft wheat adulteration in durum wheat-based foodstuffs by real-time PCR. *J. Agric. Food Chem.* **2009**, *57*, 10199–10204. [CrossRef]
13. Casazza, A.P.; Morcia, C.; Ponzoni, E.; Gavazzi, F.; Benedettelli, S.; Breviario, D. A reliable assay for the detection of soft wheat adulteration in Italian pasta is based on the use of new DNA molecular markers capable of discriminating between *Triticum aestivum* and *Triticum durum*. *J. Cereal Sci.* **2012**, *56*, 733–740. [CrossRef]
14. Silletti, S.; Morello, L.; Gavazzi, F.; Gianì, S.; Braglia, L.; Breviario, D. Untargeted DNA-based methods for the authentication of wheat species and related cereals in food products. *Food Chem.* **2019**, *271*, 410–418. [CrossRef]
15. Alary, R.; Serin, A.; Duviau, M.P.; Joudrier, P.; Gautier, M.F. Quantification of common wheat adulteration of durum wheat pasta using real-time quantitative polymerase chain reaction (PCR). *Cereal Chem.* **2002**, *79*, 553–558. [CrossRef]
16. Terzi, V.; Malnati, M.; Barbanera, M.; Stanca, A.M.; Faccioli, P. Development of analytical systems based on real-time PCR for Triticum species-specific detection and quantitation of bread wheat contamination in semolina and pasta. *J. Cereal Sci.* **2003**, *38*, 87–94. [CrossRef]
17. Matsuoka, Y.; Arami, S.; Sato, M.; Haraguchi, H.; Kurimoto, Y.; Imai, S.; Tanaka, K.; Mano, J.; Furui, S.; Kitta, K. Development of methods to distinguish between durum/common wheat and common wheat in blended flour using PCR. *Food Hyg. Saf. Sci.* **2012**, *53*, 195–202. [CrossRef] [PubMed]
18. Imai, S.; Tanaka, K.; Nishitsuji, Y.; Kikuchi, Y.; Matsuoka, Y.; Arami, S.; Sato, M.; Haraguchi, H.; Kurimoto, Y.; Mano, J.; et al. An endogenous reference gene of common and durum wheat for detection of genetically modified wheat. *Food Hyg. Saf. Sci.* **2012**, *53*, 203–210. [CrossRef] [PubMed]
19. Paternò, A.; Verginelli, D.; Bonini, P.; Misto, M.; Quarchioni, C.; Dainese, E.; Peddis, S.; Fusco, C.; Vinciguerra, D.; Marchesi, U. In-House Validation and Comparison of Two Wheat (*Triticum aestivum*) Taxon-Specific Real-Time PCR Methods for GMO Quantification Supported by Droplet Digital PCR. *Food Anal. Methods* **2018**, *11*, 1281–1290. [CrossRef]
20. Morcia, C.; Ghizzoni, R.; Gianinetti, A.; Avossa, V.; Scaramagli, S.; Carnevali, P.; Terzi, V. Validation of a DNA-based method for bread wheat contamination detection in pasta. *J. Cereal Sci.* **2017**, *76*, 69–71. [CrossRef]
21. Corbisier, P.; Barbante, A.; Berben, G.; Broothaerts, W.; De Loose, M.; Emons, H.; Georgieva, T.; Lievens, A.; Mazzara, M.; Papazova, N.; et al. *Recommendation for the Unit of Measurement and the Measuring System to Report Traceable and Comparable Results Expressing GM Content in Accordance with EU Legislation*; EUR28536 EN; Publications Office of the European Union: Luxembourg, 2017. [CrossRef]
22. Cottenet, G.; Blancpain, C.; Chuah, P.F. Performance assessment of digital PCR for the quantification of GM-maize and GM-soya events. *Anal. Bioanal. Chem.* **2019**, *411*, 2461–2469. [CrossRef]
23. Dong, X.; Gao, D.; Dong, J.; Chen, W.; Li, Z.; Wang, J.; Liu, J. Mass ratio quantitative detection for kidney bean in lotus seed paste using duplex droplet digital PCR and chip digital PCR. *Anal. Bioanal. Chem.* **2020**, *412*, 1701–1707. [CrossRef]
24. Nemoto, Y.; Kisaka, M.; Fuse, T.; Yano, M.; Ogihara, Y. Characterization and functional analysis of three wheat genes with homology to the CONSTANS flowering time gene in transgenic rice. *Plant J.* **2003**, *36*, 82–93. [CrossRef]
25. CXG 74-2010 Guidelines on Performance Criteria and Validation of Methods for Detection, Identification and Quantification of Specific DNA Sequences and Specific Proteins in Foods 2010. Available online: http://www.fao.org/fao-who-codexalimentarius/codex-texts/guidelines/en/ (accessed on 6 June 2020).
26. ENGL, European Network of GMO Laboratories. Definition of Minimum Performance Requirements for Analytical Methods of GMO Testing. 2015. Available online: http://gmo-crl.jrc.ec.europa.eu/doc/MPR%20Report%20Application%2020_10_2015.pdf (accessed on 6 June 2020).
27. Demeke, T.; Dobnik, D. Critical assessment of digital PCR for the detection and quantification of genetically modified organisms. *Anal. Bioanal. Chem.* **2018**, *410*, 4039–4050. [CrossRef] [PubMed]
28. Pecoraro, S.; Berben, G.; Burns, M.; Corbisier, P.; De Giacomo, M.; De Loose, M.; Dagand, E.; Dobnik, D.; Eriksson, R.; Holst-Jensen, A.; et al. *Overview and Recommendations for the Application of Digital PCR*; EUR 29673 EN; Publications Office of the European Union: Luxembourg, 2019; ISBN 978-92-76-00180-5, JRC115736. [CrossRef]

29. Mehle, N.; Gregur, L.; Bogožalec Košir, A.; Dobnik, D. One-Step Reverse-Transcription Digital PCR for Reliable Quantification of Different Pepino Mosaic Virus Genotypes. *Plants* **2020**, *9*, 326. [CrossRef] [PubMed]
30. Dong, L.; Meng, Y.; Wang, J.; Liu, Y. Evaluation of droplet digital PCR for characterizing plasmid reference material used for quantifying ammonia oxidizers and denitrifiers. *Anal. Bioanal. Chem.* **2014**, *406*, 1701–1712. [CrossRef] [PubMed]
31. Pavšič, J.; Žel, J.; Milavec, M. Assessment of the real-time PCR and different digital PCR platforms for DNA quantification. *Anal. Bioanal. Chem.* **2016**, *408*, 107–121. [CrossRef]
32. Debski, P.; Gewartowski, K.; Bajer, S.; Garsteki, P. Calibration-free assays on standard real-time PCR devices. *Sci. Rep.* **2017**, *7*, 44854. [CrossRef]
33. Bhave, M.; Morris, C.F. Molecular genetics of puroindolines and related genes: Allelic diversity in wheat and other grasses. *Plant Mol. Biol.* **2008**, *66*, 205–219. [CrossRef]

MDPI
St. Alban-Anlage 66
4052 Basel
Switzerland
Tel. +41 61 683 77 34
Fax +41 61 302 89 18
www.mdpi.com

Foods Editorial Office
E-mail: foods@mdpi.com
www.mdpi.com/journal/foods

www.ingramcontent.com/pod-product-compliance
Lightning Source LLC
Chambersburg PA
CBHW061830210926
51850CB00006B/852
9783036554587